Climate Change

Climate Change

Causes, Effects, and Solutions

John T. Hardy

Chair, Department of Environmental Sciences
Huxley College of the Environment
Western Washington University
Bellingham, Washington
USA

WILEY

Telephone (+44) 1243 779777

Email (for orders and customer service enquiries): cs-books@wiley.co.uk
Visit our Home Page on www.wileyeurope.com or www.wiley.com

Reprinted May 2004, June 2005, May 2006

Other Wiley Editorial Offices

John Wiley & Sons Inc., 111 River Street, Hoboken, NJ 07030, USA

Jossey-Bass, 989 Market Street, San Francisco, CA 94103-1741, USA

Wiley-VCH Verlag GmbH, Boschstr. 12, D-69469 Weinheim, Germany

John Wiley & Sons Australia Ltd, 33 Park Road, Milton, Queensland 4064, Australia

John Wiley & Sons (Asia) Pte Ltd, 2 Clementi Loop #02-01, Jin Xing Distripark, Singapore 129809

John Wiley & Sons Canada Ltd, 22 Worcester Road, Etobicoke, Ontario, Canada M9W 1L1

Wiley also publishes its books in a variety of electronic formats. Some content that appears in print may not be available in electronic books.

British Library Cataloguing in Publication Data

A catalogue record for this book is available from the British Library

ISBN-10 0-470-85019-1 (P/B)
ISBN-13 978-0-470-85019-0 (P/B)
ISBN-10 0-470-85018-3 (H/B)
ISBN-13 978-0-470-85018-3 (H/B)

Typeset in 10.5/13pt Times by Laserwords Private Limited, Chennai, India
Printed and bound in Great Britain by Antony Rowe Ltd, Chippenham, Wiltshire
This book is printed on acid-free paper responsibly manufactured from sustainable forestry in which at least two trees are planted for each one used for paper production.

Contents

Preface

"The greenhouse effect is the most significant economic, political, environmental and human problem facing the 21st Century."

Timothy Wirth, former US Senator and Undersecretary of State for Global Affairs

Unprecedented changes in climate are taking place. If we continue on our present course, life on Earth will be inextricably altered. The very sustainability of the Earth's life-support system is now in question. How did we arrive at this pivotal point in our history?

For millennia, the Earth's climate remained little changed. Early humans thrived, living on an abundance of plants and animals, some of which they domesticated for their own use. They cooked their food and warmed their dwellings largely with wood. This wood was the product of photosynthesis – the removal of carbon dioxide from the atmosphere and its conversion to living organic matter. Burning the wood returned this same quantity of carbon to the atmosphere. Human activities had little more than local impacts. Natural changes occurred in the Earth's climate, but they were gradual, occurring over tens of thousands to millions of years.

Suddenly, 200 years ago, things began to change. Modern medicine and improvements in technology led to a human population explosion. Ninety-nine percent of all human beings who ever lived are alive today. At the same time, fossil fuel (first coal, and then oil and gas) became the energy source of choice – facilitating rapid industrialization and further fossil-fuel consumption. Unlike wood, the carbon in fossil fuel was slowly formed from decaying plants millions of years ago and was stored in the Earth's crust. Its burning over the past 150 years has increased the level of atmospheric carbon dioxide by 33%. Carbon dioxide is a greenhouse gas that traps heat in the lower atmosphere, keeping our planet warm. However, like many things in nature, a little is good, but more is not necessarily better.

If we continue our heavy dependence on fossil fuel, we will double the preindustrial atmosphere's carbon dioxide level in a few decades and perhaps triple it by the end of this century. As a consequence, by most estimates, the planet will rapidly warm to a level never experienced by human beings. There will be consequences. In our hurried modern lives, we forget that our welfare is still closely linked to the health of the planet. Our health and survival depend on productive agriculture, and supplies of water, forest products, and fish. All these depend, in turn, on a favorable climate. Changes in any or all of these, as a result of climate change, will affect the economy.

Human-induced climate change is now a recognized phenomenon. Our ability to predict how climate will change and how those changes will impact ecosystems and humans improved markedly during the last decade. Debate continues about the exact degree of future change, and there are many uncertainties. Some argue that immediate and drastic measures must be taken to stem greenhouse gas emissions before it is too late. The precautionary principle (better safe than sorry) is invoked. Others argue that action will be costly and should be delayed until more research is completed. Many nations have joined together in an international treaty to limit greenhouse gas emissions. However, nations that emit the greatest share of global greenhouse gas emissions remain reluctant to join this effort, and the treaty remains ineffective.

Thirteen years ago, I developed a university course titled "Effects of Global Climate Change." Every year since, I have attempted to convey to students, and on occasion to the general public, the importance of this global problem. I truly believe that, unaltered, our present course will cause hardship for millions of humans, particularly the poor. In 1988, a conference of over 300 scientists and policymakers from 46 countries declared that "*humanity is conducting an unintended, uncontrolled, globally pervasive experiment whose ultimate consequences could be second only to a global nuclear war.*"

Greenhouse warming is a problem in search of a solution. It is often difficult to avoid a sense of hopelessness as individuals and governments continue consuming fossil fuel at record rates and fight to ensure an uninterrupted supply of oil. I am a short-term pessimist, but a long-term optimist. Humans often react only in the face of a crisis. In human lifetimes, climate change is slow, barely perceptible, and its potential impacts are understood by few. However, in the not too distant future, it will be impossible to deny the impacts as the changes reach a crisis level. Alternative fuels will be adopted, fossil fuel will be conserved, and humans will adapt. However, the longer the denial continues, the more severe the ultimate crisis will be. My purpose here is to describe how humans are causing the climate to change, what effects we can expect from that change, and the variety of actions that can be taken to minimize climate change and its impacts. If this book contributes in any way to a more informed public or a more climate-friendly energy policy, it will have served its purpose.

Many individuals have made this book possible. Over 400 climate-change students have inspired me with their hope for the future. Their probing questions sharpened my insight into many aspects of climate change that I might have otherwise overlooked. Knowledge stands on the shoulders of others, and I attempted to distill the most important points from hundreds of detailed reports by respected scientists around the world into one readable volume. For their work, we must all be grateful. Kevin Short and Ray Mutchler provided invaluable graphics assistance in preparation of figures and illustrations. Diane Peterson kept my administrative job manageable, allowing me to focus on this work. Katie Frankhauser very competently checked the many citation details. My thanks to the editorial team at John Wiley & Sons, including Lyn Roberts, Keily Larkins, Susan Barclay, and the staff at Laserwords for ably facilitating the timely completion of this work. I am grateful to my parents who instilled persistence to see a job to completion. To my wife Kathie, my appreciation goes far beyond the typical author's appreciation of patience. Her insightful comments helped turn a technical treatise into (hopefully) a readable book. Without her inspiration and support, none of this would have been written. Finally, to Kevin, Amy, and Tanya – its your planet now.

SECTION I

Climate Change – Past, Present, and Future

Chapter 1

Earth and the Greenhouse Effect

"... if the carbonic acid content of the air [atmospheric CO_2] rises to 2 [i.e. doubles] the average value of the temperature change will be... +5.7 degrees C"

Svante Arrhenius 1896

Introduction

The physics and chemistry of the Earth's atmosphere largely determines our climate (Lockwood 1979). Although the atmosphere seems like a huge reservoir capable of absorbing almost limitless quantities of our industrial emissions, it is really only a thin film. Indeed, if the Earth were shrunk to the size of a grapefruit, its atmosphere would be thinner than the skin of the grapefruit. Our understanding of how the chemistry and physics of the atmosphere affect climate developed over many centuries, but has greatly accelerated during the past few decades (Box 1.1).

The Earth's atmosphere is layered. In the lower atmosphere, from the surface up to about 11-km altitude (troposphere), temperature decreases with increasing altitude. This layer is only about 1/1,200 of the diameter of the globe, but its physics and chemistry are crucial to sustaining life on the planet. Because cold dense air on top of warm less dense air is unstable, the layer is fairly turbulent and well mixed. It contains 99% of the atmospheric mass. From 15 to 50 km, the temperature increases with altitude, resulting in a stable upper atmosphere (stratosphere) with almost 1% of the atmospheric mass. Above 50 km are the mesosphere and the thermosphere, which have little effect on climate (Figure 1.1).

During the past 100 years we humans, as a result of burning coal, oil, and gas and clearing forests, have greatly changed the chemical composition of this thin atmospheric layer. These changes in chemistry, as described in subsequent chapters, have far-reaching consequences for the climate of the Earth, the ecosystems that are sustained by our climate, and our own human health and economy.

The Greenhouse Effect

Three primary gases make up 99.9% by volume of the Earth's atmosphere – nitrogen (78.09%), oxygen (20.95%), and argon (0.93%). However, it is the rare trace gases, that is, carbon dioxide (CO_2), methane (CH_4), carbon monoxide (CO), nitrogen

Climate Change: Causes, Effects, and Solutions John T. Hardy
 ISBNs: 0-470-85018-3 (HB); 0-470-85019-1 (PB)

Box 1.1 The history of atmospheric science

Our current understanding of the chemistry and physics of the atmosphere has a long and fascinating history (Crutzen and Ramanathan 2000). Some highlights of this history include the following:

340 B.C. – The Greek philosopher Aristotle publishes Meteorologica; its theories remain unchallenged for nearly 2000 years.

1686 – Edmond Halley shows that low latitudes receive more solar radiation than higher ones and proposes that this heat gradient drives the major atmospheric circulation.

1750s – Joseph Black identifies CO_2 in the air.

1781 – Henry Cavendish measures the percentage composition of nitrogen and oxygen in air.

1859 – John Tyndall suggests that water vapor, CO_2, and other radiatively active ingredients could contribute to keeping the Earth warm.

1896 – Svante Arrhenius publishes a climate model demonstrating the sensitivity of surface temperature to atmospheric CO_2 levels.

1938 – GS Callendar calculates that 150 billion tons of CO_2 was added to the atmosphere during the past half century, increasing the Earth's temperature by 0.005 °C per year during that period.

1920 – Milutin Milankovitch publishes his theory of ice ages based on variations in the Earth's orbit.

1957 – Roger Revelle and Hans Suess, renowned oceanographers, proclaim that "human beings are now carrying out a large-scale geophysical experiment," that is, altering the chemistry of the atmosphere without knowing the result.

1959 – Explorer Satellites provide images of cloud cover. Verner Suomi estimates the global radiation heat budget.

1967 – Syukuro Manabe and Richard Wetherald develop the one-dimensional radiative-convective atmospheric model and show that a doubling of atmospheric CO_2 can warm the planet about 3 °C.

1970–1974 – Destruction of stratospheric ozone by man-made chlorofluorocarbons is described through the work of several researchers.

1985 – The British Antarctic Survey reports a 40% drop in springtime stratospheric ozone between 1956 and 1985.

1986 – Many countries sign the Montreal Protocol on Substances that Deplete the Ozone Layer.

1990s – Researchers discover the cooling effect of atmospheric aerosols and their importance in offsetting the greenhouse effect. The global warming trend continues and record temperatures are repeatedly set.

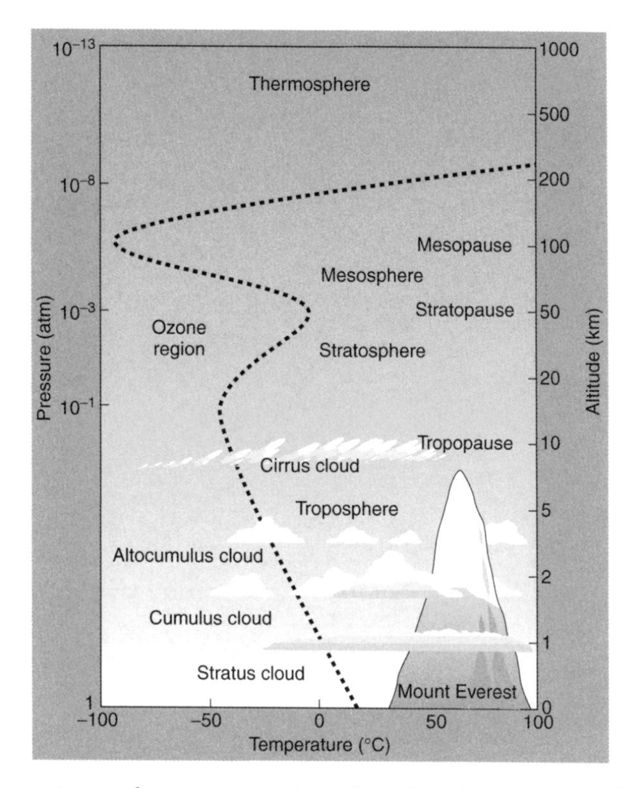

Fig. 1.1 Vertical temperature and pressure structure of the Earth's atmosphere (From Graedel TE and Crutzen PJ 1997. *Atmosphere, Climate, and Change*. Scientific American Library, New York: W.H. Freeman and Co, p. 3. Henry Holt & Co.).

oxides (NO_x), chlorofluorocarbons (CFCs), and ozone (O_3) that have the greatest effect on our climate. Water vapor, with highly variable abundance (0.5–4%), also has a strong influence on climate. These trace gases are known as *greenhouse gases* or *radiatively important trace species* (RITS). They are radiatively important because they influence the radiation balance or net heat balance of the Earth.

Thermonuclear reactions taking place on our nearest star, the Sun, produce huge quantities of radiation that travel through space at the speed of light. This solar radiation includes energy distributed across a wide band of the electromagnetic spectrum from short-wavelength X rays to medium-wavelength visible light, to longer-wavelength infrared. The greatest amount of energy (44%) is in the spectral region, visible to the human eye from 0.4 (violet) to 0.7 μm (red) (Figure 1.2a).

As incoming solar radiation passes through the atmosphere, particles and gases absorb energy. Owing to its physical or chemical structure, each particle or gas has specific wavelength regions that transmit energy and other regions that absorb energy. For example, ozone in the stratosphere absorbs short- and middle-wavelength ultraviolet radiation. A large percentage of incoming solar radiation is in the visible region. Atmospheric water vapor, carbon dioxide, and methane have low absorption in this region and allow most of the visible light to reach the Earth's surface.

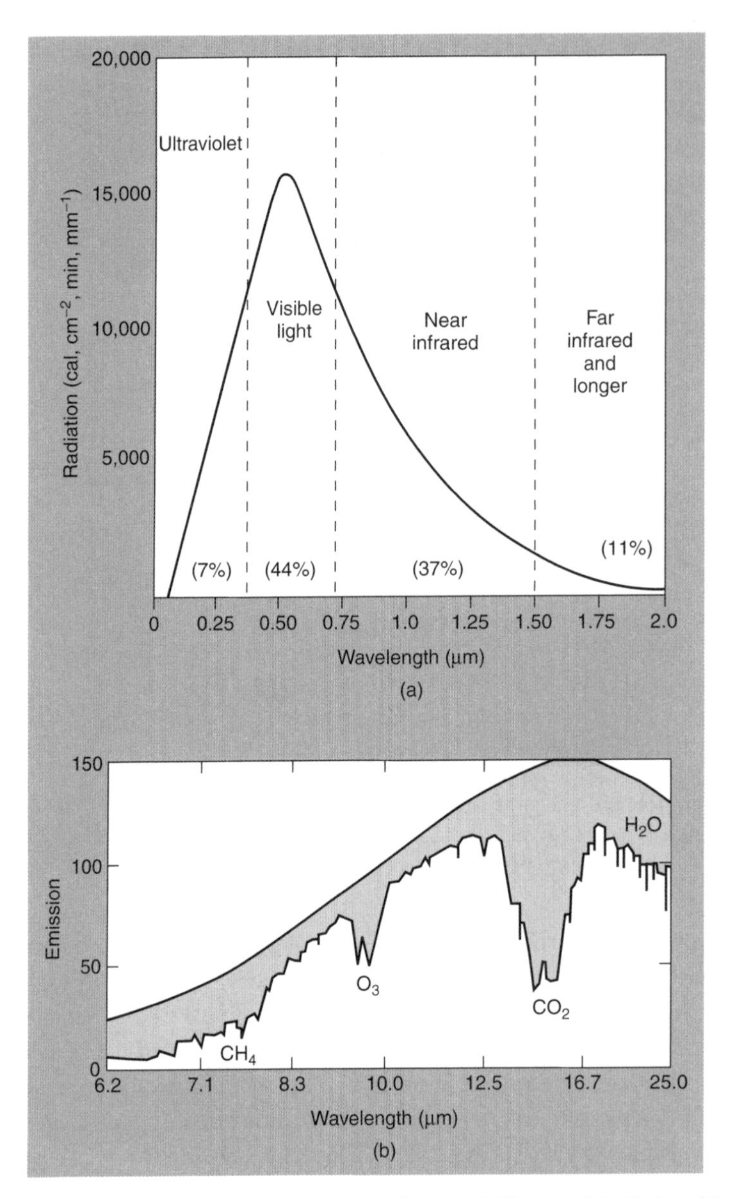

Fig. 1.2 (a) Spectral distribution of incoming solar radiation; 44% is visible light with a maximum at a wavelength of 0.49 micrometers (μm) in the blue-green part of the spectrum (From Oliver JE and Hidore JJ 2002. *Climatology: An Atmospheric Science*. Upper Saddle River, NJ: Prentice Hall, p. 410). (b) Spectral distribution of outgoing far-infrared radiation emitted by the Earth. Upper line indicates energy distribution emitted by warm Earth in the absence of an atmosphere. The light area is the actual energy escaping through the top of the atmosphere (as measured by satellite). Water vapor and clouds absorb and reduce heat loss over a wide band of wavelengths (light shaded area). CO_2 absorbs strongly at wavelengths from 12 to 18 μm and methane (not shown) at about 3.5 μm (Adapted from Ramanathan V 1988. The greenhouse theory of climate change: a test by an inadvertent global experiment. *Science* **240**: 293–299. Reproduced/modified by permission of the American Geophysical Union).

After absorption by the Earth's surface, visible energy is transformed and radiated back in the far-infrared (heat) region of the spectrum at wavelengths greater than 1.5 μm (Figure 1.2b). The transparency of the atmosphere to outgoing far-infrared radiation (heat) determines how much heat can escape from the Earth back into space and how much is trapped. The important feature of greenhouse gases is that they absorb (are opaque to) certain infrared wavelengths. Water vapor, carbon dioxide, and methane, the same gases that transmit visible wavelengths, absorb strongly in the far infrared (Figure 1.2b). Thus, they trap heat in the troposphere and stop it from escaping to space. Window glass used to trap heat in a greenhouse has similar absorption and transmission properties; hence, the term "greenhouse gases."

Of the total amount of solar energy reaching the Earth's atmosphere (342 W m^{-2}), an average 31% is reflected back to space by the upper surface of clouds, the particles in the atmosphere (dust and aerosols), or the surface of the Earth (Figure 1.3). While the overall average reflectivity (also called *albedo*) of the

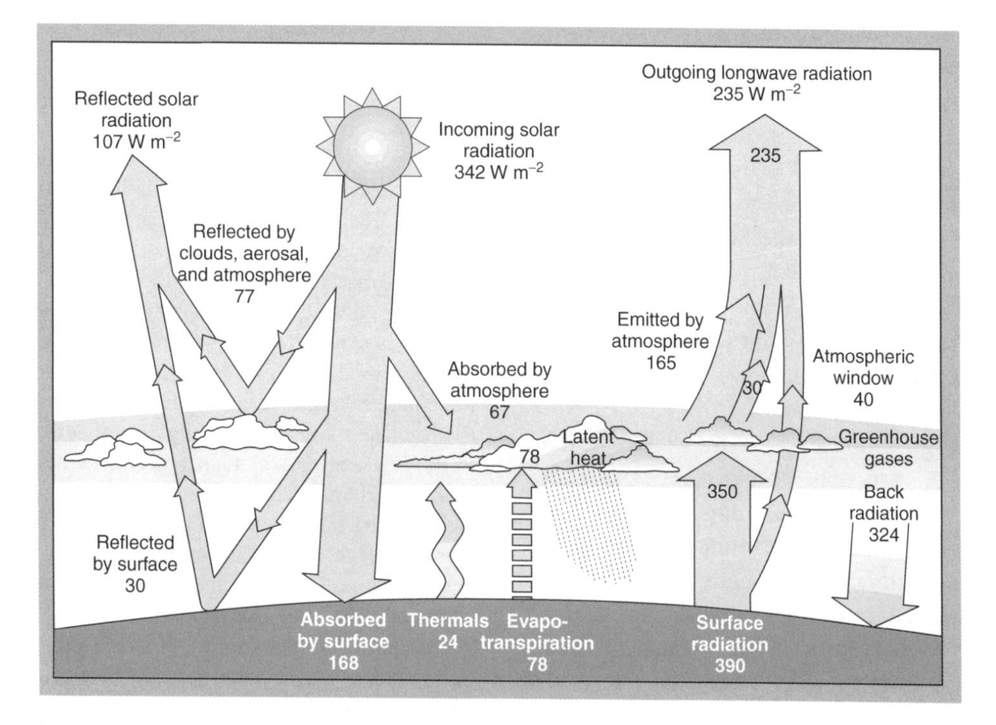

Fig. 1.3 The Earth's radiation and heat balance. Greenhouse gases are transparent to visible and near-infrared wavelengths of sunlight, but they absorb and reradiate downward a large fraction of the longer far-infrared wavelengths (heat). The net incoming solar radiation of 342 W m^{-2} is partially reflected by clouds, the atmosphere, and the Earth's surface, but 49% is absorbed by the surface. Some of that heat is returned to the atmosphere as sensible heat – most as evapotranspiration that is released as latent heat in precipitation. The rest is radiated as thermal infrared radiation. Most of that is absorbed by the atmosphere, which in turn, emits radiation both up and down, producing a greenhouse effect (From Kiehl JT and Trenberth KE 1997. Earth's annual global mean energy budget. *Bulletin of the American Meteorological Society* **78**(2): 197–208. Reproduced by permission of the American Meteorological Society).

Earth is 31%, albedo differs greatly between surfaces. Clouds, with an albedo of 40 to 90%, are by far the most important reflectors of incoming solar radiation. The albedo of fresh snow is 75 to 90%, forests 5 to 15%, and that of water, which depends on the angle of inclination of the Sun, ranges from 2 to >99%. Incoming energy that is not reflected (the remaining 69%) is absorbed by the troposphere and the Earth's surface (Figure 1.3). Evaporation of water requires a considerable amount of energy. This energy is essentially stored as "latent heat" in water vapor and released back into the air as heat when water vapor condenses (Figure 1.3).

Of the total far-infrared (heat) energy reradiated from the Earth's surface, 83% is back-radiated and does not directly escape the atmosphere. This back radiation is about double the amount of energy absorbed by the surface directly from the Sun. This additional atmospheric energy warms the Earth to its present temperature. Without the greenhouse warming effect of the atmosphere, the Earth's average surface temperature would be about −20 °C (−4 °F) instead of 15 °C (59 °F). Simply put, greenhouse gases trap solar heat in the lower atmosphere and keep the Earth warm. As long as the amount of incoming solar energy and the amount of greenhouse gas in the atmosphere remain fairly constant, the Earth's temperature remains in balance (Figure 1.3). However, the greater the concentration of greenhouse gases, the greater the amount of long-wave (heat) radiation trapped in the lower atmosphere.

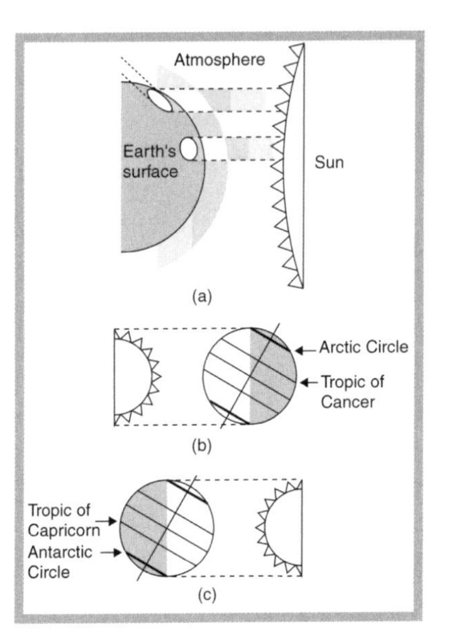

Fig. 1.4 (a) The low-latitude tropics are warm because they receive greater amounts of heat per unit surface area than the high-latitude temperate regions. The amount of heat received at higher latitudes is less than that at lower latitudes for three reasons: (1) because of the greater angle of incidence, a unit of solar energy striking a surface is spread over a greater area, (2) because it is absorbed through a greater thickness of atmosphere before reaching the surface, and (3) because more of the energy is reflected owing to the low angle of incidence. (b) In winter in the Northern Hemisphere, areas north of the Arctic Circle receive no direct solar radiation, while areas south of the Antarctic Circle receive continuous radiation for months. (c) The reverse of the above is true during the Northern Hemisphere summer. Throughout the year, the Sun is directly over some latitude between the tropics (From Thurman HV 1991. *Introductory Oceanography* (6th Edition). New York: Macmillan Publishing Co., p. 169).

Large-Scale Heat Redistribution

The Earth's temperature is not uniform but differs greatly – geographically (horizontally), by elevation (vertically), and over time (seasons and decades). To understand climate and how it might change over decades to centuries in the future, we need to understand how heat is distributed over the Earth. The tropics are warmer than temperate or polar regions because incoming solar radiation strikes the Earth's surface at a greater angle of incidence, resulting in more radiation per unit surface area (Figure 1.4). The same energy

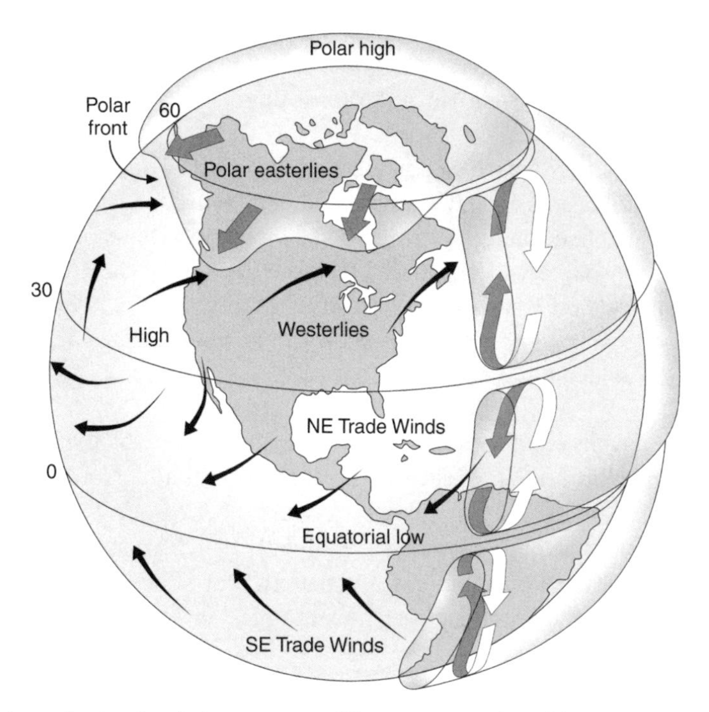

Fig. 1.5 Major atmospheric circulation patterns. The great quantity of heat received in the tropics moves poleward driving large-scale atmospheric currents (From Thurman HV 1991. *Introductory Oceanography* (6th Edition). New York: Macmillan Publishing Co., p. 170).

striking higher-latitude areas is distributed over a greater surface area. Thus, the radiation balance of the Earth is unevenly distributed and the excess heat from the tropics will (according to the laws of thermodynamics) tend to redistribute itself poleward to cooler areas. This occurs as direct heat transfer by the poleward movement of warm air and water masses. Also, significant heat energy, used in evaporation of surface water in the tropics, is carried as latent heat and released at higher latitudes when water vapor cools and condenses into rainfall.

The initial uneven distribution of heat and density in the atmosphere and ocean results in important large-scale circulation patterns (Figure 1.5). Warm air near the equator rises, and cools as it flows poleward at high altitude.

Box 1.2 The Coriolis force

Because of the Earth's rotation, moving objects (including air and water masses) are diverted to the right or left in the Northern and Southern Hemispheres, respectively. This apparent force is named after the French mathematician Gustave Gaspard Coriolis (1792–1843), who first described it. Imagine watching from outer space as an object moves from the North Pole toward the equator. The object appears to move straight toward the south. Now, as observers on the Earth's surface, we watch the same object. However, since we rotate along with the Earth (in an easterly direction), under the path of the object, it appears to us to veer in a curve toward the right (in a westerly direction), with respect to its direction of movement.

The Coriolis force manifests itself in a number of ways, from riverbanks that erode deeper on one side than the other to winds that rotate counterclockwise around low-pressure areas in the Northern Hemisphere and clockwise in the Southern Hemisphere. The large-scale circulation of the atmosphere and ocean is strongly affected by the Coriolis force. Useful graphical animations of the Coriolis force have been presented by a number of authors (e.g. *http://www.windpower.dk* or *http://satftp.soest.hawaii.edu/ocn620/coriolis/*).

At about 30° latitude, it sinks and flows southward again at lower altitudes. To replace the sinking air mass, air is drawn from polar latitudes, flows south at high altitudes, sinks, and then flows poleward near the surface. These patterns are modified by the Coriolis force, which pushes the circulation clockwise in the Northern Hemisphere and counterclockwise in the Southern Hemisphere (Box 1.2). This results in major surface wind patterns at lower altitudes including, for example, in the Northern Hemisphere, the

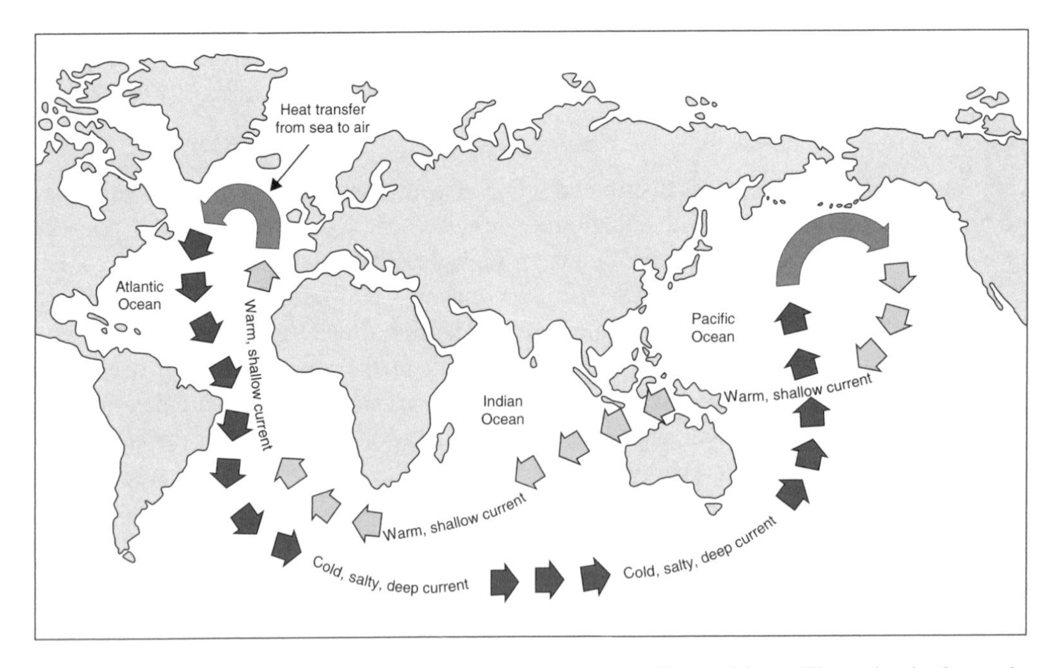

Fig. 1.6 The oceanic conveyor belt transports huge quantities of stored heat. The Atlantic Ocean loses more water by evaporation than it receives from runoff. The resulting saline dense water sinks in the North Atlantic and flows southward near the bottom. Carrying 20 times more water than all the world's rivers combined, it flows near the bottom, back to the Pacific, where it rises to the surface (upwells). The upwelled water then flows along the surface as a warm shallow current back to the Atlantic (From Hileman B 1989. Global warming. *Chemical and Engineering News* **19**: 25–40).

temperate westerly and northeast trade winds (Figure 1.5).

In the ocean, wind patterns, along with density (salinity) differences drive the major circulation currents such as the Gulf Stream in the North Atlantic and the Koroshio Current in the North Pacific. Huge quantities of heat are transported with the surface currents from south to north in the Western Atlantic and Pacific Oceans by what has been termed the *oceanic conveyor belt* (Figure 1.6). Thus, major ocean and atmosphere circulation patterns are closely tied to the Earth's heat balance and any disruption of these patterns could cause rapid changes in global climate.

Greenhouse Gases

Historically, greenhouse gas concentrations in the Earth's atmosphere have undergone natural changes over time and those changes have been closely followed by changes in climate. Warmer periods were associated with higher atmospheric greenhouse gas concentrations and cooler periods with lower greenhouse gas concentrations. However, those changes were part of natural cycles and occurred over periods of tens of thousands to millions of years (Chapter 2). Recent human-induced changes in atmospheric chemistry have occurred over decades (Ramanathan 1988). When referring to the postindustrial era, scientists generally use the term *climate change* in the way defined by The UN Framework Convention on Climate Change. Thus, "climate change" is a change of climate that is attributed directly or indirectly to human activity that alters the composition of the global atmosphere and which is, in addition to natural climate variability, observed over comparable time periods.

Human activities generate several different greenhouse gases that contribute to climatic change. To determine the individual and cumulative effects of these gases on the Earth's climate, we need to examine their total quantity, their natural and human sources to the atmosphere, their rates of loss to natural sinks, their past and projected rates of increase, and their individual and cumulative heating capacities (Table 1.1).

Water vapor traps heat in the atmosphere and makes the greatest contribution to the greenhouse effect. Its level in the atmosphere is not directly the result of human activities. However, because warmer air can hold more water vapor, an increase in the Earth's temperature resulting from other greenhouse gases produces a "positive feedback," that is, more warming means more water vapor in the atmosphere, which in turn contributes to further warming (Chapter 3).

Carbon dioxide is a natural component of the atmosphere and is very biologically reactive. It can be reduced to organic carbon biomass through photosynthetic uptake in plants and, through biological oxidation (respiration), converted back to gaseous CO_2 and returned to the atmosphere. Major natural sources to the atmosphere are animal respiration, microbial breakdown of dead organic matter and soil carbon, and ocean to atmosphere exchange (flux). Sinks include photosynthetic uptake by plants and atmosphere to ocean flux. These natural cycles maintained the atmospheric concentration of CO_2 at about 280 ± 10 ppmv (parts per million by volume) for several thousand years prior to industrialization in the mid-nineteenth century.

During the past 150 years, and especially during the last few decades, humans greatly increased the concentration of atmospheric CO_2. Huge reservoirs of carbon, stored for millions of years as fossilized organic carbon (coal, oil, and gas) in the Earth's crust, have been removed and burned for fuel. When carbon fuels burn, they combine with

Table 1.1 The main greenhouse gases (From UNEP 2001. *United Nations Environment Programme: Introduction to Climate Change*. Accessed April 17, 2001 from *www.grida.no/climate/vital/intro.htm*).

Greenhouse gases	Chemical formula	Preindustrial concentration (ppbv)	Concentration in 1994 (ppbv)	Atmospheric lifetime (years)[a]	Anthropogenic sources	Global Warming Potential (GWP)[b]
Carbon dioxide	CO_2	278,000	358,000	Variable	Fossil-fuel combustion Land-use conversion Cement production	1
Methane	CH_4	700	1,721	12.2 ± 3	Fossil fuels Rice paddies Waste dumps Livestock	21[c]
Nitrous oxide	N_2O	275	311	120	Fertilizer Industrial processes Combustion	310
CFC-12	CCl_2F_2	0	0.503	102	Liquid coolants Foams	6,200–7,100[d]
HCFC-22	$CHClF_2$	0	0.105	12.1	Liquid coolants	1,300–1,400[d]
Perfluoro-methane	CF_4	0	0.070	50,000	Production of aluminum	6,500
Sulfur hexa-fluoride	SF_6	0	0.032	3,200	Dielectric fluid	23,900

Note: ppbv = parts per billion volume; 1 ppbv of CO_2 in the Earth's atmosphere is equivalent to 2.13 million metric tons of carbon (*www.cdiac.esd.ornl.gov*, accessed on December 10, 2000).

[a] No single lifetime for CO_2 can be defined because of the different rates of uptake by different sink processes.

[b] Global Warming Potential (GWP) for 100-year time horizon.

[c] Includes indirect effects of tropospheric ozone production and stratospheric water vapor production.

[d] Net GWP (i.e. including the indirect effect due to ozone depletion).

atmospheric oxygen to produce carbon dioxide, which enters the atmosphere (Figure 1.7). Globally, more than 80% of human CO_2 emissions come from transportation and industrial sources. The remaining 20% comes primarily from deforestation and biomass burning. A forest stores about 100 tons of carbon per acre and about half the world's forest was destroyed in the last half of the twentieth century. Carbonate minerals used in cement production also release CO_2 to the atmosphere. These sources altogether contribute 6.5 billion tons or gigatons of carbon (GtC) to the atmosphere each year (Figure 1.7). The largest emitters of CO_2 are the United States, China, and the Russian Federation (Figure 1.8). Furthermore, the rate of addition to the atmosphere from these sources exceeds the rate of loss to major CO_2 sinks by about 3.3 GtC per year (Box 1.3, Figure 1.9). Thus, the atmospheric concentration of CO_2 continues to increase.

The GWP of atmospheric carbon dioxide is not a new concept, but dates back to the nineteenth century. In 1896, the Swedish chemist Svante Arrhenius estimated that fossil-fuel burning would result in a doubling of the atmospheric CO_2 concentration 3,000 years in the future (Arrhenius 1896). He correctly described the process, but severely underestimated the time frame in which humans could double the atmospheric CO_2 concentration. By 1938 some scientists concluded that human combustion of fossil fuel

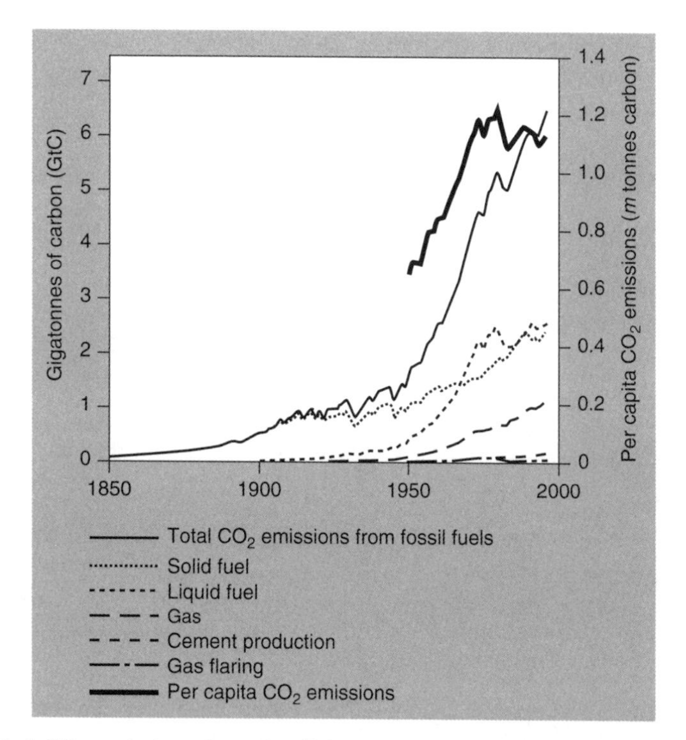

Fig. 1.7 Total global CO_2 emissions from fossil-fuel combustion and cement production and average global per capita emissions (Adapted from Marland G, Boden TA and Andres RJ 2002. Global, regional, and national fossil fuel CO_2 emissions. *In Trends: A Compendium of Data on Global Change*. Carbon Dioxide Information Analysis Center, Oak Ridge National Laboratory, US Department of Energy, Oak Ridge, Tenn., USA).

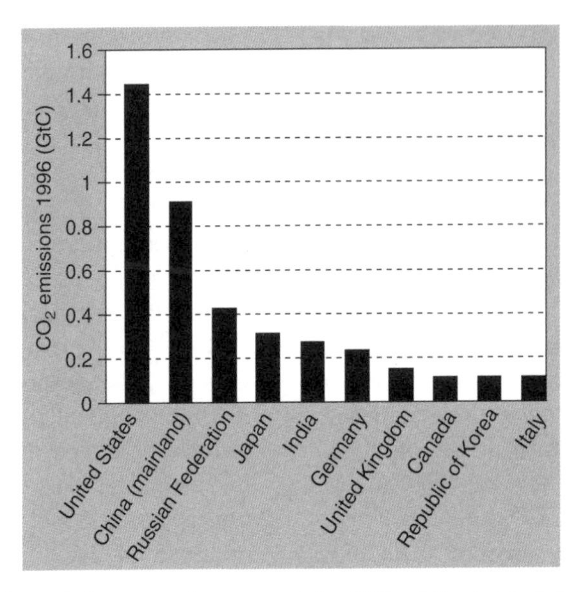

Fig. 1.8 The top 10 CO_2-emitting countries account for 66% of the total global CO_2 emissions. The US is number one accounting for 23% of the global total (Data from Marland G, Boden TA and Andres RJ 2002. Global, regional, and national fossil fuel CO_2 emissions. *In Trends: A Compendium of Data on Global Change*. Carbon Dioxide Information Analysis Center, Oak Ridge National Laboratory, US Department of Energy, Oak Ridge, Tenn., USA).

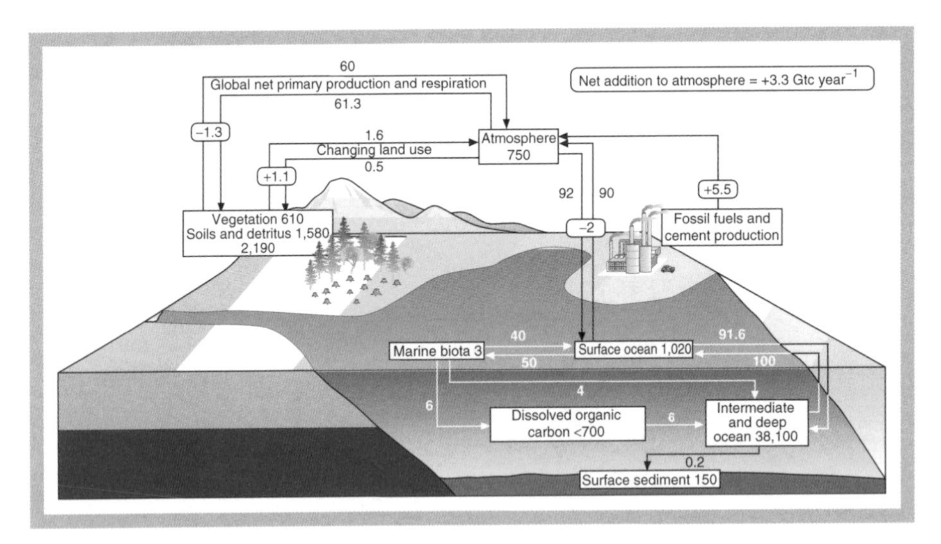

Fig. 1.9 The global carbon cycle showing reservoirs (boxes) in gigatons of carbon (GtC) and fluxes (arrows) in GtC year^{-1}. Anthropogenic emissions are adding about 6.6 GtC year^{-1} to the atmospheric reservoir of 750 GtC. Because of natural sinks, the net contribution to the atmosphere is about 3.3 GtC year^{-2} (From Schimel D, Alves D, Enting I, Heimann M, Joos F, et al. 1996. Radiative forcing of climate change. In: Houghton JT, Filho LGM, Callander BA, Harris N, Kattenberg A, Maskell K, eds *Climate Change 1995: The Science of Climate Change*. Intergovernmental Panel on Climate Change. Cambridge: Cambridge University Press, p. 77. Reproduced by permission of the Intergovernmental Panel on Climate Change).

Box 1.3 The global carbon cycle

The Earth's carbon exists in a number of reservoirs. The largest is the intermediate and deep water of the ocean that contains about 38,100 gigatons of carbon (GtC) (Figure 1.9). Terrestrial vegetation and soils represent the second largest reservoir, totaling 2,190 GtC. The atmosphere contains about 750 GtC. Carbon is the chemical building block of life. Through the process of photosynthesis, terrestrial plants remove carbon dioxide (CO_2) gas from the atmosphere and marine phytoplankton take up dissolved carbon from seawater. Both reduce it to organic carbon, a temporary living carbon reservoir. Animals, and most microbes, derive their energy from the breakdown of organic carbon and, through respiration, release CO_2 back to the atmosphere. Because the solubility of CO_2 in water is inversely proportional to temperature, the cold high-latitude regions of the ocean absorb atmospheric CO_2, while warm tropical waters release it back to the atmosphere. This movement of carbon between atmosphere, ocean, and land constitutes the natural carbon cycle. Undisturbed, this cycle has maintained atmospheric CO_2 levels at a relatively stable level for millennia.

However, there is another large carbon reservoir – one being exploited by humans. Dense forests and swamps in past eras left their stored carbon as deposits preserved in the Earth's crust. The industrial era, beginning in the nineteenth century, was and continues to be driven by the burning of huge quantities of this fossil carbon (coal, oil, and gas). As the carbon is burned (oxidized), it is converted to CO_2, water, and energy. Thus, carbon that remained undisturbed for millions of years in the Earth's crust is being injected into the atmosphere as CO_2 in what represents an instant in geologic time. In addition, carbon stored as biomass in trees and vegetation is being transferred to the atmosphere by biomass burning and land clearing. The land and the ocean store about half of the carbon emitted annually by fossil-fuel combustion and industrial activity; the other half is accumulating in the atmosphere. The net result is that human activities are increasing the concentration of heat-trapping CO_2 in the atmosphere by 3.3 GtC each year (Figure 1.9).

Although the quantities of carbon described here are generally accurate, some uncertainty remains and research continues in an effort to refine carbon cycle models. For example, some research suggests that more carbon is taken up from the atmosphere than can be accounted for, implicating a "missing sink." Satellites and other techniques are being used to map the global distribution of terrestrial and oceanic plant productivity (USGCRP 2002). Studies using NASA's Sea-Viewing Wide Field-of-view Sensor (SeaWiFS) indicate that global net photosynthetic carbon uptake may be slightly greater than previously thought (111 to 117 GtC per year) with about half being accounted for by marine phytoplankton (Behrenfeld et al. 2001).

was already leading to a significant increase in atmospheric CO_2 and global average temperature (Callendar 1938).

By the end of the twentieth century, the atmospheric concentration of CO_2 had risen to over 367 ppmv – an increase of 31% above its preindustrial level – and it continues to increase exponentially at about 0.5% per year (Figure 1.10). The present atmospheric CO_2 concentration has not been exceeded during the past 420,000 or perhaps even 20 million years (Houghton et al. 2001). Even if current

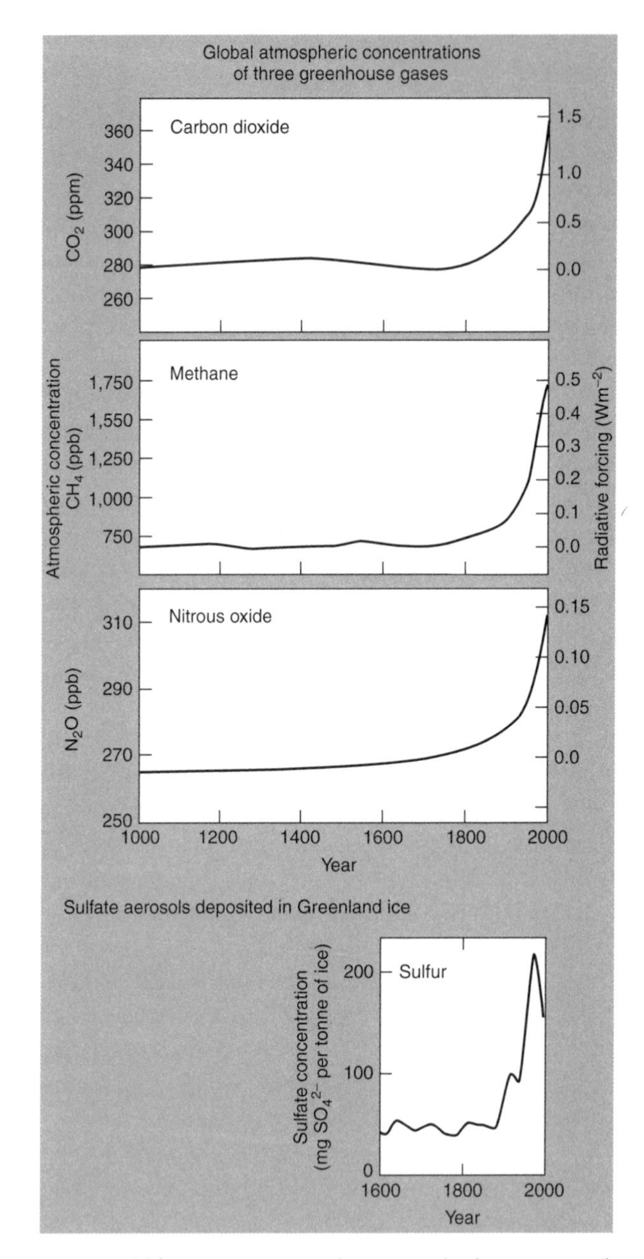

Fig. 1.10 (a) During the past 1,000 years or more, the atmospheric concentrations of greenhouse gases remained relatively constant. However, since the beginning of the industrial era in the nineteenth century, human activities have led to an exponential increase in greenhouse gases. (b) Anthropogenic sulfate concentrations in Greenland ice cores. Sulfates cause negative radiative forcing (cooling). For both (a) and (b), earlier data are based on polar ice core and other paleoclimatological evidence (Chapter 2). Later, mostly twentieth-century, values are from actual chemical analysis (From IPCC 2000. *Summary for Policymakers: The Science of Climate Change.* Intergovernmental Panel on Climate Change. IPCC Working Group I, 26 Feb. Available from: *http://www.ipcc.ch/pub/* Reproduced by permission of Intergovernmental Panel on Climate Change).

CO_2 emissions are reduced and maintained at or near 1994 rates, the atmospheric concentration will reach 500 ppmv (nearly double the preindustrial level) by the end of this century – far sooner than Arrhenius could have imagined. The major long-term reservoir (sink) for CO_2 is the deep ocean. Carbon dioxide produced today will take more than 100 years to be absorbed by this reservoir. Thus, even if all emissions ceased today, atmospheric CO_2 would remain above its preindustrial level for the next 100 to 300 years.

Fossil fuels are nonrenewable and their supply is finite. However, current supplies are abundant, relatively inexpensive, and could last for another 40 to 200 years (Table 1.2). If we continue to burn the carbon remaining in tropical rainforests, oil, gas, and coal reserves, we could more than quadruple the concentration of atmospheric CO_2 in the next few centuries.

Methane gas (CH_4) is produced by the microbial breakdown of organic matter in the absence of oxygen. Natural wetland soils, swamps, and some coastal sediments release significant quantities of CH_4 to the atmosphere. In the atmosphere it can combine with hydroxyl radicals (OH^-) to form carbon monoxide (CO). Its atmospheric concentration has increased by 150% since 1750 and is increasing rapidly by about 1.1% per year (Figure 1.10). About half the current methane emissions are from anthropogenic (human-produced) sources. These sources include livestock production (incomplete digestion of food), wetland rice cultivation, solid waste landfills, and coal, oil, and gas production (Figure 1.11). However, global emission rates appear variable and are difficult to quantify exactly (Houghton et al. 2001).

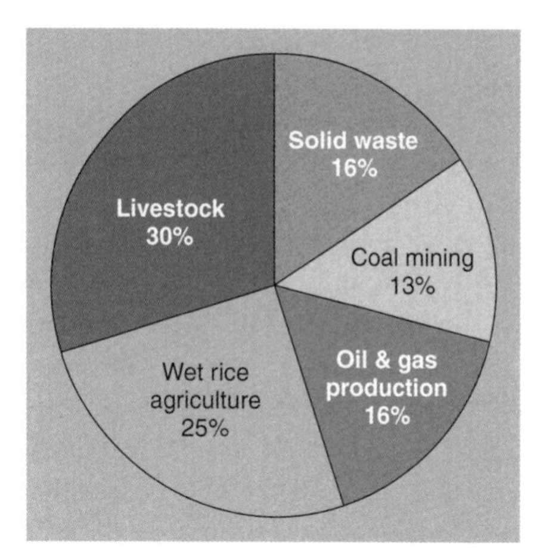

Fig. 1.11 Anthropogenic sources of methane. Total = 270 million tones per year (Based on 1996–1997 data) (WRI 2002).

Nitrous oxide (N_2O) originates from the microbial breakdown of agricultural fertilizers, fossil-fuel combustion, and biomass burning. Coal combustion is a major contributor of N_2O to the atmosphere. N_2O has a long atmospheric lifetime (170 years). Its atmospheric

Table 1.2 World reserves of fossil fuel (Data from Dresselhaus MS and Thomas IL 1991. Alternative energy technologies. *Nature* **414**: 332–337).

Source	World reserves	World consumption rate	Approximate lifetime (years) at current consumption rate[a]
Oil	1.6×10^{14} L	1.2×10^{10} L day^{-1}	37
Natural gas	1.4×10^{14} m^3	2.4×10^{12} m^3 year^{-1}	58
Coal	9.1×10^{11} tonnes	4.5×10^9 tonnes year^{-1}	202

[a]Assuming no new discoveries.

concentration has increased since the preindustrial era by 16%, and it continues to increase by about 0.25% per year. It makes a significant contribution to the overall global warming (Figure 1.10).

Chlorofluorocarbons (CFCs) and *hydrochlorofluorocarbons* (HCFCs) are a relatively inert class of manufactured industrial compounds containing carbon, fluorine, and chlorine atoms. They are used as coolants in refrigerators and air conditioners, and in foam insulation, aerosol sprays, and industrial cleaning solvents. These compounds escape to the atmosphere where they destroy the stratospheric ozone layer that shields the Earth from harmful ultraviolet radiation. Their role in ozone depletion led to the first comprehensive international environmental treaty – the Montreal Protocol – to phase out the use of chlorofluorocarbons. However, CFCs and HCFCs are also greenhouse gases. The atmospheric concentration of CFCs has increased rapidly since the 1960s. Although they are involved in the destruction of the stratospheric ozone layer (Box 1.4), which leads to some cooling, they still make an overall positive contribution to greenhouse warming. The Montreal Protocol now restricts their use. However, because of their long lifetimes in the atmosphere (60 to >100 years), they must

Box 1.4 Stratospheric ozone

The stratospheric ozone layer extends upward from about 10 to 30 miles and protects life on Earth from the Sun's harmful ultraviolet-b radiation (UV-b, 280- to 320-nm wavelength). Ozone occurs naturally in the stratosphere and is produced and destroyed at a constant rate. But, man-made chemicals, CFCs, and halons (used in coolants, foaming agents, fire extinguishers, and solvents) are gradually destroying this "good" ozone. These ozone-depleting substances degrade slowly and can remain intact for many years as they move through the troposphere until they reach the stratosphere. There they are broken down by the intensity of the Sun's ultraviolet rays and release chlorine and bromine molecules, which destroy "good" ozone. One chlorine or bromine molecule can destroy 100,000 ozone molecules, causing ozone to disappear much faster than nature can replace it. It can take years for ozone-depleting chemicals to reach the stratosphere, and even though we have reduced the use of many CFCs, their impact from years past is just starting to affect the ozone layer. Substances released into the air today will contribute to ozone destruction well into the future. Satellite observations indicate a worldwide thinning of the protective ozone layer. The most noticeable losses occur over the North and South Poles because ozone depletion accelerates in extremely cold weather conditions. As the stratospheric ozone layer is depleted, higher UV-b levels reach the Earth's surface. Increased UV-b can lead to more cases of skin cancer, cataracts, and impaired immune systems. Damage to UV-b-sensitive crops, such as soybeans, reduces yield. High-altitude ozone depletion is suspected to cause decreases in phytoplankton, a plant that grows in the ocean. Phytoplankton are an important link in the marine food chain and, therefore, food populations could decline. Because plants "breathe in" carbon dioxide and "breathe out" oxygen, carbon dioxide levels in the air could also increase. Increased UV-b radiation can be instrumental in forming more ground-level (tropospheric) or "bad" ozone pollution (Box 10.1).

be considered as significant greenhouse gases. Also, hydrofluorocarbons (HFCs), a CFC substitute, and related chemicals [perfluorocarbon (PFC) and sulfur hexafluoride (SF6)] currently contribute little to warming, but their increasing use could contribute several percent to warming during the twenty-first century (IPCC 2000).

Tropospheric ozone (O_3) – motor vehicle emissions are the major source of this greenhouse gas. On clear warm days with a stable atmosphere, vehicle combustion hydrocarbons and nitrogen oxides undergo a photochemical reaction to produce a hazy air pollution condition (smog) with high concentrations of O_3 (Box 10.1). The atmospheric concentration increased an estimated 20 to 50% during the twentieth century and continues to increase at about 1% per year (Beardsley 1992). In the atmosphere, chemical reaction with hydroxyl radicals OH^- results in loss of O_3; however, as a result of other reactions, increasing atmospheric CO_2 will probably decrease this removal process. Globally, the degree of warming due to O_3 is not well known, but believed to be on the order of 15% of the total warming. Tropospheric ozone (bad ozone) is not to be confused with the natural stratospheric ozone layer (good ozone) that protects the Earth from excess damaging ultraviolet radiation (Box 1.4).

Aerosols are small microscopic particles resulting from fossil-fuel and biomass combustion, and ore smelting. They are formed largely from sulfur, a constituent of some fuels, particularly some high-sulfur coal and oil. Sulfate aerosols increase the acidity of the atmosphere and form acid rain. They also reflect solar energy over a broadband, including the infrared, and thus have a negative radiative forcing or cooling effect on the atmosphere. Sulfate aerosols, unlike the greenhouse gases discussed above, have a short lifetime in the atmosphere (days to weeks). Globally, sulfate aerosols may be responsible for counteracting 20 to 30% of human-induced greenhouse warming. In some regions of the industrialized Northern Hemisphere, the sulfate-induced cooling appears to be great enough to completely offset the current warming effect of greenhouse gases. Natural sources of aerosols such as volcanic eruptions can also inject particles into the atmosphere, resulting in temporary global-scale cooling events, lasting months to several years.

Other greenhouse gases in total account for approximately 9% the total net warming. These include carbon monoxide (CO) and nitrogen oxides (NO_x) – both largely from fossil-fuel and biomass combustion.

Black carbon (soot), from the incomplete combustion of fossil fuel, may contribute substantially to greenhouse warming, at least on a regional scale (Chameides and Bergin 2002). It is not a gas, but the black particles making up soot absorb solar radiation. Its lifetime in the atmosphere is short compared to most greenhouse gases and its warming potential depends on the source and its fate on the atmosphere. Recent studies are assessing the contribution of black carbon to global warming.

Warming Potentials

The postindustrial increases in greenhouse gases have resulted in an increase in global radiative forcing (warming) of 2.45 watts per square meter ($W\,m^{-2}$). This is only about 1% of the net incoming solar radiation, but it amounts to the energy content of about 1.8 billion tonnes of oil every minute or more than 100 times the world's current rate of commercial energy consumption (UNFCC 2002). Each greenhouse gas contributes to

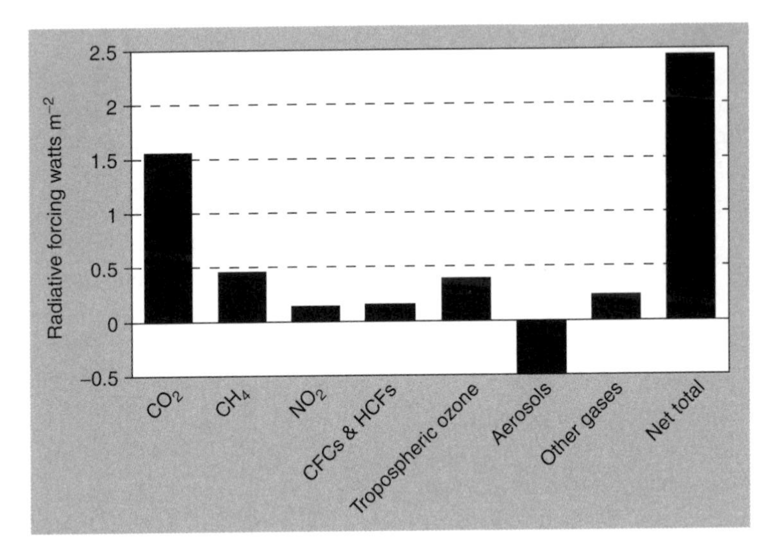

Fig. 1.12 Relative contribution of anthropogenic increases in atmospheric greenhouse gas concentrations to global radiative forcing (warming) (Data from IPCC 2000. *Summary for Policymakers: The Science of Climate Change*. Intergovernmental Panel on Climate Change. IPCC Working Group I, 26 Feb. Available from: *http://www.ipcc.ch/pub/*).

this warming. Equal quantities of different greenhouse gases have widely different warming potentials (Table 1.1). Also, the lifetime of the gas in the atmosphere affects its resultant concentration and warming potential. For example, carbon dioxide, nitrous oxide, and CFCs have average lifetimes of 100 years or more, whereas methane has a lifetime of 5 to 10 years and carbon monoxide only 5 months. Each greenhouse gas has a characteristic per molecule greenhouse effect or warming potential. For example, one molecule of CFC-11 or CFC-12 traps 6 to 7 thousand times more heat than one molecule of CO_2 and one molecule of methane traps 21 times more heat than one molecule of CO_2. However, because CO_2 is much more abundant, about 60% of the current human-induced greenhouse warming results from CO_2, 15 to 20% from methane, and the remaining 20% or so from nitrous oxide, chlorofluorocarbons, and tropospheric ozone (Figure 1.12).

Summary

A thin layer of mixed gases surrounds the Earth. The greenhouse gases (CO_2, CH_4, N_2O, CFCs, and O_3), although less than 0.1% of the atmospheric volume, have a profound influence on the Earth's climate. These gases, most importantly carbon dioxide and methane, allow sunlight to penetrate, but trap outgoing heat. A large quantity of heat, received in the tropics, is redistributed to higher latitudes by major atmospheric and oceanic currents. During the past 150 years, human activities have led to an exponential growth in greenhouse gas emissions. These activities include extracting and burning fossilized carbon (coal, oil, and gas) for fuel, forest clearing and burning, wetland rice cultivation, livestock rearing, solid waste landfilling, and nitrogen fertilization of agriculture. The result has been a major increase in the concentrations of greenhouse gases, with a consequent increase in the warming potential (heat-trapping ability) of the atmosphere.

References

Arrhenius S 1896 On the influence of carbonic acid in the air upon the temperature of the ground. *Philosophical Magazine and Journal of Science, Series 5* **41**(251): 237–276.

Beardsley T 1992 Add ozone to the global warming equation. *Scientific American* **266**(3): 29.

Behrenfeld MJ, Randerson JT, McClain CR, Feldman GC, Los SO, Tucker CJ, et al. 2001 Biospheric primary production during an ENSO transition. *Science* **291**: 2594–2597.

Callendar GS 1938 The artificial production of carbon dioxide and its influence on temperature. *Quarterly Journal of the Royal Meteorological Society* **64**: 223–240.

CDIAC 2000 Carbon Dioxide Information Analysis Center, US Department of Energy, Oak Ridge National Laboratory, *http://cdiac.esd.ornl.gov*.

Chameides WL and Bergin M 2002. Soot takes center stage. *Science* **297**: 2214, 2215.

Crutzen PJ and Ramanathan V 2000 The ascent of atmospheric sciences. *Science* **290**: 299–304.

Graedel TE and Crutzen PJ 1997 *Atmosphere, Climate, and Change*. Scientific American Library, New York: W.H. Freeman, p. 3.

Hileman B 1989 Global warming. *Chemical and Engineering News* **19**: 25–40.

IPCC 2000 *Summary for Policymakers: The Science of Climate Change*. Intergovernmental Panel on Climate Change. IPCC Working Group I, 26 Feb. Available from: *http://www.ipcc.ch/pub/*.

Houghton JT, Ding Y, Griggs DJ, Noguer M, van der Linden PJ, Dai X, et al. 2001 *Climate change 2001: The Scientific Basis*. Intergovernmental Panel on Climate Change. IPCC. Cambridge: Cambridge University Press, p. 39.

Lockwood JG 1979 *Causes of Climate*. New York: Halsted Press, John Wiley & Sons.

Oliver JE and Hidore JJ 2002 *Climatology: An Atmospheric Science*. Upper Saddle River, NJ: Prentice Hall, p. 23.

Ramanathan V 1988 The greenhouse theory of climate change: a test by an inadvertent global experiment. *Science* **240**: 293–299.

Schimel D, Alves D, Enting I, Heimann M, Joos F, et al. 1996 Radiative forcing of climate change. In: Houghton JT, Filho LGM, Callander BA, Harris N, Kattenberg A, Maskell K, eds *Climate Change 1995: The Science of Climate Change*. Intergovernmental Panel on Climate Change. Cambridge: Cambridge University Press, p. 77.

Thurman HV 1991 *Introductory Oceanography* (6th Edition). New York: Macmillan Publishing Co.

Trenberth KE, Houghton JT and Filho LGM 1995 The climate system: overview. In: Houghton JT ed *Climate Change 1995: The Science of Climate Change*. Intergovernmental Panel on Climate Change. Cambridge: Cambridge University Press, pp. 50–64.

UNEP 2001 *United Nations Environment Programme: Introduction to Climate Change*. Accessed April 17, 2001 from: *www.grida.no/climate/vital/intro.htm*.

UNFCC 2002 *United Nations Framework Convention on Climate Change*, *http://unfcc.int/resource*.

USGCRP 2002 *U. S. Global Change Research Program Carbon Cycle Program: An Interagency Partnership*, *http://www.carboncyclescience.gov*.

WRI 2002. World Resources Institute, Washington, DC, *http://wri.igc.org/wri/*.

Chapter 2

Past Climate Change: Lessons from History

"The farther backward you can look,
the farther forward you are likely to see."

Winston Churchill

"From the experience of the past we derive instructive lessons for the future."

John Quincy Adams: US Presidential Inaugural Addresses, 1789

Introduction

Since the Earth formed more than four billion years ago, its climate has periodically shifted from warm to cool and back again – sometimes dramatically. Fossils, preserved in ancient sedimentary rocks, provide evidence that populations of tropical plants and animals once thrived in Europe and elsewhere, where today's climate is cool and temperate. Sheets of glacial ice, a mile thick, covered much of North America only 20,000 years ago. About 8,000 years ago Saharan North Africa, now an arid desert, was home to numerous wetlands and lakes dotted with shoreline human settlements.

More recently, very small climate changes over the past few thousand years have greatly impacted human civilization. Only 1,100 years ago, Viking settlers took advantage of a particularly mild and warm period to colonize a temperate area that they named "Greenland." From there they explored, and for some time settled in, North America. At that same time, the great Mayan civilization of Central America collapsed. Climate change and prolonged drought is one of several competing theories attempting to explain the sudden and mysterious collapse of the Maya (NOAA 2002). Europe in 1816 experienced "the year without a summer," and widespread crop failures resulted in food shortages and political unrest (Gore 1993). In New England in that same year, it snowed in June. The immediate cause of this global cold spell was a series of massive volcanic eruptions in Indonesia, which released huge quantities of dust into

Climate Change: Causes, Effects, and Solutions John T. Hardy
 ISBNs: 0-470-85018-3 (HB); 0-470-85019-1 (PB)

the atmosphere and reduced the amount of sunlight reaching the Earth. Although cooling probably only averaged a degree or so globally, the effects were dramatic. In East Africa, written documents and oral histories covering the past 1,000 years suggest that local civilizations prospered during periods of greater precipitation and suffered during periods of drought (Verschuren et al. 2000).

Past climate changes, more than 150 years before the present (BP), occurred prior to human emissions of greenhouse gases. We can learn much about the future of our climate by examining past climate change. What were the extremes of past climate? How rapidly did past changes occur? What triggered such changes? How did past change affect populations of plants and animals including humans? Can we use past changes to validate and test our predictive models of future climate (Chapter 4)? In this chapter we examine these questions and evaluate the lessons we might learn from the field of paleoclimatology – the study of past climates (e.g. Crowley and North 1991).

Past Climate Change – Six Historic Periods

Climate change occurs over time, and the degree of change depends on the timescale we choose to examine. Changes in the last decade seem insignificant when compared to those over the past million years. Computer models are used to predict how human emissions of greenhouse gases will change our climate over the next century or more (Chapter 4). To put these predictions into context, we must first examine how past climate changes have altered the chemistry of the atmosphere and influenced the plants and animals of the Earth. Climate change can be described in terms of six major time periods (Kutzbach 1989).

First, a major cooling trend occurred more than one billion years ago with the beginning of photosynthetic organisms. The atmosphere had a relatively high concentration of CO_2, and owing to the greenhouse effect, the Earth was correspondingly warm. However, with the appearance of photosynthetic plants, CO_2 was removed from the atmosphere and stored as organic carbon. This reduced the heat-trapping capacity of the atmosphere and led to a major cooling trend.

Second, several hundred million years ago, the Earth experienced a period of intense tectonic activity involving crustal movements, continental drift, and volcanic eruptions. Massive outgassing of CO_2 from the Earth's crust led to an enhanced greenhouse effect with temperatures on average 5 °C warmer than now. There was a general rise in the diversity of life forms, but at least five periods of major species extinctions have occurred since.

Third, beginning about 100 million years ago, tectonic activity subsided. Outgassing of CO_2 decreased, lessening the greenhouse effect of the atmosphere, and the climate cooled once more.

Fourth, during the past million years, shorter-term alternating cool and warm periods occurred on a scale of tens of thousands of years. These are the glacial–interglacial cycles resulting from a natural fluctuating pattern in the orbital configuration of the Earth with respect to the Sun. The elliptical path of the Earth around the Sun (eccentricity) brings it closer to or farther from the Sun every 100,000 years (Figure 2.1a). Also, the Earth like a spinning top wobbles as it rotates on its axis, exposing more or less of each hemisphere to the direct rays of the Sun. It does this, in a process called *precession*, with a periodicity of 20,000 years (Figure 2.1b). Finally, the tilt of the Earth's axis with respect to the Sun (obliquity) changes over a period

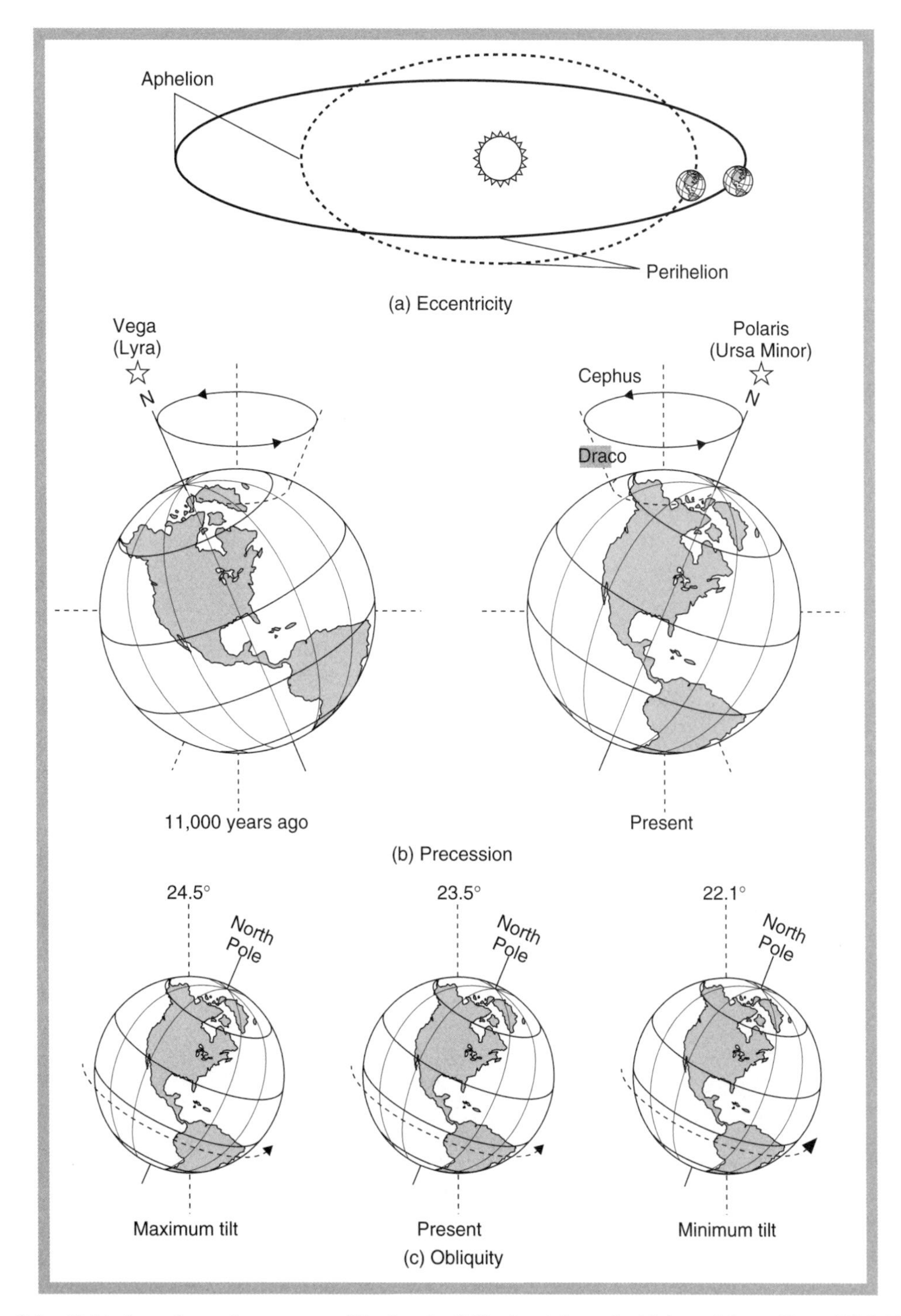

Fig. 2.1 Orbital configurations responsible for the Milankovich cycle (Adapted from Gates DM 1993. *Climate Change and its Biological Consequences*. Sunderland, MA: Sinauer Associates, p. 280).

of 40,000 years (Figure 2.1c). The summation of these three periodicities (Figure 2.2) determines the amount of solar radiation reaching the Earth at a particular time. The resulting cold and warm periods (Figure 2.3a) and glacial retreats and advances are known as Milankovich Cycles after the Serbian mathematician who first proposed the relationship (Box 2.1). The most recent glaciation peaked about 18,000 years ago, and between then and 6,000 years ago the Earth's climate warmed by an average 5 °C (Figure 2.3b).

Fifth, smaller magnitude cycles of 1,000 years or less occur. These cycles may be related to changes in solar activity, but are not well understood (Struiver and Quay 1980). Although small, they probably have significant effects on human civilization. For example, during the "Medieval Warm Epoch," peaking about 1,100 years ago, vineyards thrived in southern England and Vikings crossed through ice-free seas to North America (Figure 2.3c). From about 200 to 600 years ago (1400 to 1800 AD), the "Little Ice Age" brought frequent bitterly cold winters to the temperate regions of the Earth (Figure 2.4). Cold summers led to crop failures and starvation in parts of Europe.

Finally, during just the last 150 years, the Earth's global average temperature has increased by about 0.8 °C, and at higher

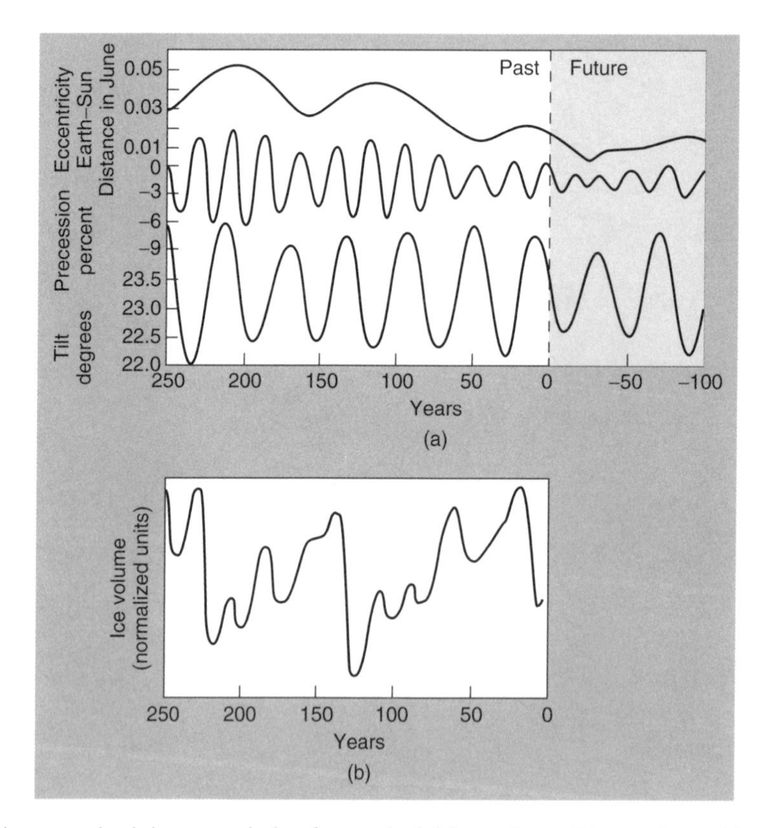

Fig. 2.2 Maximum and minimum periods of past glacial ice volume (b) correlate with periods of low and high solar insolation, resulting from the summation of the three cycles (a) (Adapted from Gates DM 1993. *Climate Change and its Biological Consequences*. Sunderland, MA: Sinauer Associates, p. 280).

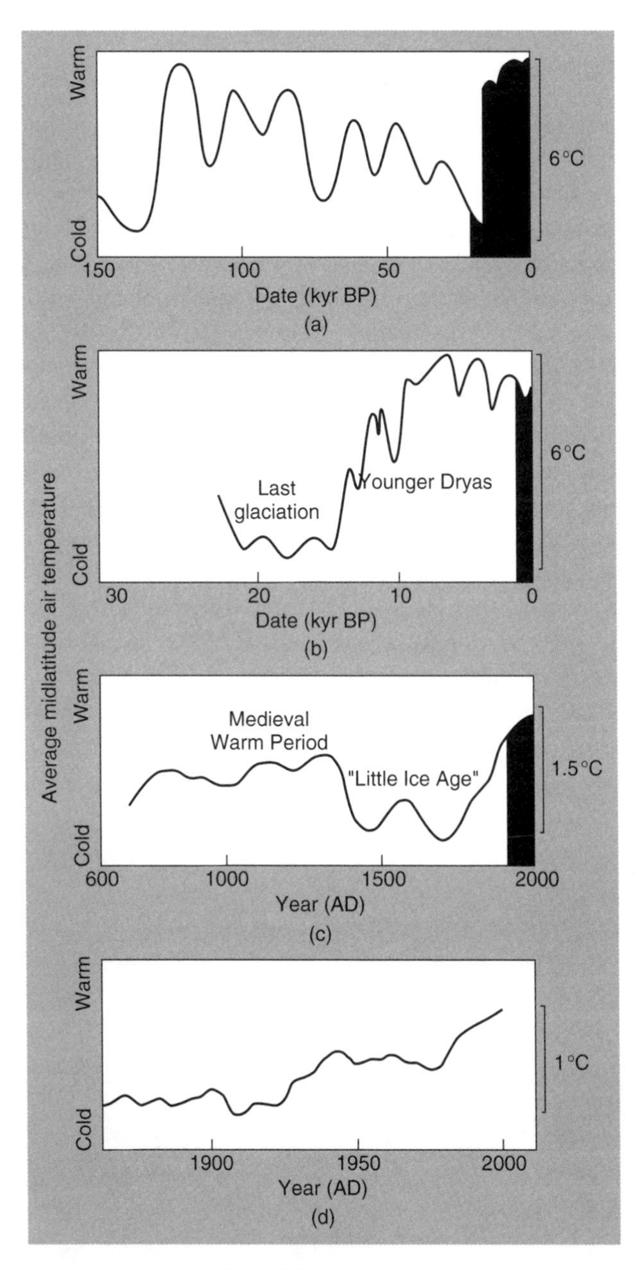

Fig. 2.3 Global temperature changes on four different timescales from decades to hundreds of thousands of years. From the bottom, the shaded area indicates the time segment that is expanded in the display immediately above. Data are derived primarily from (a) isotope ratios in marine plankton and sea-level marine terraces; (b) pollen data and alpine glacier volume; (c) historical reports; (d) instrument measurements (Adapted from Webb III T, Kutzbach J and Street-Perrott FA 1985. 2,000 years of global climate change: palaeoclimatic research plan. In: Malone TF and Roederer JG, eds *Global Change: The Proceedings of a Symposium Sponsored by the ICSU. ICSU (QE1 G51)*. Cambridge: Cambridge University Press, pp. 182–218. Reproduced by permission of the ICSU).

Box 2.1 The Milankovich Cycle

In the 1910s, the Serbian mathematician Milutin Milankovich developed a theory that would eventually explain natural fluctuations in the Earth's climate. He used equations that predict the cyclical variations in the Earth's eccentricity and precession, but went further by incorporating astronomical calculations of the German scientist Ludwig Pilgrim on the obliquity or tilt of the Earth. He also reasoned that summer, rather than winter, temperatures were the main contributors to the growth and decline of the polar icecaps. Finally, he calculated summer radiation curves for key latitudes of 55, 60, and 65 °N that correlated well with evidence from the geologic record. The Earth's orbital position relative to the three cycles of eccentricity, precession, and tilt determine the quantity of solar energy received by the Earth (solar insolation). Periods of low insolation and high insolation correspond to glacial and interglacial periods, respectively (NOAA 2002).

The last ice age, the "Wisconsin," and recent Holocene warming beginning about 18,000 years ago correspond well to the Milankovich Cycle. However, the Cycle does not always correspond to warm and cold periods exactly. For example, 135,000 years ago summer insolation values in the Northern Hemisphere were apparently too low to be responsible for termination of the then prevalent ice age that led to the last warm period, the Eemian. The insolation cycle may be more complex than previously thought or other interacting and complicating factors may be interacting to modify the warming–cooling trends predicted by Milankovich (Karner and Muller 2000).

Fig. 2.4 From 1607 to 1814, during the European "Little Ice Age Period," "frost fairs" were regularly held on London's Thames River, which froze over each winter. Today the Thames is ice-free. Woodcut depicting winter of 1683–1684 (From The Granger Collection, New York).

latitudes has increased by several degrees Celsius. Although small in magnitude, this is a very rapid rate of increase, unprecedented in the Earth's long history.

Thus, the Earth has undergone periodic natural fluctuations in climate of about ± 1 to $6\,^{\circ}$C. We are currently in a warm interglacial period and the Earth is about as warm as it has been for 140,000 years.

Methods of Determining Past Climates and Ecosystems

One million or even ten thousand years ago, humans were not collecting climatological data. So, how do we know what the climate was like long ago? Scientists use a number of techniques, each appropriate for different periods of the past. Fossilized remains of ancient life

Box 2.2 Isotopic temperature and age determinations

Oxygen – Three isotopes of oxygen occur naturally: ^{16}O, ^{17}O, and ^{18}O. Water (H_2O) contains both the light isotope (^{16}O) and the much rarer heavy isotope (^{18}O). These oxygen isotopes can be used to indicate past temperature and water evaporation patterns. When water evaporates, the lighter isotope ($H_2^{16}O$) evaporates at a faster rate. Therefore, the ratio of ^{18}O to ^{16}O in rain, snow, and ice decreases as the air temperature, and thus evaporation, increases. The reverse happens when water condenses and the heavier $H_2^{18}O$ preferentially condenses compared to $H_2^{16}O$. Thus, the ratio of ^{18}O to ^{16}O in lake water is controlled mainly by the balance between evaporation and precipitation.

Carbonates – Foraminifera are tiny animals that form part of the marine plankton community. When living, they deposit calcium carbonate ($CaCO_3$) from seawater to form their cell walls. When they die, they settle and form ocean floor deposits. In core samples from the ocean floor, the deeper ocean sediments contain the oldest deposits. The $^{18}O/^{16}O$ ratio in the $CaCO_3$ shells of the foraminifera indicates the isotopic composition and hence seawater temperature at the time they lived. Similarly, reef-building corals deposit calcium carbonate skeletons, and cores from old reefs can reveal the temperature of the ocean at the time the animals lived.

Alkenones – some marine phytoplankton produce straight chain hydrocarbons called alkenones. The colder the temperature where the phytoplankton lives, the greater the number of double bonds in the alkenone chains of their cell membranes. When the plankton die, they settle to the ocean bottom and their alkenones are incorporated in the sediment. Thus, analysis of the ratio of different alkenones in ocean sediments indicates the past temperature of the seawater.

Carbon – Carbon-14 is a radioactive isotope that occurs naturally in the atmosphere in very low concentrations, and through photosynthesis, is incorporated in plants along with the much more abundant and stable ^{12}C form of carbon. When an organism dies and is no longer accumulating carbon, the ^{14}C slowly decays to ^{12}C over time with a half-life (the time for 1/2 of the original amount to decay) of 5,730 years. Thus, the ratio of $^{14}C/^{12}C$ declines over time and can be used to date the age of the animal or plant remains.

forms, preserved in rock formations, can indicate the types of species inhabiting a region millions of years ago. Their relationship to present-day tropical or temperate species, or to desert or rainforest species, along with information on continental drift, can suggest the type of climate that existed at that time.

Oxygen isotope ratios are used to determine past temperatures and rates of precipitation and evaporation that occurred tens of thousands to a million years ago (Box 2.2). Oxygen bubbles trapped in ancient polar ice deposits indicate past climate conditions. Samples of calcium carbonate, deposited by living organisms such as corals and marine plankton, and subsequently preserved in reef structures or sediments, can reveal past ocean temperatures during the life of the organism (Figure 2.5). For example, estimates of past sea-surface temperatures from fossil coral indicate that 10,200 years ago, waters of the tropical southwestern Pacific Ocean were 5 °C colder than today (Beck et al. 1992).

Fig. 2.5 Scientists drilling a core from a large colony of the coral *Porites lobata* at Clipperton Atoll in the Pacific. The core will be sectioned, age-dated, and the oxygen isotope ratios preserved in the $CaCO_3$ skeleton at the time of live deposition, which is used to construct a record of past ocean temperatures (From NOAA 2002. National Oceanic and Atmospheric Administration Paleoclimatology Program, *http://www.ngdc.noaa.gov/paleo/slides*).

Polar ice cores have provided invaluable insights to climate over the past 150,000 years. Each year, snow deposits to form surface ice, which is then buried the next year. The ice thus provides a stratigraphic record from the recent (shallow) to the distant past (deep). In 1982, Russian scientists, using techniques similar to drilling an oil well, removed cores of ice from the Antarctic ice sheet down to a depth of 2,083 m. Sections of the "Vostok" core and from subsequent deep ice cores provide material for several types of analyses (Figure 2.6) (Raynaud et al. 1993). Analysis of the concentrations of CO_2 and methane (CH_4) in the air bubbles trapped within the ice indicated that of these the greenhouse gases showed similar parallel fluctuations in atmospheric concentration over different depths (past times) (Figure 2.7).

Fig. 2.6 Removing ice from a core just recovered from 90 m deep at Siple Dome, Antarctica. The drill is on the sled beside the core. In the background is the support tower for a larger drill that can recover cores to a depth of 1,000 m (From Taylor K 1999. Rapid climate change. *American Scientist* **87**: 320–327).

Oxygen isotope ratios in the ice indicate past temperature variations of about ±2 to 3 °C. During the warming period of a Milankovich cycle, greenhouse gases are slowly released from natural reservoirs into the atmosphere. For example, CO_2 solubility in water decreases with increasing temperature, so in warmer waters more remains in the atmosphere. Also, as soils warm, the rate of microbial breakdown of soil organic matter increases, releasing CO_2 and N_2O into the atmosphere. Melting of methane clathrates

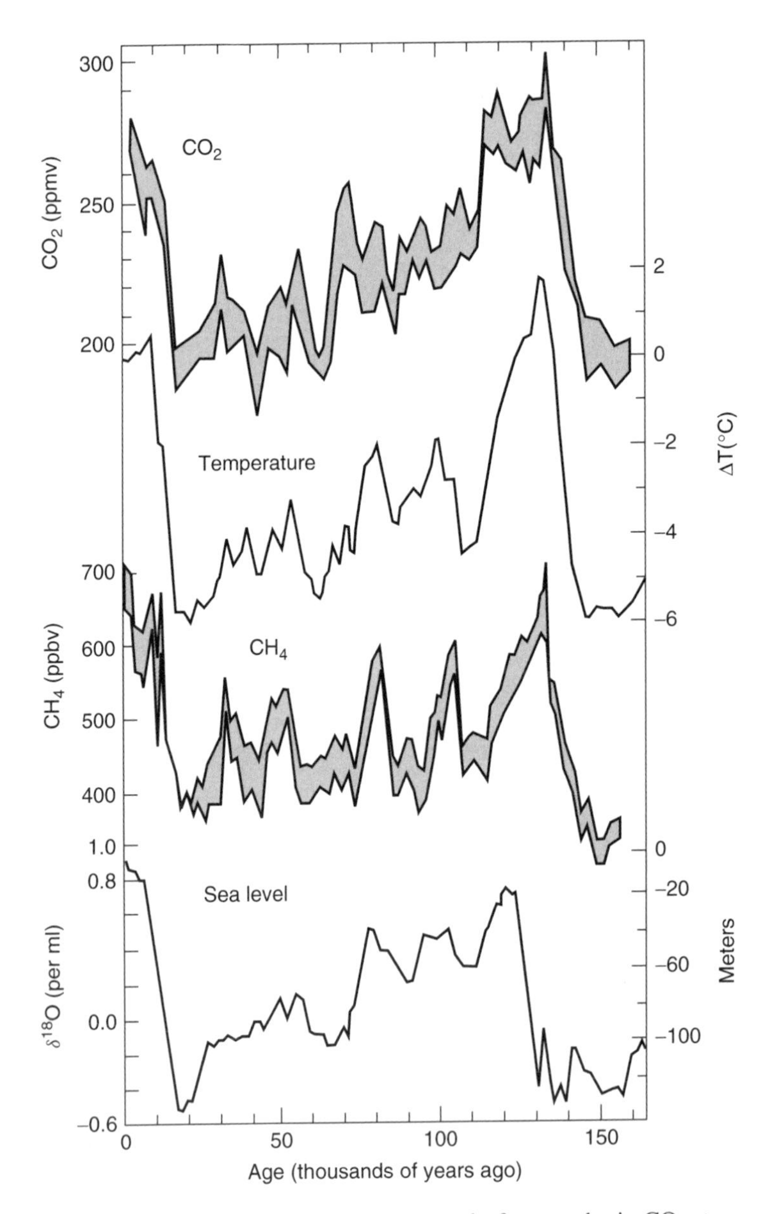

Fig. 2.7 Vostok Ice Core data showing 160,000-year record of atmospheric CO_2, temperature, CH_4, and sea level.

(solid methane) found in wetland sediments and permafrost, released gaseous methane to the atmosphere and contributed further to the warming. Thus, concentrations of the greenhouse gases CO_2, CH_4, and N_2O were lower during past ice ages and increased during deglaciation. These processes lag behind the increase in temperature so that peak atmospheric concentrations of greenhouse gases follow temperature maxima by about 1,000 years. Also, during warmer ice-free periods, the darker heat-absorbing surface of the Earth is revealed, and reflection of solar energy decreases. This reduction in reflectivity (reduced albedo) may account for up to half the interglacial warming.

Other deep core projects in Antarctica and Greenland confirm that global temperatures vary by ± 2 to $3\,^{\circ}C$ over periods of tens of thousands of years. Generally, past warm periods correspond to periods of high atmospheric CO_2 and CH_4 levels and vice versa. However, the cycles are complex and the relationship between past climates and CO_2 alone is sometimes weak (Velzer et al. 2000).

For recent records (less than 75,000 years), remains of once-living plants or animals can be dated using isotopes of carbon (Box 2.2). The growth rings of tree species that live to very old ages can be used to infer climatic conditions of a region over a thousand years ago. For example, in cores collected from live Huon pine trees in Tasmania, more recent outer rings deposited from 1896 to 1988 were correlated with actual measured temperature records. These correlations were then used to reconstruct the temperature records at which older inner rings were deposited. Results indicate that no period over the past 1,089 years has been as warm as that since 1965 (Cook et al. 1991). In another study annual temperature patterns revealed by a 2,000-year tree-ring record in the Southern Sierra Nevada Mountains of California correlate well with the Medieval Warm Epoch (800 to 1200 AD) and the Little Ice Age (1400 to 1800 AD) (Scuderi 1993).

Samples of plant pollen can reveal the nature of past climates. Plant pollen is generally resistant to decay. Pollen falling on the surface of lakes and bogs becomes buried in the sediments. When cores are extracted from the bottom sediments of lakes and sectioned, the older pollen deposits are deep and the recent deposits are shallow. Isotope dating, combined with microscopic enumeration of the abundance of pollen of different species, can provide a detailed picture of the plants inhabiting a region for hundreds or, in some cases, tens of thousands of years. Assuming the species preferred the same climate then as they do now, one can infer much about the regional climate over time (Figure 2.8). For example, the abundance of tree pollen of different tree species found in lake sediments in Southern Ontario matches model predictions of changes in forest composition during a $2\,^{\circ}C$ cooling of the "Little Ice Age" (1200 to 1850 AD) (Campbell and McAndrews 1993).

For recent historical periods, human records often describe the prevailing climatic conditions. Useful written records extend back hundreds to thousands of years. For example, study of private correspondence of the Jesuit religious order in Spain during the period 1634 to 1648 suggests intense rainfall and cold – a pattern confirming the "Little Ice Age" climate pattern (Rodrigo et al. 1998). Instrumental measurements of rainfall and temperature data for some areas (such as Central England) date back several hundred years, although instrumental records from more than 100 years ago may be significantly less accurate than more recent measurements. In summary, different methods are used to reveal

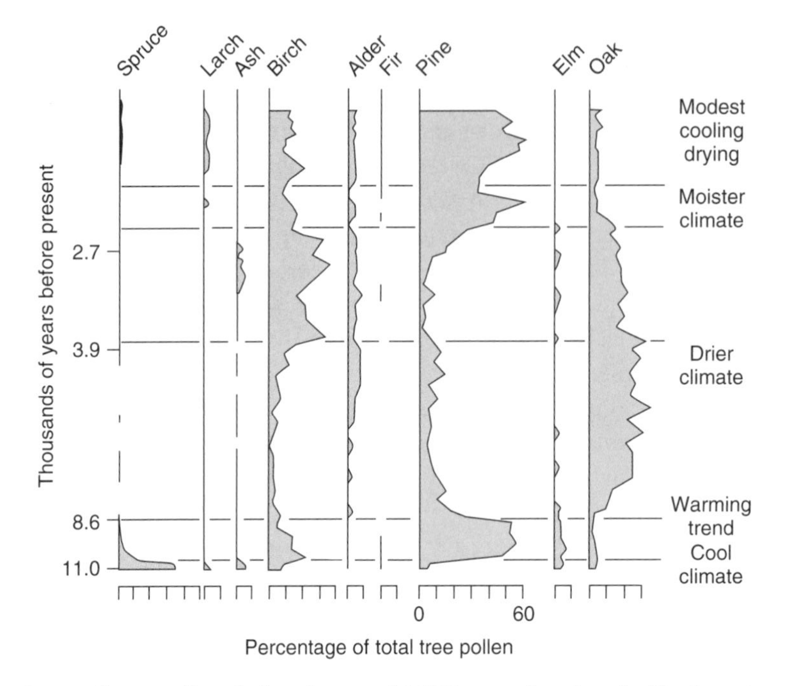

Fig. 2.8 Abundance of tree pollen during the past 11,000 years in a bog in Northwestern Minnesota, USA (Adapted from McAndrews JH 1967. Pollen analysis and vegetation history of the Itasca region, Minnesota. In: Cushing EJ and Wright Jr HE, eds *Quaternary Paleoecology. Volume 7 of the Proceedings of the 7th Congress of the International Association for Quaternary Research*. University of Minnesota Press, pp. 218–236).

climatic conditions over different past time periods (Figure 2.9).

Rapid Climate Change

Paleoclimatologists, using oxygen isotope ratios from ice cores in Greenland to indicate past temperatures, discovered an intriguing puzzle. Following the last ice age, the Earth's climate was warming and polar ice was retreating. Then, about 14,000 years ago, the warming suddenly stopped. Within 1,500 years, the climate cooled by 6 °C and the glaciers returned. Then, 11,650 years ago, the Earth experienced an unprecedented rapid warming. What could possibly cause such a rapid climatic change? Further investigation finally linked this event to changes in circulation of the oceanic conveyor belt.

As the polar ice retreated, following the Wisconsin glaciation, the melting ice and freshwater inflow lowered the salinity and hence the density of the North Atlantic. With a drop in ocean salinity, the less dense North Atlantic waters stopped sinking. The oceanic conveyer belt (Chapter 1) that transfers heat from the tropical Atlantic northward suddenly stopped, and North America and Europe experienced a reappearance of cold temperatures. This cooling is known as the "Younger Dryas Event" after a cold-thriving *Dryas* plant that invaded Europe at that time. The cooling and resultant decrease in oceanic evaporation lowered precipitation (Kerr 1993) and led to prolonged droughts in the African Sahel

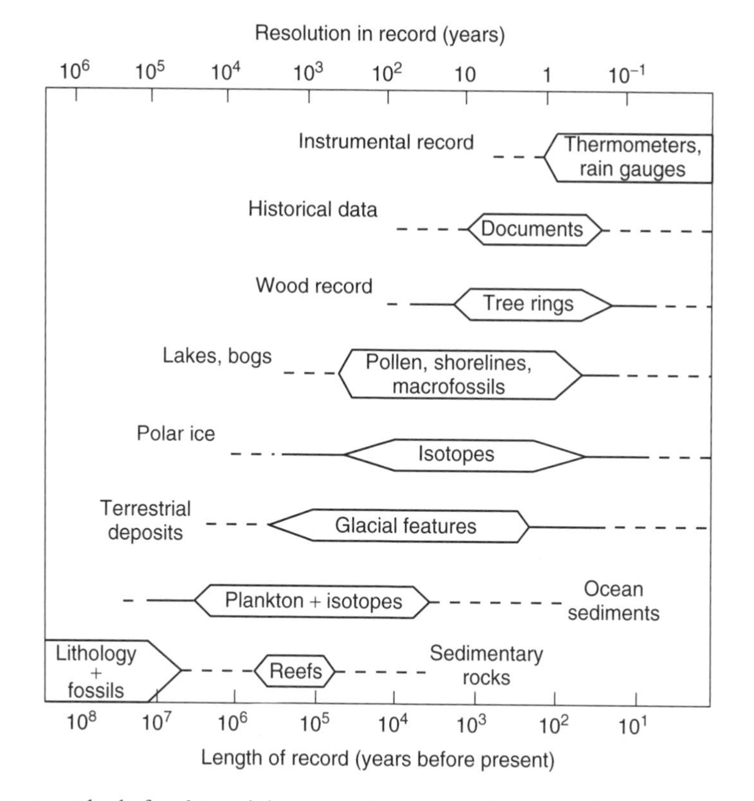

Fig. 2.9 Different methods for determining past climates on timescales from 1 to 100 million years (bottom scale). Top scale indicates approximate time resolution of the techniques (Adapted from Webb III T, Kutzbach J and Street-Perrott FA 1985. 2,000 years of global climate change: palaeoclimatic research plan. In: Malone TF and Roederer JG, eds *Global Change: The Proceedings of a Symposium Sponsored by the ICSU. ICSU (QE1 G51)*. Cambridge: Cambridge University Press, pp. 182–218. Reproduced by permission of the ICSU).

and Mexico at that time (Street-Perrot and Perott 1990). Finally, as the low salinity water mixed and dispersed, the oceanic conveyor belt was restored and the interglacial warming rapidly reappeared. In fact, within 20 years, air temperatures increased by 5 to 10 °C and increased precipitation led to a doubling of snow accumulation in Greenland (Taylor 1999).

Another rapid climate event occurred about 8,200 years ago. Earlier, as glaciers advanced, they pushed ahead massive quantities of earth and rock, forming a large "terminal moraine" (a natural dam). As the Earth warmed, the melting ice formed a huge lake over Eastern Canada. When the lake finally filled and overflowed the moraine, it collapsed sending huge quantities of freshwater into the North Atlantic Ocean. The lowered salinity again interrupted the oceanic conveyor belt and halted the warming trend for about a 400-year period. These events demonstrate how rapidly climate can change.

Lessons of Past Climate Change

Life and climate are inextricably linked. What lessons can we learn from past climate changes? For the most part, prior to the industrial era, climate change occurred slowly in human time frames. Natural changes on the order of 5 °C occurred over periods of tens of thousands of years – slowly enough to allow many animals and plants to migrate to more favorable climates. The migration of plant and animal species to higher latitudes during interglacial and to lower latitudes during glacial periods is well documented.

However, evidence also shows that small perturbations of the global system can lead to dramatic and rapid changes in regional or global climate and result in species extinctions. An estimated 95% of all species that have ever lived are now extinct. Many of these extinctions are linked to climatic change. For example, one of the strongest theories for extinction of the dinosaurs 65 million years ago relates to a massive meteor impact and dust cloud that resulted in a dramatic global cooling. Also, plant pollen records show rapid changes in plant species during the Younger Dryas Event. Some scientists, after examining past climatic fluctuations, warn that if human-induced climate change continues, the Earth's climate could become unstable, resulting in rapid unpredictable change (Lorius and Oeschger 1994).

To examine what a warmer world would be like, scientists often look at the last warm interglacial period (about 120,000 years ago). However, an older interglacial period (423,000 to 362,000 years ago) known as "MIS 11" may provide more insight into the future than the last interglacial period. Study of MIS 11 indicates that the sea level may have risen to 20 m above its present level, and increasing temperature, as well as complex changes in ocean biogeochemistry, lowered the ocean's ability to hold CO_2. These past trends lead one to ask, "Are we setting in motion a chain of biogeochemical feedbacks that will add an oceanic contribution to the anthropogenic greenhouse-gas enrichment?" (Howard 1997).

In conclusion, biota may migrate or adapt to slow climatic change, but rapid change could have far-reaching consequences, including extinction. Human-caused global warming will be similar in magnitude to some of the largest changes of the past (6 °C), but will occur 20 to 50 times faster (see Chapter 4).

Summary

The Earth's climate has changed over periods of millions of years. Long-term natural changes resulted from volcanic activity releasing huge quantities of heat-trapping CO_2 into the atmosphere and from the evolution of plants that removed CO_2 from the atmosphere through photosynthesis. Climate also undergoes natural cooling and warming cycles as a result of periodic fluctuations in the amount of solar energy reaching the Earth's surface. These fluctuations occur on a scale of tens of thousands of years and result from periodic changes in the orbital alignment of the Earth with respect to the Sun, known as Milankovich Cycles.

Paleoclimatologists use many methods to determine past climates. These include (from long past to recent) fossil rocks, isotope ratios in plankton and coral reefs, polar ice cores, plant pollen, tree rings, historical documents, and scientific instrumental records.

About 11,000 years ago, melting glaciers lowered the density of seawater in the North Atlantic. This slowed the oceanic conveyor belt carrying heat to Europe and North America. This event, "The Younger Dryas," demonstrates that climate can change rapidly – in decades – in response to an environmental disturbance. Thus, natural fluctuations in the global average Earth temperature of ± 2 to

3 °C occur on a timescale of tens of thousands of years. Smaller fluctuations (perhaps due to fluctuations in the energy output of the Sun) of about ±0.8 °C occur over thousands of years. These long-term natural fluctuations have been accompanied by major changes in ecosystems and in recorded history have influenced human civilizations in a number of ways. Now, human emissions of greenhouse gases threaten to raise the Earth's global average temperature by 2 to 6 °C or even more during this century (Chapter 4).

References

Beck JW, Edwards RL, Ito E, Taylor FW, Recy J, Rougerie F, et al. 1992 Sea-surface temperature from coral skeletal strontium/calcium ratios. *Science* **257**: 644–647.

Campbell ID and McAndrews JH 1993 Forest disequilibrium caused by rapid little ice age cooling. *Nature* **366**: 336–338.

Cook E, Bird T, Peterson M, Barbetti M, Buckley B, D'Arrigo R, et al. 1991 Climatic change in Tasmania inferred from a 1089 year tree-ring chronology of Huon pine. *Science* **253**: 1266–1268.

Crowley TJ and North GR 1991 *Paleoclimatology*. Oxford: Oxford University Press.

Gates DM 1993 *Climate Change and its Biological Consequences*. Sunderland, MA: Sinauer Associates, p. 42.

Gore A 1993 Chapter 3, Climate and Civilization *Earth in the Balance: Ecology and the Human Spirit*. New York: Plume Publications, pp. 56–80.

Howard WR 1997 A warm future in the past. *Nature* **388**: 418, 419.

Karner DB and Muller RA 2000 A causality problem for Milankovitch. *Science* **288**: 2143, 2144.

Kerr RA 1993 How ice age climate got the shakes. *Science* **260**: 890–892.

Kutzbach J 1989 Historical perspectives: climatic changes throughout the millennia. In: DeFries RS and Malone TF, eds *Global Change and our Common Future*. Washington, DC: National Academy Press, pp. 50–61.

Lorius C and Oeschger H 1994 Paleoperspectives: reducing uncertainties in global climate change? *Ambio* **23**(1): 30–36.

McAndrews JH 1967 Pollen analysis and vegetation history of the Itasca region, Minnesota. In: Cushing EJ and Wright Jr HE, eds *Quaternary Paleoecology. Volume 7 of the Proceedings of the 7th Congress of the International Association for Quaternary Research*. University of Minnesota Press, Minneapolis, USA, pp. 218–236.

NOAA 2002 National Oceanic and Atmospheric Administration Paleoclimatology Program, *http://www.ngdc.noaa.gov/paleo/slides*.

Raynaud D, Jouzel J, Barnola JM, Chappellaz J, Delmas RJ and Lorius C 1993 The ice record of greenhouse gases. *Science* **259**: 926–933.

Rodrigo FS, Esteban-Parra MJ and Castro-Diez Y 1998 On the use of the Jesuit order private correspondence records in climate reconstructions: a case study from Castille (Spain) for 1634–1648 AD. *Climate Change* **40**: 625–645.

Scuderi LA 1993 A 2000-year tree ring record of annual temperatures in the Sierra Nevada Mountains. *Science* **259**: 1433–1436.

Street-Perrot FA and Perrot RA 1990 Abrupt climate fluctuations in the tropics: the influence of Atlantic Ocean circulation. *Nature* **343**: 607–612.

Struiver M and Quay PD 1980 Changes in atmospheric carbon-14 attributed to a variable Sun. *Science* **207**: 11–19.

Taylor K 1999 Rapid climate change. *American Scientist* **87**: 320–327.

Velzer J, Godderis Y and François LM 2000 Evidence for decoupling of atmospheric CO_2 and global climate during the Phanerozoic eon. *Nature* **408**: 698–701.

Verschuren D, Laird KR and Cummings BF 2000 Rainfall and drought in equatorial east Africa during the past 1100 years. *Nature* **403**: 410–414.

Webb III T, Kutzbach J and Street-Perrott FA 1985 2000 years of global climate change: paleoclimatic research plan. In: Malone TF and Roederer JG, eds *Global Change: The Proceedings of a Symposium Sponsored by the ICSU. ICSU(QE1 G51)*. Cambridge: Cambridge University Press, pp. 182–218.

Chapter 3

Recent Climate Change: The Earth Responds

"The body of statistical evidence ... now points towards a discernible human influence on global climate"

Intergovernmental Panel on Climate Change, 1995

Introduction

The Earth's climate changed substantially over periods of thousands to millions of years (Chapter 2). How does recent climate change during the past 100 to 150 years compare with the natural long-term fluctuations of the past? More importantly, are recent climate changes caused by human emissions of greenhouse gases?

Glaciers in the European Alps are melting rapidly and have lost more than half their volume since 1850. The major water reservoir supplying Athens, Greece, suffered nearly a decade of drought in the 1990s that left the lake reservoir low and threatened the water supply of four million people. Globally, the 1998 average annual temperature was the greatest and that for 2001 the second greatest since 1860. Of the 15 warmest years recorded during the past 150 years, 10 were in the 1990s. In fact, the last decade of the twentieth century was the warmest in the entire global instrumental temperature record.

Anecdotal examples of recent unusually warm, dry, or wet climate extremes like these abound (Box 3.1). They *suggest* that the climate is undergoing unnatural change. However, scientific detection of an unnatural change requires demonstrating that the observed change is significantly (in a statistical sense) different from the natural pattern of variation. Even then, establishing a cause–effect relationship (attribution) between human greenhouse gas emissions and global greenhouse warming would require additional evidence. Perhaps, human responsibility for recent climatic trends can be "proven," as in some criminal jury trials, only by examining a large number of individual circumstantial pieces of evidence, the preponderance of which indicate unusual,

Climate Change: Causes, Effects, and Solutions John T. Hardy
 ISBNs: 0-470-85018-3 (HB); 0-470-85019-1 (PB)

Box 3.1 Anecdotal evidence of recent unusual climate change

1995–February – Los Angeles, USA, has a record high temperature of 35 °C (95 °F), which is repeated the following year.
1996–April 4 – Seattle, USA, temperature is 5.6 °C (10 °F) greater than average and mountain snow level elevation is much greater than normal.
1998–Summer – Uncontrolled wildfires sweep through large areas of Florida, USA.
1998–August – Forty-one people die of heat-related stress in Louisiana – more than double the previous record since the state began keeping such statistics in 1986.
1999–July – Heat-related deaths in the US midwest climb to 250.
2000–2001 – Record rainfall in the United Kingdom causes widespread flooding.
2001 – The second warmest year (next to 1998) globally, since records began (142 years).
2001–2002 – November, December, and January is the warmest US winter on record. Drought conditions prevail in 15 states from Georgia to Maine and water reservoirs in New York State are 30% below normal.
2002–Spring – A new lake formed at the base of the rapidly melting Belvedere Glacier in the Italian Alps.
On April 15 and 16, 70 record high temperatures are set in the United States, and in New York City the temperature is 16 °C (29 °F) above normal. On April 25, five Western US states have drought emergencies, mountain snow packs are 80% below normal, and moisture levels are at record lows. April 29 – severe tornadoes hit the Eastern United States, leaving six dead.
2002–August – The worst floods in 200 years hit central Europe with four times the average rainfall, leaving over 100 dead and more than $20 billion in damage.

that is, unnatural, trends. Researchers have recently uncovered such evidence by identifying numerous alarming changes in the Earth's land, air, oceans, and biota that occurred during the past 150 years. Furthermore, these recent changes have occurred over decades, rather than over hundreds of thousands of years, proving human-induced climate change "beyond a reasonable doubt."

Atmospheric Temperatures

The twentieth century was the warmest century, and 1990 to 2000 was the warmest decade, of the past millennium (Figure 3.1a). During the past 150 years, the global average annual temperature of the Earth's lower atmosphere, the troposphere, has warmed about $0.6 \pm 0.2\,^{\circ}\mathrm{C}$ (Figure 3.1b). The recent temperature increase is evident as a time-series anomaly (Box 3.2). Thousands of individual measurements of near-ground air temperatures using standard thermometers at numerous sites located around the world have supported the existence of this warming trend (Hansen and Lebedoff 1987, Jones et al. 1999, Nicholls et al. 1996).

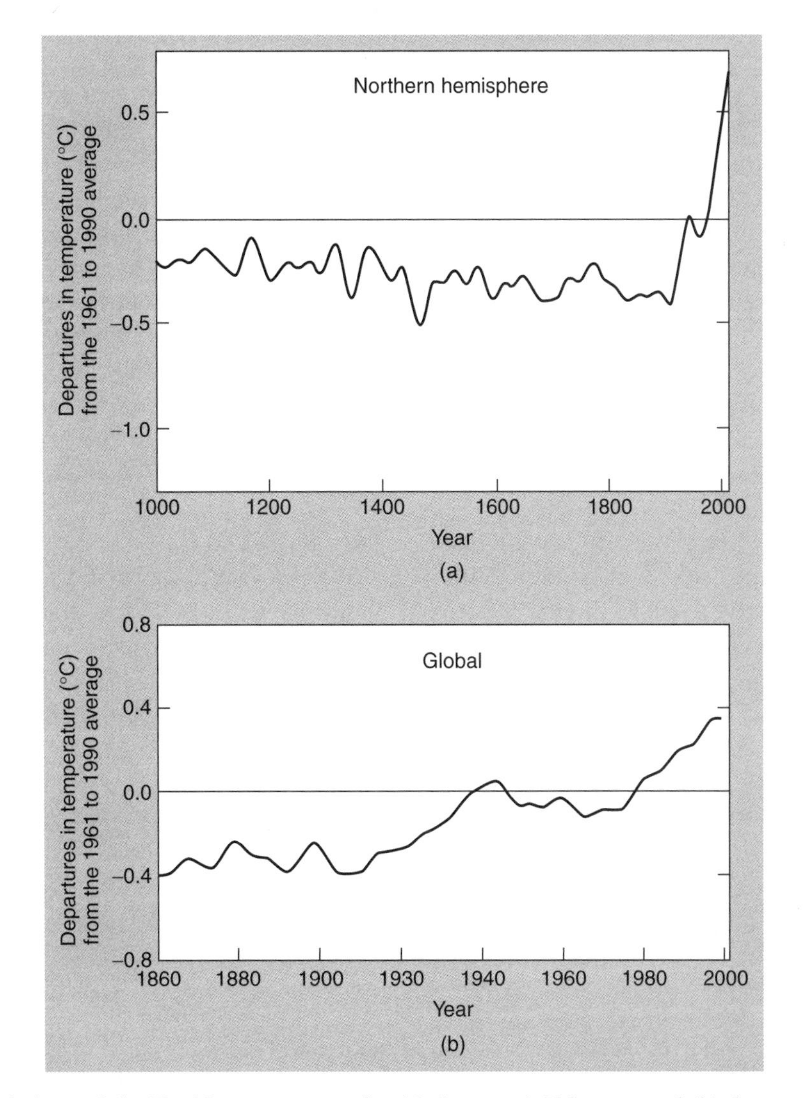

Fig. 3.1 Variations of the Earth's temperature for (a) the past 1,000 years and (b) the past 140 years. Data previous to accurate thermometer records in the mid-nineteenth century are derived from proxy measurements based on tree rings, corals, ice cores, and historical records (see Chapter 2) (From IPCC 2001. Houghton JT, Ding Y, Griggs DJ, Noguer M, van der Linden PJ, Dai X, et al., eds *Climate Change 2001: The Scientific Basis*. Intergovernmental Panel on Climate Change, Working Group I. Cambridge: Cambridge University Press, p. 881. Reproduced by permission of Intergovernmental Panel on Climate Change).

Warming in recent decades has generally been greater at higher latitudes and in some midcontinental regions (Plate 1). In addition to global changes, many local and regional decade-long climate trends are consistent with predictions of anthropogenic climate change. In the United States, for example, the average February temperature for Bellingham, Washington, increased by 2.2 °C between 1920 and 1997. In Europe,

Box 3.2 Time-series anomalies

Data such as temperature, precipitation, frost-free days, and glacial volume generally vary widely over time. Thus, the average Earth temperature 18,000 years ago, during the Wisconsin glaciation, was considerably less than the average temperature today. To compare change over time, it is useful to select a mean value for a reference time period, and then subtract all values over the entire time period from that mean. For example, in Figure 3.1, global temperatures each year from 1861 to 2000 are subtracted from the mean value for the period 1961 to 1990. Thus, in this example, the reference mean temperature becomes zero, and values from earlier times are generally negative, while values after the reference period are positive. The overall change over the past 140 years is slightly less than 0.8 °C (Figure 3.1b), although, because of variability, the best statistical estimate is $0.6 \pm 0.2\,°C$.

changes in temperature from 1946 to 1999 are significant and consistent with a human influence on climate (ECA 2002) (Figure 3.2).

While the lower atmosphere (the troposphere) is warming, the upper atmosphere (the stratosphere) is cooling. As greenhouse gas concentrations increase, theory predicts that more heat will be trapped in the lower atmosphere instead of escaping to the upper atmosphere (stratosphere) and space. Also, depletion of stratospheric ozone (a greenhouse gas) by man-made chlorofluorocarbons may

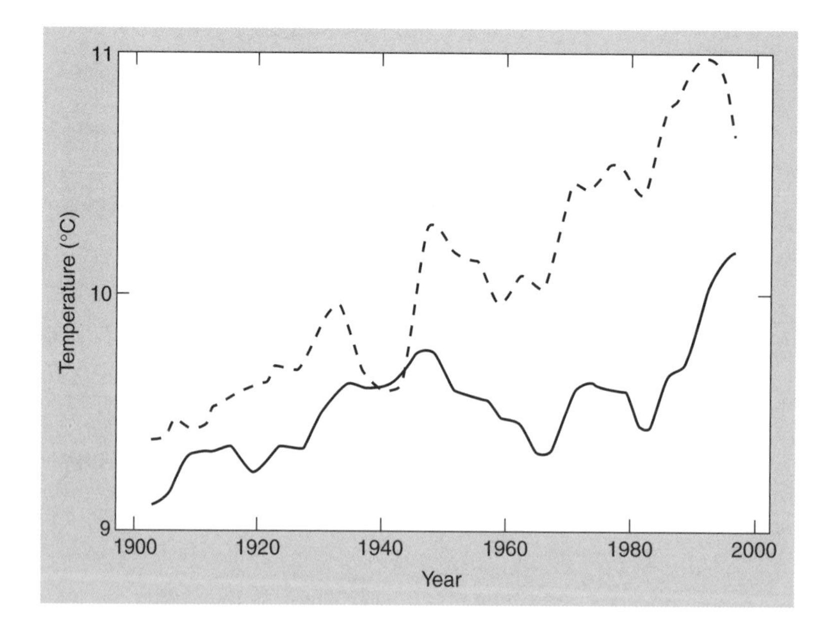

Fig. 3.2 Examples of two European regional average annual temperature trends for the twentieth century. Solid line = Central England. Dashed line = Wein, Austria (Adapted from ECA 2002. Tank AK, Wijngarrd J and van Engelen A, eds *Climate of Europe. European Climate Assessment.* (Working Group of 36 institutions in 34 countries) *http://www.knmi.nl/voorl/*).

contribute to greater heat loss from the stratosphere to space. Vertical profiles of temperature from the Earth's surface upward to the stratosphere do indeed show such a temperature increase at lower elevations accompanied by a temperature decrease in the upper atmosphere (Lambeth and Callis 1994, IPCC 2001).

Ocean temperatures are rising. Since about 1980, infrared sensors aboard satellites have measured day and night sea-surface temperatures (Strong 1989). These measurements provide more than two million temperature readings each month over the entire globe. Such a large data set allows trends to be examined statistically in a meaningful way. The surface temperature of the ocean increased between 1976 and 2000 at a rate of 0.14 °C per decade (IPCC 2001). Deeper ocean temperatures, at least in some areas, are also increasing. Atlantic deep (800 to 2,500 m) water along the 24-degree latitude line has warmed remarkably since 1957. Warming at 1,100 m is occurring at a rate of 1 °C per century (Parrilla et al. 1994).

Water Vapor and Precipitation

A warmer troposphere will increase evaporation of water from the oceans leading to a general global average increase in atmospheric water vapor and rainfall. Overall, global land precipitation has increased by about 2% since 1900 (Folland and Karl 2001). Evaporation is greatest in warm waters of the low-latitude tropical ocean where solar heat input is greatest. Global atmospheric circulation patterns (Chapter 1) tend to distribute this moisture poleward to higher latitudes, where it cools and condenses into rain. This condensation releases large amounts of latent heat.

In response to the additional heat energy from global warming, models predict that water, evaporated from the ocean surface in tropical latitudes, will be carried further poleward before precipitating out. Thus, as the greenhouse effect intensifies, precipitation should increase poleward of 30° latitude and decrease between latitudes 5° and 30°. An examination of rainfall data between 1900 and 1999 demonstrates just such a global trend (Figure 3.3). This pattern also holds in Europe, where between 1946 and 1999, the number of days and quantity of precipitation generally increased at northern sites and decreased in the south (ECA 2002). At Boulder, Colorado, USA, a temperate site at 40 °N latitude, concentrations of water vapor in the lower atmosphere significantly increased from 1981 to 1994 (Ottmans and Hofmann 1995).

Finally, since the 1980s the natural El Niño Southern Oscillation (ENSO) (Box 8.2) has occurred more frequently and has lingered longer than previously. Some researchers suggest that this intensification of the ENSO may, at least partially, result from greenhouse warming (Trenberth and Hoar 1996).

Clouds and Temperature Ranges

These temperature and ocean evaporation increases should lead to an increase in clouds, at least in temperate regions. Global average cloud cover has increased in recent decades (Nicholls et al. 1996). Furthermore, increased cloud cover should lead to warmer winters (when clouds trap heat) and cooler summers (when clouds tend to reflect the more intense solar energy). Just as predicted, measurements in the United States show that cloud cover has increased more than 10%, while the summer–winter temperature difference has decreased (Figure 3.4).

With overcast cloudy conditions, the Earth's surface loses less heat at night, resulting in a decrease in the day–night temperature difference. Indeed, since the 1940s

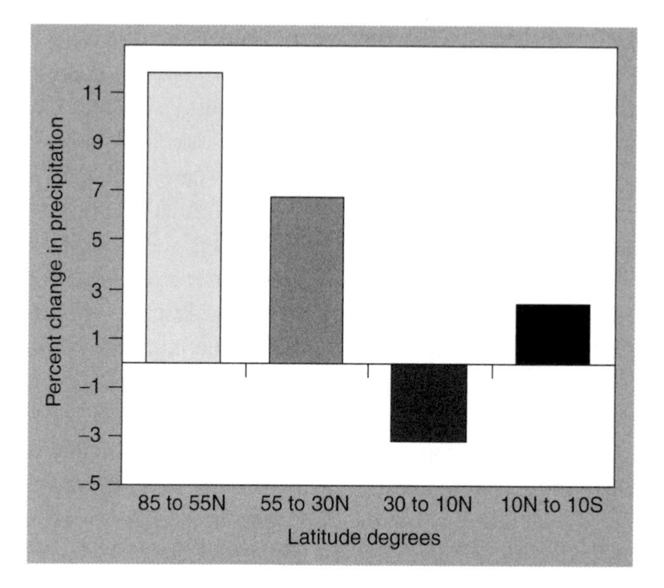

Fig. 3.3 Annual precipitation between 1900 and 1999 increased at temperate and decreased at subtropical latitudes (Adapted from Folland CK and Karl TR 2001. Observed climate variability and change. In: Houghton JT, Ding Y, Griggs DJ, Noguer M, van der Linden PJ, Dai X, et al., eds *Climate Change 2001: The Scientific Basis*. Intergovernmental Panel on Climate Change, Working Group I. Cambridge: Cambridge University Press, p. 144. Reproduced by permission of Intergovernmental Panel on Climate Change).

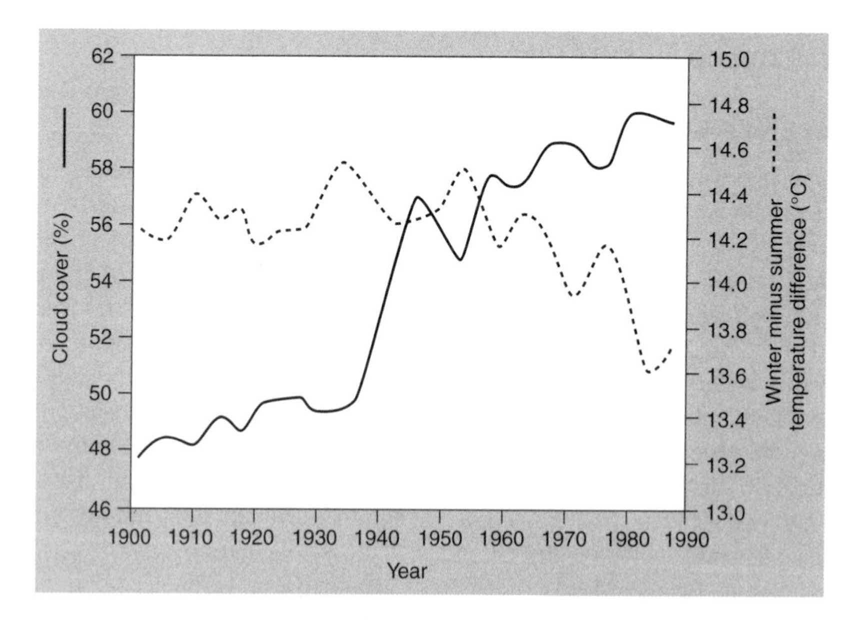

Fig. 3.4 Annual temperature range and cloud cover for the United States. Increasing cloudiness is leading to warmer winter and cooler summer temperatures (From Pearce F 1994. Not warming, but cooling. *New Scientist* **143**: 37–41).

the difference in the daytime and nighttime average temperature has decreased in both Europe and the United States (ECA 2002, Kukla and Karl 1993). Globally, the daily range of surface air temperature has decreased since the 1950s, with an increase in the night-time minimum temperature exceeding the increase in daytime maximum temperature. Increasing cloud cover is the most probable cause (Dai et al. 1997).

Ocean Circulation Patterns

Global patterns of ocean circulation are changing. For example, upwelling of deep water along some coastal areas is increasing. During summer, coastal areas on several continents develop significant atmospheric pressure differences between the low pressure over the warm land and the higher pressure over the cool ocean. This natural atmospheric pressure gradient drives vigorous alongshore winds. The wind blowing parallel to the coast pushes surface water ahead, and because of the Coriolis force (Box 1.2), the surface water moves to the right (Northern Hemisphere) or left (Southern Hemisphere) of the wind direction. This is called *Ekman transport*, and on the western

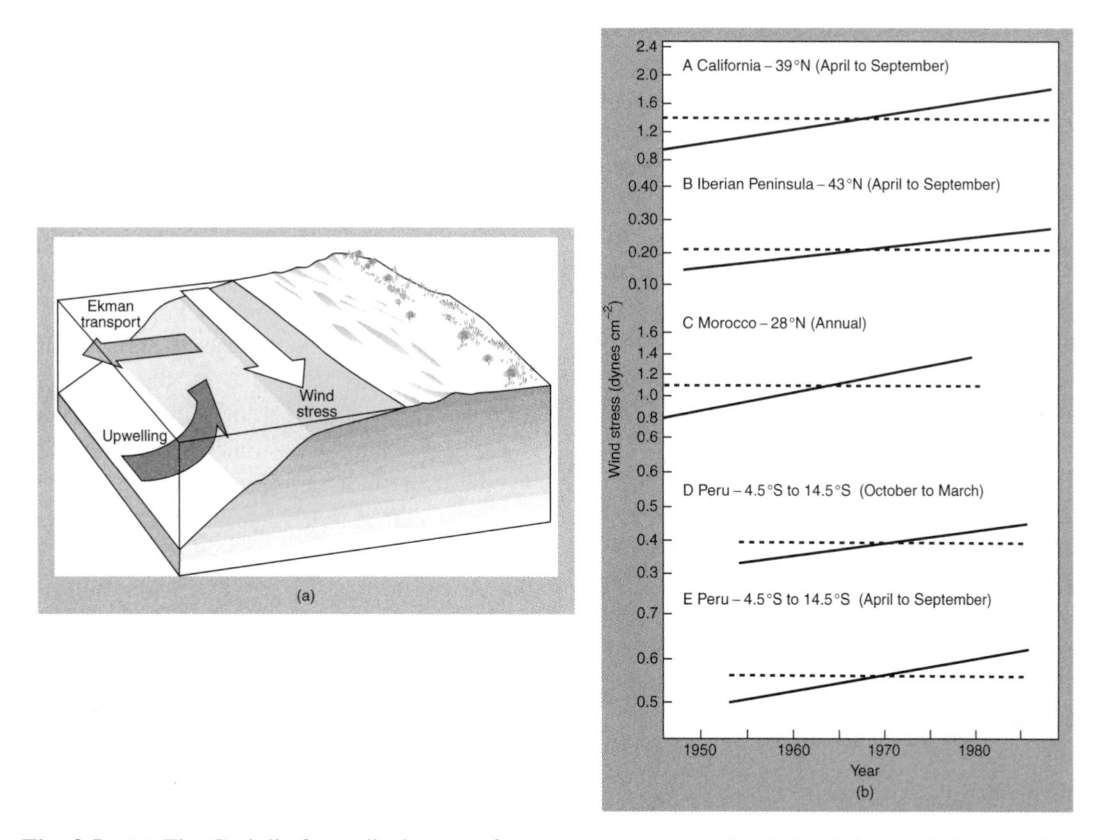

Fig. 3.5 (a) The Coriolis force displaces surface ocean currents to the right of the wind direction (in the Northern Hemisphere). This is called *Ekman transport* and results in upwelling of deep water to the surface. (b) Alongshore wind strengths have intensified at widely separated coastal locations since 1946. Horizontal dashed line indicates long-term mean (Reprinted from Bakun A 1990. Global climate change and intensification of coastal ocean upwelling. *Science* **247**: 198–201).

margins of continents the direction is generally offshore (Figure 3.5a). The surface water is replaced by cold nutrient-rich water that moves upward from depth (upwells).

If surface air becomes warmer year after year, the land should heat more rapidly than the coastal water (which has a higher heat capacity) and theoretically, the pressure gradient and alongshore wind strength should intensify. Evidence supports such a trend. In five widely spaced geographic areas where coastal upwelling is known to occur, wind strengths increased significantly between 1946 and 1989 (Figure 3.5b).

If this trend continues, the implications go beyond upwelled deep-ocean water. The regional decrease in the ocean surface temperature, compared to the land, might increase the frequency and intensity of coastal fog. In addition, with warmer temperatures inland, the increased pressure difference could intensify winds through the passes to the interior. At the same time, cooler coastal water would mean less evaporation and hence a decrease in rainfall. For an area like Southern California, the decreased rainfall and increased wind could spell severe fire danger.

The Younger Dryas Period 11,650 to 14,000 years ago (Chapter 2) exemplifies another possible change in ocean circulation patterns. The increased freshwater inflow from melting ice slowed currents that bring heat to the North Atlantic. Some evidence suggests this pattern might again be under way, although the longevity of the trend remains uncertain (Schlosser et al. 1991).

Snow and Ice

Alpine glaciers around the world are shrinking dramatically. Quantitative measurements of alpine glacial mass from many global regions verify this trend (Oerlemans 1994). Recent glacial retreat is well documented in the European Alps and Western North America where data, as well as photographs taken since the nineteenth century, document the retreat of glaciers (Figure 3.6). In Africa, the glaciers of Mount Kenya lost 75% of their area between 1899 and 1987, with 40% of this loss occurring between 1963 and 1987 (Hastenrath and Krus 1992). The rate of ice loss from Alaskan Glaciers has more than doubled in a decade and now provides half the current global contribution of glaciers to sea-level rise (Arendt et al. 2002, Meier and Dyurgerov 2002).

Less visible, but still important, is the 10% decline in annual snow cover over the Northern Hemisphere during the past 20 years (Groisman et al. 1994). Spring snowmelt in the Arctic occurred two weeks earlier in the 1980s than it did in the 1940s and 1950s (Walsh 1991). In Canada, in response to warmer temperatures, the southern boundary of the permafrost (permanently frozen layer of soil) migrated northward 120 km between 1964 and 1990 (Kwong et al. 1994).

Polar sea ice has decreased significantly during the last 50 years. In the Northern Hemisphere, the extent of summer sea ice decreased about 15% during the last half of the twentieth century (Figure 3.7). The decrease between 1978 and 1987 alone was 2.1% (Gloersen and Campbell 1991). In the Antarctic, sea-ice cover in the Bellingshausen Sea decreased in the late 1980s and 1990s compared to the 1970s (Jacobs and Comiso 1993). During the 1990s, large segments of the ice shelf of Antarctica broke loose and drifted to sea, and in January 1995 the northernmost section of the Larsen Ice Shelf collapsed abruptly. One study suggests that, "unless the situation changes dramatically and ice-front retreat ceases almost immediately, it seems almost certain that another ice shelf will

(a)

(b)

Fig. 3.6 Typical example of a shrinking alpine glacier. The South Cascade Glacier, Washington State US photographed in the years (a) 1928 and (b) 2000. Photos courtesy of US Geological Survey.

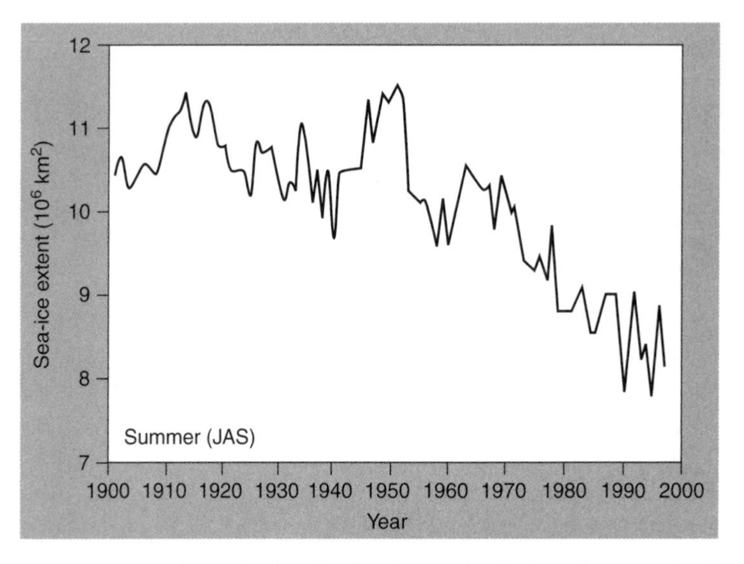

Fig. 3.7 Decline in summer sea-ice extent in the Northern Hemisphere between 1901 and 1999 (Adapted from IPCC 2001. Houghton JT, Ding Y, Griggs DJ, Noguer M, van der Linden PJ, Dai X, et al., eds *Climate Change 2001: The Scientific Basis*. Intergovernmental Panel on Climate Change, Working Group I. Cambridge: Cambridge University Press, p. 125. Reproduced by permission of Intergovernmental Panel on Climate Change).

disappear..." (Doake et al. 1998). Finally, on the basis of records of whaling ships since 1931, averaged over October to April, the Antarctic summer sea-ice edge moved southwards by 2.8° latitude between the mid-1950s and the early 1970s – a decline in ice-covered area of about 25% (Mare 1997).

Thickening of the central Greenland ice sheet (Zwally 1989) seems at first to run counter to the thinning trend of ice in alpine glaciers. However, even with greenhouse warming, temperatures at high latitudes such as Greenland generally remain below freezing point. Thus, greater precipitation at such northern latitudes comes as snow. Increased precipitation in Greenland may also be related to changes in large-scale atmospheric circulation and resultant spatial shifts in the tracks of storms over the North Atlantic (Kapsner et al. 1995). In fact, a transect across Greenland suggests that between 1954 and 1995, although the ice thickened somewhat in the east, it thinned in Western Greenland at a rate of about 31 cm year^{-1} (Paterson and Reeh 2001).

Sea-Level Rise

The sea level is rising at an unprecedented rate. As water warms, its volume expands. In the case of the ocean, this can only result in a rise in sea level relative to the land. About two-thirds of the twentieth-century sea-level rise results from thermal expansion of ocean water and one-third from melting glaciers and ice caps that add freshwater to the sea. When corrected for land movement, historical tide-gauge records, in some cases covering the last 100 years, show a general increase in sea level. One of the most extensive studies of recent trends in sea level analyzed 500 tide-gauge records that had data for more than a 10-year period. Results indicate a global average sea-level rise of ±2.4 mm per year (Peltier and Tushingham 1989). Consideration of both

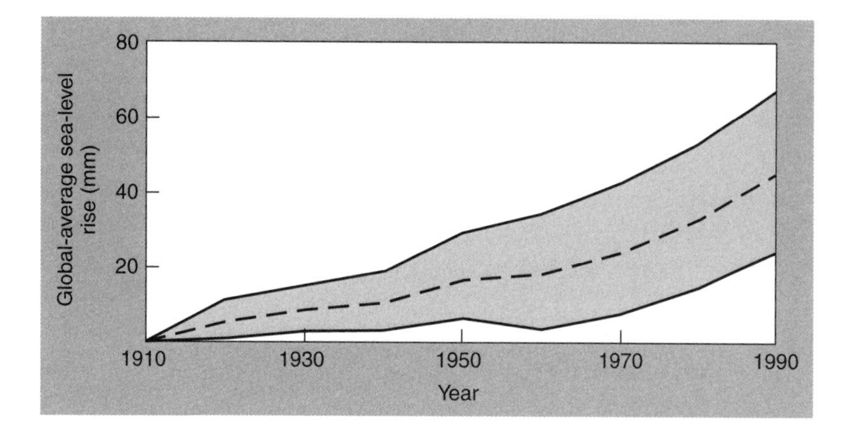

Fig. 3.8 Midrange and upper- and lower-bound estimates for the response of sea level to climate change from 1910 to 1990 (From Church JA and Gregory JM 2001. Changes in sea level. In: Houghton JT, Ding Y, Griggs DJ, Noguer M, van der Linden PJ, Dai X, et al., eds *Climate Change 2001: The Scientific Basis*. Intergovernmental Panel on Climate Change, Working Group 1. Cambridge: Cambridge University Press, p. 666. Reproduced by permission of Intergovernmental Panel on Climate Change).

tide-gauge and satellite altimeter data together indicates that during the twentieth century the sea level rose at a rate of 0.3 to 0.8 mm year^{-1} (Church and Gregory 2001) (Figure 3.8).

Animal Populations

Changes in populations of plants and animals may be among the most sensitive indicators of climate change. As Professor Donella Meadows of Dartmouth College says, "... there are people in touch with the planet's living systems and leading indicators, to whom it seems that the earth could hardly be sending a clearer signal if it were jumping up and down and yelling, Hey, I'm changing!"

In the marine environment the growth and reproduction of most organisms are closely linked to a specific optimum temperature for that species. But, only in a few instances do we have long-term records of the abundance of marine species. The abundance of animal plankton in offshore waters of Southern California decreased 80% between 1951 and 1993. This decrease occurred concurrently with a rise of surface water temperatures of 1.2 to 1.6 °C (Roemmich and McGowan 1995). The intertidal community of Monterey Bay, California, was described in detail between 1931 and 1933 and then examined again in 1993 to 1994. There was a significant shift in 32 of 45 invertebrate species with an increase in more southern species and a decrease in more northern species (Barry et al. 1995).

Warmer ocean temperatures threaten tropical coral reefs, which are among the world's most biologically diverse communities. In tropical areas, reefs shelter coastal areas from storms and are economically important as areas for fishing and tourism. Reef-building corals are symbiotic organisms; that is, microscopic photosynthetic algae in the tissues of the coral are vital to their survival. If stressed by higher-than-normal temperature, the coral lose their algae or "bleach." The coral may recover or, if the stress is prolonged, may die. Since the 1980s the frequency and extent of "coral bleaching" has grown alarmingly. In some areas large expanses of reef have died. Satellite sea-surface temperature maps demonstrate that the large-scale bleaching events are almost always associated with anomalous "oceanic hotspots" – areas of the ocean that exceed the long-term mean monthly maximum temperature by 1 °C or more (Goreau and Hayes 1994).

In terrestrial populations, evidence suggests that birds, at least in some temperate regions, are laying their eggs earlier in the season. Specifically, studies in the United Kingdom reveal that the onset of egg laying in 20 out of 65 bird species studied has shifted 4 to 17 (average 8.8) days earlier in the season over the 25-year period from 1971 to 1995 (Crick et al. 1997). Amphibians show a similar trend toward earlier spawning in response to earlier spring temperature increases (Beebee 1995). The extent of annual sea ice in the Arctic decreased 6% during the 1980s and the 1990s. For polar bears, the larger ice-free areas and longer ice-free periods restrict their hunting area and threaten their survival (Stefan et al. 2002).

Vegetation

Unusual shifts in terrestrial vegetation are another indicator of recent climate change. In high-latitude (boreal) regions, temperatures have increased about 2 °C since 1880 – more than double the global average. In West Central Mongolia, the growth rate of Siberian pine trees, based on a 450-year tree-ring chronology, increased dramatically in response to this recent warming (Jacoby et al. 1996). In Alaska annual growth of white spruce trees increased in the 1930s in response to warming. However, as fall–winter–spring temperatures continued increasing, other factors such as availability of sunlight and increased water loss from leaves probably became limiting. By the 1970s, water stress and increasing attacks from pests such as the bark beetle slowed tree growth markedly (Jacoby and D'Arrigo 1995).

In the North Eastern United States, red spruce *Picea rubens* made up 45% of the forest cover in some areas of New Hampshire in the early nineteenth century. By 1984, in the same areas (not previously logged), red spruce had declined to only 5% of the forest. Although factors such as acid precipitation from air pollution might play a role, the most likely explanation for the shift is the strong correlation with increasing summer temperatures over the same period (Hamburg and Cogbill 1988).

Satellite data suggest that from 1981 to 1991 the photosynthetic activity of terrestrial vegetation has increased, especially between latitudes 45 °N and 70 °N (Myneni et al. 1997), where marked warming has led to disappearance of snow cover earlier in the season. The increased summertime photosynthesis and CO_2 uptake is reflected by a 20% increase in the seasonal (winter–summer) difference in concentrations of atmospheric CO_2.

Finally, warming is altering the community ecology of phytoplankton in high-latitude aquatic habitats. Sediment cores of high-arctic freshwater ponds in Canada indicate recent changes in microfossil diatoms – unicellular photosynthetic plankton. Assemblages, whose species composition had been stable over the last few millennia, changed dramatically in composition beginning in the nineteenth century (Douglas et al. 1994). Also, extreme ecological changes have altered the phytoplankton community inhabiting the lakes of Signy Island near the Antarctic Peninsula. Between 1980 and 1995 lake temperatures increased an average 1 °C – four times the global mean average temperature increase. Permanent ice cover receded 45%

since 1951 and the ice-free period increased by 63 days between 1980 and 1993 (Quayle et al. 2002).

Attribution

The recent changes described above are consistent with a global-warming trend, but can the warming trend be attributed to human emissions of greenhouse gases? Human-induced changes are superimposed on natural changes. How can we sort out the human-induced variation (the signal) from the natural background variation (the noise)? Actual attribution is most likely to come from an examination of simultaneous patterns of change over both space and time.

One approach is to ask if an observed change, for example, an increase in global temperature, over some recent period, is statistically different from any long-term natural (preindustrial) variation. Global mean temperature shows considerable variability on all timescales and may show natural trends of up to 0.3 °C over intervals of up to 100 years. Although this natural variability is quite large, it is insufficient to explain the observed global warming during the twentieth century (Wigley and Raper 1990). Several noted scientists, after examining such data, conclude that there is an 80 to 95% probability that the global average temperature increase over the past 100 years is outside the range of natural variation. The increase, instead, is due to human emissions of greenhouse gases (Tol 1994, Schneider 1994).

A second approach to establishing a link between global warming and human emissions of greenhouse gases is used by Cynthia Kuo and others at the Bell Laboratories in New Jersey. Using time-series statistics they show a strong positive correlation between the increasing atmospheric concentration of carbon dioxide and the increasing temperature over the 30-year period that they examined from 1958 to 1988. The probability that the level of coherence between the two variables (CO_2 concentration and temperature) is due to chance alone is about 2 out of 1 million (Kuo et al. 1990).

A third approach to determining a human link to climate change is to compare model (expected owing to greenhouse gas increases) and observed patterns of change for a single variable (e.g. temperature) in four dimensions – over latitude, longitude, altitude, and time. For example, the observed latitude versus altitude changes in temperature over time for the last few decades agree well with patterns predicted by global climate models. Also, models predict less warming in areas of high sulfate aerosols. Observed spatial and temporal patterns of high aerosol distribution are consistent with model predictions of areas of reduced warming (see Chapter 1).

Climate change involves much more than temperature change. For example, in addition to overall changes in temperature or precipitation, at least five measures of climatic change are thought to be sensitive to increased atmospheric concentrations of greenhouse gases. Multivariate statistics are used to correlate simultaneous changes in multiple variables. Such changes include unequal increases in maximum and minimum temperature, increases in cold season precipitation, severe summertime drought, and the proportion of total precipitation that is derived from extreme daily precipitation events, and decreases in day-to-day temperature variations. For example, the Greenhouse Climate Response Index, an analysis of such changes in the United States over time since about 1910, indicates that the probability that the increase in the index since 1910 is due purely to natural causes is only 1 to 9% (Karl et al. 1996).

Summary

A great deal of circumstantial events during the past decade or two suggests that the Earth's climate is changing. More importantly, scientific research on climate trends and the Earth's responses during the past 150 years indicate that change is now occurring much more rapidly than during past historical periods. There can be little doubt that during the twentieth century, humans altered the Earth's climate by emitting huge quantities of greenhouse gases. Recent changes include

- an historically rapid tropospheric warming of almost 1 °C in global average surface temperature and increased average sea-surface temperature;
- a stratospheric cooling, resulting from increased retention of heat trapped in the troposphere due to higher greenhouse gas concentrations;
- increased atmospheric water vapor at temperate latitudes, general increases in precipitation poleward of 30° latitude, and decreases in precipitation at lower latitudes;
- increased cloudiness and decreased day–night temperature differences at temperate locations;
- increased coastal winds, ocean surface currents, and upwelling of cold deep-ocean water along the western margins of continents;
- decreases in snow and ice cover, shrinking alpine glaciers, and increases in the frost-free season in temperate regions;
- an average global sea-level rise of about 2.4 cm per decade;
- changes in animal populations including geographic shifts (poleward) of coastal marine species populations, and earlier spring breeding of birds and amphibians;
- a dramatic increase in "bleaching" and mass mortality of coral reefs;
- changes in terrestrial and aquatic vegetation, including changing growth patterns in trees and altered species composition in aquatic phytoplankton.

Statistical analyses indicate that many of these changes are outside the range of natural variation and are consistent with model predictions of human-induced greenhouse warming. Predicting the magnitude of future human-induced climate change remains difficult and will probably come slowly as computer models improve (Chapter 4).

References

Arendt AA, Echelmeyer KA, Harrison WD, Lingle CS and Valentine VB 2002 Rapid wastage of Alaska glaciers and their contribution to rising sea level. *Science* **297**: 382–386.

Bakun A 1990 Global climate change and intensification of coastal ocean upwelling. *Science* **247**: 198–201.

Barry JP, Baxter CH, Sagarin RD and Gilman SE 1995 Climate-related, long-term faunal changes in a California rock intertidal community. *Science* **267**: 672–674.

Beebee TJC 1995 Amphibian breeding and climate. *Nature* **374**: 219–220.

Church JA and Gregory JM 2001 Changes in sea level. In: Houghton JT, Ding Y, Griggs DJ, Noguer M, van der Linden PJ, Dai X, et al., eds. *Climate Change 2001: The Scientific Basis*. Intergovernmental Panel on Climate Change, Working Group 1. Cambridge: Cambridge University Press, p. 666.

Crick HQP, Dudley C, Glue DE and Thompson DL 1997 UK birds are laying eggs earlier. *Nature* **388**: 526.

Dai A, Del Genio AD and Fung IY 1997 Clouds, precipitation and temperature range. *Nature* **386**: 665, 666.

de la Mare WK 1997 Abrupt mid-twentieth-century decline in Antarctic sea-ice extent from whaling records. *Nature* **389**: 57–59.

Doake CS, Corr HFJ, Rott H, Skvarca P and Young NW 1998 Breakup and conditions for stability of the northern Larsen Ice Shelf, Antarctica. *Nature* **391**: 778–780.

Douglas MSV, Smol JP and Blake Jr W 1994 Marked post-18th century environmental change in high-arctic ecosystems. *Science* **266**: 416–419.

ECA 2002 Tank AK, Wijngarrd J and van Engelen A, eds *Climate of Europe. European Climate Assessment*. (Working Group of 36 institutions in 34 countries) *http://www.knmi.nl/voorl/*.

Folland CK and Karl TR 2001 Observed climate variability and change. In: Houghton JT, Ding Y, Griggs DJ, Noguer M, van der Linden PJ, Dai X, et al., eds *Climate Change 2001: The Scientific Basis*. Intergovernmental Panel on Climate Change. Working Group I. Cambridge: Cambridge University Press, pp. 99–181.

Gloersen P and Campbell WJ 1991 Recent variations in Arctic and Antarctic sea-ice covers. *Nature* **352**: 33–36.

Goreau TJ and Hayes RL 1994 Coral bleaching and ocean "Hot Spots". *Ambio* **23**(3): 176–180.

Groisman PY, Karl TR and Knight RW 1994 Observed impact of snow cover on the heat balance and the rise of continental spring temperatures. *Science* **263**: 198–200.

Hamburg SP and Cogbill CV 1988 Historical decline of red spruce populations and climatic warming. *Nature* **331**: 428–430.

Hansen J and Lebedoff S 1987 Global trends of measured surface air temperature. *Journal of Geophysical Research* **92**(D11): 13,345–13,372.

Hastenrath S and Krus P 1992 The dramatic retreat of Mount Kenya's glaciers between 1963 and 1987: greenhouse forcing. *Annals of Glaciology* **16**: 127–133.

IPCC 2001 Houghton JT, Ding Y, Griggs DJ, Noguer M, van der Linden PJ, Dai X, et al., eds *Climate Change 2001: The Scientific Basis*. Intergovernmental Panel on Climate Change. Working Group I. Cambridge: Cambridge University Press.

Jacobs SS and Comiso JC 1993 A recent sea-ice retreat west of the Antarctic peninsula. *Geophysical Research Letters* **20**(12): 1171–1174.

Jacoby GC, D'Arrigo RD and Davaajamts T 1996 Mongolian tree rings and 20th century warming. *Science* **273**: 771–773.

Jacoby GC and D'Arrigo RD 1995 Tree-ring width and density evidence of climatic and potential forest change in Alaska. *Global Biogeochemical Cycles* **9**(2): 227–234.

Jones PD, New M, Parker DE, Martin S and Rigor IG 1999 Surface air temperature and its changes over the past 150 years. *Reviews of Geophysics* **37**: 173–199.

Kapsner WR, Aley RB, Schuman CA, Anandakrishnan S and Grootes PM 1995 Dominant influence of atmospheric circulation on snow accumulation in Greenland over the past 18,000 years. *Nature* **373**: 52–54.

Karl TR, Knight RW, Easterling DR and Quayle RG 1996 Indices of climate change for the United States. *Bulletin of the American Meteorological Society* **77**(2): 279–292.

Kukla G and Karl TR 1993 Nighttime warming and the greenhouse effect. *Environmental Science and Technology* **27**(8): 1468–1474.

Kuo C, Lindberg C and Thompson DJ 1990 Coherence established between atmospheric carbon dioxide and global temperature. *Nature* **343**: 709–713.

Kwong Y, John T and Gan TY 1994 Northward migration of permafrost along the Mackenzie highway and climatic warming. *Climatic Change* **26**: 399–419.

Lambeth JD and Callis LB 1994 Temperature variations in the middle and upper stratosphere: 1979–1992. *Journal of Geophysical Research* **99**(D10): 20,701–20,712.

Meier MF and Dyurgerov MB 2002 How Alaska affects the world. *Science*. **297**: 350, 351.

Myneni RB, Keeling CD, Tucker CJ, Asrar G and Nemani RR 1997 Increased plant growth in the northern high latitudes from 1981 to 1991. *Nature* **386**: 698–702.

Nicholls N, Gruza GV, Jouzel J, Karl TR, Ogallo, LA, and Parker DE 1996 Observed climate variability and change. In: Houghton JT, Filho, LGM, Callander BA, Harris N, Kattenberg A and Maskell K, eds *Climate Change 1995: The Science of Climate Change*. Intergovernmental Panel on Climate Change. Cambridge: Cambridge University Press, p. 132–192.

Oerlemans J 1994 Quantifying global warming from the retreat of glaciers. *Science* **264**: 243–245.

Ottmans SJ and Hofmann DJ 1995 Increase in lower-stratospheric water vapour at a mid-latitude

Northern Hemisphere site from 1981 to 1994. *Nature* **374**: 146–149.

Parrilla G, Lavin A, Bryden H, Garcia M and Millard R 1994 Rising temperatures in the subtropical North Atlantic Ocean over the past 35 years. *Nature* **369**: 48–51.

Paterson WSB and Reeh N 2001 Thinning of the ice sheet in northwest Greenland over the past forty years. *Nature* **414**: 60–62.

Pearce F 1994 Not warming, but cooling. *New Scientist* **143**: 37–41.

Peltier WR and Tushingham AM 1989 Global sea level rise and the greenhouse effect: might they be connected? *Science* **244**: 806–810.

Quayle WC, Peck LS, Peat H, Ellis-Evans JC and Harrigan PR 2002 Extreme responses to climate change in Antarctic lakes. *Science* **295**: 645.

Revkin A 1992 *Global Warming: Understanding the Forecast. American Museum of Natural History and Environmental Defense Fund*. New York: Abbeville Press, p. 180.

Roemmich D and McGowan J 1995 Climatic warming and the decline of Zooplankton in the California current. *Science* **367**: 1324–1326.

Schlosser P, Bönisch PG, Rhein M and Bayer R 1991 Reduction of deepwater formation in the Greenland sea during the 1980s: evidence from tracer data. *Science* **251**: 1054–1056.

Schneider S 1994 Detecting climatic change signals: Are there any "fingerprints"? *Science* **263**: 341–347.

Stefan N, Rosentrater L and Eid PM 2002 *Polar Bears at Risk*. World Wildlife Fund International Arctic Programme. Accessed May, 2002 from: *http://www.panda.org/climate/pubs.cfm*.

Strong AE 1989 Greater global warming revealed by satellite-derived sea-surface temperature trends. *Nature* **338**: 642–645.

Tol RSJ 1994 *Greenhouse Statistics*. Time series analyses, Pt II. Institute for Environmental Studies, Amsterdam (ISBN 90-5383-302-1) *http://www.vu.nl/ivm/*.

Trenberth KE and Hoar TJ 1996 The 1990–1995 El Nino-Southern oscillation event: longest on record. *Geophysical Research Letters* **23**(1): 57–60.

Walsh J 1991 The Arctic as a bellweather. *Nature* **352**: 19–20.

Wigley TML and Raper SCB 1990 Natural variability of the climate system and detection of the greenhouse effect. *Nature* **344**: 324–327.

Zwally HJ 1989 Growth of Greenland ice sheet: interpretation. *Science* **246**: 1589–1591.

Chapter 4

Future Climate Change: The Twenty-First Century and Beyond

"Prediction is very difficult, especially about the future."

Niels Bohr (1885–1962)

Introduction

In 1896, the Swedish scientist Svante Arrhenius predicted that fossil-fuel burning (coal) would double atmospheric CO_2 over the next 3,000 years, leading to an increase in average global temperature of about 5 °C (Arrhenius 1896). Although his theory remains sound today, he underestimated the rate of this increase. The concentration of atmospheric CO_2 has actually increased 30% in 100 years – or about 18 times faster than Arrhenius predicted, and will double before the end of this century. Furthermore, this increase has already resulted in an historically unprecedented rapid increase in global average temperature. Recent changes in sea level and precipitation patterns also agree with those expected from an enhanced greenhouse effect (Chapter 3).

Predictions of future climate, based on computer models, are increasing in accuracy and precision. However, the Earth-climate system is complicated and our knowledge of some of the factors affecting climate remains uncertain. Positive and negative feedbacks of the climate system could either increase or decrease climatic change. Also, there is the question of sensitivity, that is, how much will the Earth's temperature rise in response to a given increase in greenhouse gases? Finally, future rates of economic growth and fossil-fuel combustion are major uncertainties.

Most recent estimates indicate a 2 to 5 °C global average rise in temperature for a doubling of atmospheric CO_2 but suggest that warming at mid to high latitudes could be much greater. A 3 °C increase in global average annual temperature would be unprecedented in the history of human civilization and would have serious consequences for humans and the ecosystems on which we depend. A 5 °C increase at midlatitudes would

Climate Change: Causes, Effects, and Solutions John T. Hardy
© 2003 John Wiley & Sons, Ltd ISBNs: 0-470-85018-3 (HB); 0-470-85019-1 (PB)

shift terrestrial habitats about 500 to 750 km northward. Such a shift could change the climate of Washington DC to that of Charleston, South Carolina, or that of Southern France to Algeria. This rate of temperature increase is 10 to 60 times faster than the natural increase from the end of the last ice age to the present.

Global Climate Models

Predictions of future climate rely on numerical computer models, referred to as *General Circulation Models* (GCMs), which simulate the Earth's climate system. Climate-modeling studies are under way at universities and research institutes around the world and many are large collaborative international efforts. They involve hundreds of scientists at centers such as the Hadley Centre for Climate Prediction and Research in the United Kingdom; the Max Planck Institute in Germany; the Laboratory for Modeling of Climate and Environment in France; the Canadian Climate Center; the National Oceanic and Atmospheric Administration (NOAA), Geophysical Fluid Dynamics Laboratory (GFDL), the National Center for Atmospheric Research (NCAR), the National Aeronautics and Space Administration (NASA), and Goddard Institute for Space Studies (GISS) in the United States; and the World Climate Research Program (WCRP) – a joint program of the United Nations, International Union for the Conservation of Nature, and the Intergovernmental Oceanographic Commission. The Intergovernmental Panel on Climate Change (IPCC), an international cooperative effort of hundreds of scientists sponsored by the World Meteorological Organization and the United Nations Environment Program, is in the forefront of evaluating and summarizing worldwide studies on climate change, including modeling efforts.

Typically, GCMs represent the atmosphere and ocean (fluids) on a grid of 1 to 4° latitude by longitude with 10 to greater than 200 vertical layers in each fluid (Figure 4.1). Inputs to each box within the model generally include the physical factors that determine the climate, such as solar radiation gain and loss rates, humidity, barometric pressure, ocean temperature, and salinity (which determines density), and atmospheric gas concentrations (including the concentration of the greenhouse gases) (Box 4.1). The models can be run with past, current, or assumed future greenhouse gas concentrations. After the models are "set in motion," that is, given the initial conditions, they compute changes in temperature or precipitation over time in intervals (time steps) for each box in the global grid. At each time step, the model recomputes the position and properties of the air and water as it mixes in response to wind and density differences.

Models are generally coupled, meaning they include interactions between separate submodels of several systems such as the atmosphere, land surface, ocean, and cryosphere (snow cover, glaciers, and polar caps). Submodels may use different timescales. The fast climate system adjusts to changes in the atmosphere in days or to the upper ocean in months. The slow climate system includes the deep-ocean and perennial land ice with responses of decades to centuries (Manabe 1998).

In "equilibrium simulations" the models are integrated for several decades – first with present and then with increased (often doubled) greenhouse gas concentrations. "Transient simulations" incorporate external forcing over time, for example, a scenario might include a CO_2 concentration increasing gradually over time. This is obviously more realistic, but also more complex.

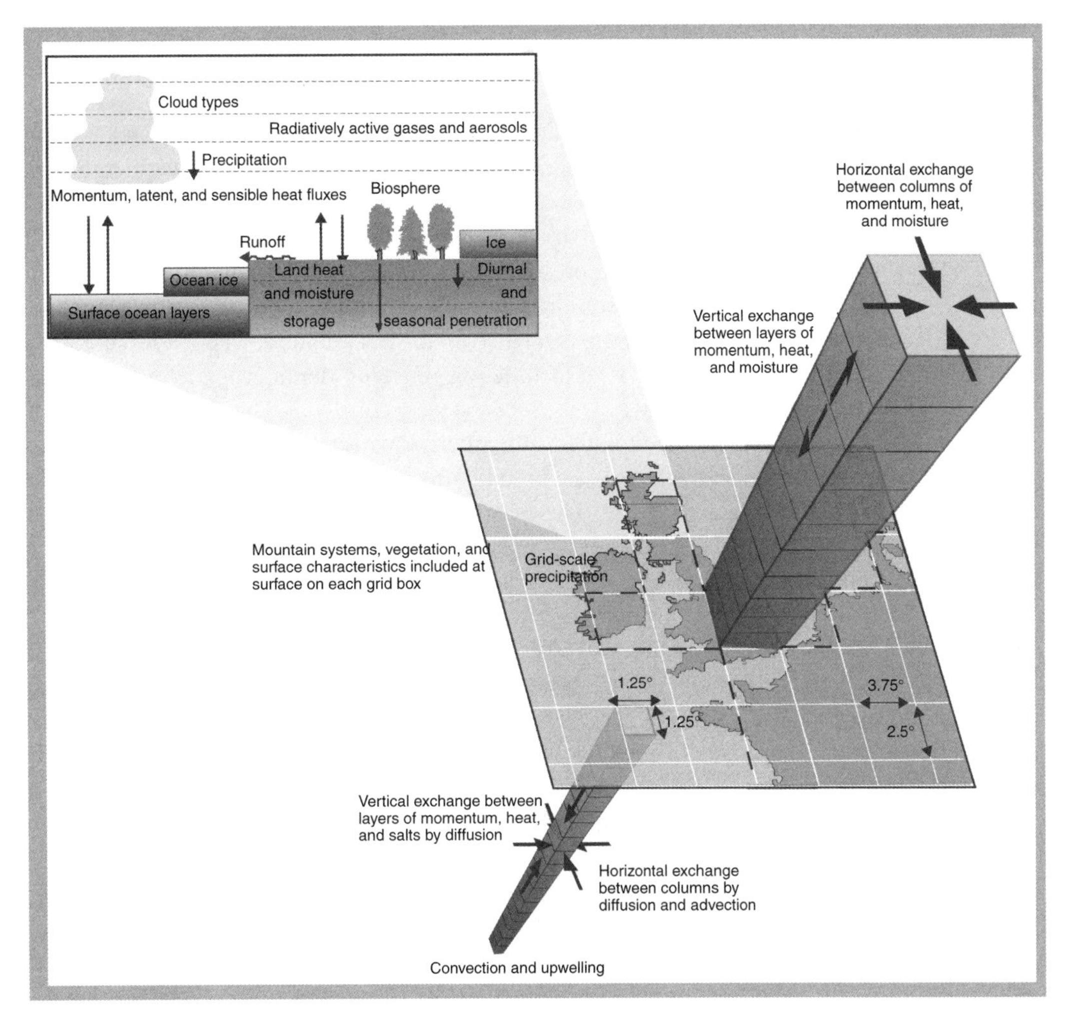

Fig. 4.1 Elements of a global climate model (Courtesy of David Viner 2002. Climatic Research Unit, University of East Anglia, UK).

Box 4.1 Example of a global climate model (Adapted from Hadley Centre 2002. Hadley Centre for Climate Prediction and Research, Meteorological Office, UK, *http://www.meto.gov.uk/research/hadleycentre/index.html*)

The Hadley Centre for Climate Prediction and Research of the UK Meteorological Office is one of the leading global centers for climate study and modeling. One of the most recent coupled atmosphere-ocean general circulation models (AOGCMs) is HadCM3 (Figure 4.1). In a simulated projection of over a thousand years, it showed little drift in known surface climate.

The atmospheric component of HadCM3 has 19 levels with a horizontal resolution of 2.5° of latitude by 3.75° of longitude, which produces a global grid of 7,008 grid cells. This is equivalent to a surface resolution of about 417 km × 278 km at the equator, reducing to 295 km × 278 km at 45° of latitude.

A radiation scheme includes six and eight spectral bands in the solar (short wave) and terrestrial thermal (long wave) wavelengths, respectively. The model represents the radiative effects of minor greenhouse gases as well as CO_2, water vapor, and ozone and also includes a simple parameterization of background aerosol.

A land surface scheme includes a representation of the freezing and melting of soil moisture, as well as surface runoff and soil drainage; the formulation of evaporation includes the dependence of stomatal resistance in plant leaves on temperature, vapor pressure, and CO_2 concentration. The surface albedo is a function of snow depth, vegetation type, and also of temperature over snow and ice.

A penetrative convective scheme includes an explicit down-draught and the direct impact of convection on momentum. The large-scale precipitation and cloud scheme is formulated in terms of an explicit cloud water variable. The effective radius of cloud droplets is a function of cloud water content and droplet number concentration. The atmospheric component of the model also permits the direct and indirect forcing effects of sulfate aerosols to be modeled, given scenarios for sulfur emissions and oxidants.

The oceanic component of HadCM3 has 20 levels with a horizontal resolution of 1.25° × 1.25°. At this resolution, it is possible to represent important details in oceanic current structures. Because of its higher ocean resolution, it does not need flux adjustment (additional "artificial" heat and freshwater fluxes at the ocean surface) to produce a good simulation. Horizontal mixing is included and horizontal momentum varies with latitude. Regional adjustments in circulation are included in some areas such as the Denmark Straits, and the Iceland–Scotland, and Mediterranean–Atlantic connections.

The sea-ice model uses a simple thermodynamic scheme including snow cover. The surface ocean current advects ice, preventing convergence, when the depth exceeds 4 m. There is no explicit representation of iceberg calving, so a prescribed water flux is returned to the ocean at a rate calibrated to balance the net snowfall accumulation on the ice sheets, geographically distributed within regions where icebergs are found. In order to avoid a global average salinity drift, surface water fluxes are converted to surface salinity fluxes using a constant reference salinity of 35 parts per thousand.

The model is initialized directly from the observed ocean state at rest, with a suitable atmospheric and sea-ice state. The atmosphere and ocean exchange information once per day. Heat and water fluxes are conserved in the transfer between their different grids.

Model simulations can be compared to recent observed climate. Agreement between modeled and observed change depends on the relationship between CO_2 concentration and temperature (temperature sensitivity) assumed in the model. For example, a doubling of CO_2 could result in a temperature increase of as little as 1.5 °C or as great as 4.5 °C (Figure 4.2a). Also, correlation between observations and model predictions increases when the cooling effect of sulfate aerosols is included (Figure 4.2b).

GCMs have evolved and improved over the past 40 years. Early GCMs treated the

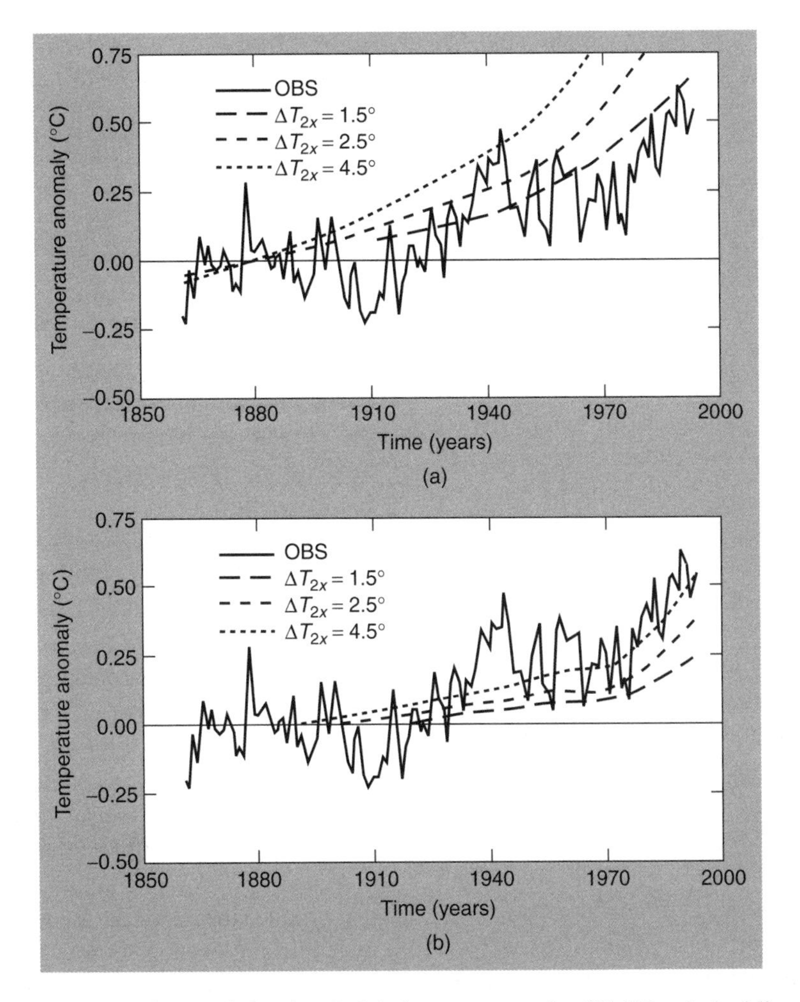

Fig. 4.2 Observed (solid line) and simulated global temperature for 2X CO_2, dashed lines = three different assumptions of temperature sensitivity to doubling of atmospheric CO_2. (a) Model forcing with greenhouse gases only; (b) greenhouse gases and cooling effect of sulfate aerosols (From IPCC 1996. *Climate Change 1995: The Science of Climate Change*. Intergovernmental Panel on Climate Change, World Meteorological Organization and United Nations Environment Program. Cambridge: Cambridge University Press, p. 37. Reproduced by permission of Intergovernmental Panel on Climate Change).

ocean heat sink as a static "swamp," without mixing. By 1987, GCMs incorporated clouds, seasonally changing sea ice, and a three-layered ocean with a surface-mixed layer that served as the primary heat exchange with the atmosphere (Schlesinger and Mitchell 1987). By 1993 scientists began to look at the important ocean component with more sophisticated models. A model developed by Albert Semtner of the US Naval Post Graduate School and Robert Chervon of the NCAR in the early 1990s divided the ocean into one-quarter degree blocks and calculated the average properties within each. It used eight supercomputers in parallel, capable of trillions of calculations per second. Even so, given the complexity of the system, it took more than 100 h of computer time to simulate just one year of detailed ocean circulation. This "Transient Ocean Model" correctly simulated the vast looping oceanic conveyer belt current linking the Pacific, Indian, and Atlantic Oceans (see Chapter 1 and Kerr 1993).

Dozens of newer coupled Atmosphere-Ocean General Circulation Models (AOGCMs) have now been developed and assessed. An accurate prediction of climate must incorporate not only the physical climate system but the biosphere as well. For example, the way in which the biosphere (living matter) reacts to climate change may either dampen or magnify climate change. Photosynthesis is a major sink for atmospheric CO_2, the most important greenhouse gas. Microbes break down organic matter (e.g. dead vegetation), and their release of CO_2 from respiration represents an important source to the atmosphere. Microorganisms are also active in the biogeochemical cycles of several radiatively important trace gases including methane, sulfur, and nitrogen. Some models now incorporate the interactions of major biomes, for example, desert, savanna, rain forest, and so on, with the atmosphere (Baskin 1993). Linked climate/biosphere models such as the Dynamic Global Phytogeography Model of the UK Meteorological Office demonstrate that predictions of climate change including life may be quite different from, and even more severe than, climate-change predictions in the absence of life.

Several approaches are used to evaluate the accuracy of climate models. Models are generally tested for their ability to simulate the current global climate including seasonal cycles. They may also be tested to see if they accurately "predict" changes that occurred thousands of years ago (Chapter 2) or recent trends in regional precipitation measured over the past few decades (Figure 4.3). In general, most models do a good job in predicting atmospheric temperature, precipitation, surface heat flux, and percent cloudiness. For example, 11 different GCMs give similar predictions of mean precipitation patterns in relation to latitude over the globe (Figure 4.4). However, models are less accurate at predicting the slow processes of changing ice cover or large-scale ocean circulation.

Feedback Loops and Uncertainties

Any attempt at predicting future conditions will contain uncertainties. Given the complexity of the climate system, it is not surprising that model predictions of global average warming for the period 1990 to 2100 range from 1.4 to 5.8 °C (IPCC 2001). One variable is that of setting the initial conditions, that is, how much anthropogenic change has occurred prior to the initial year of the model run. Also, there are errors in the data against which the model is compared. For example, models can be tested for their ability to replicate past climates under conditions present long ago, but the methods used to construct the past climates have inherent errors of their own.

A feedback is an interaction in which factors that produce the result are themselves modified by that result. Feedbacks are a common feature of the natural and the human-engineered world. For example, a household thermostat senses the room temperature. When the room temperature drops below the thermostat setting, the thermostat triggers the furnace to start. As the room heats up, the rising temperature, acting as a negative feedback, switches the heat supply off and ends the heating cycle. In a positive feedback loop, a malfunctioning thermostat would respond to increased temperatures by increasing its setting each time a higher temperature was reached. Positive feedbacks are of particular concern because of the possible "runaway" behavior they produce. For example, warming of wetlands could lead to increased microbiological activity in the sediments and increased

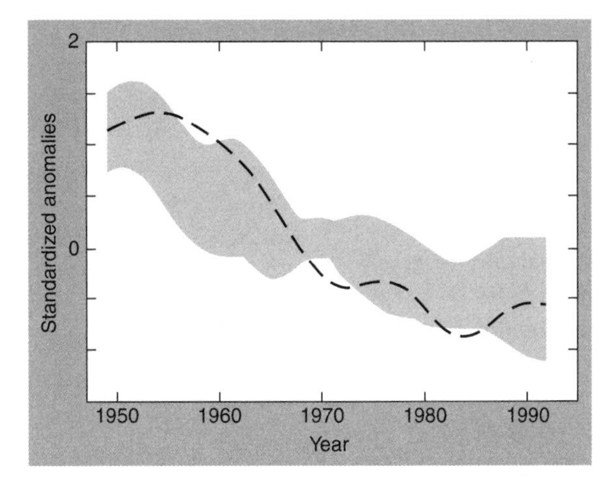

Fig. 4.3 Observed (dashed line) and modeled (shaded area) July to September rainfall from 1949 to 1993 for the Sahel Africa, using seven different climate models. 0 = reference period 1955 to 1988 (From Gates WL, Henderson-Sellers A, Boer GJ, Folland CK, Kitoh A, et al. 1996. In: Houghton JT, Filho LGM, Callender BA, Harris N, Kattenberg A, Maskell K, eds *Climate Change 1995: The Science of Climate Change*. Intergovernmental Panel on Climate Change, World Meterological Organization and United Nations Environment Program. Cambridge: Cambridge University Press, p. 258).

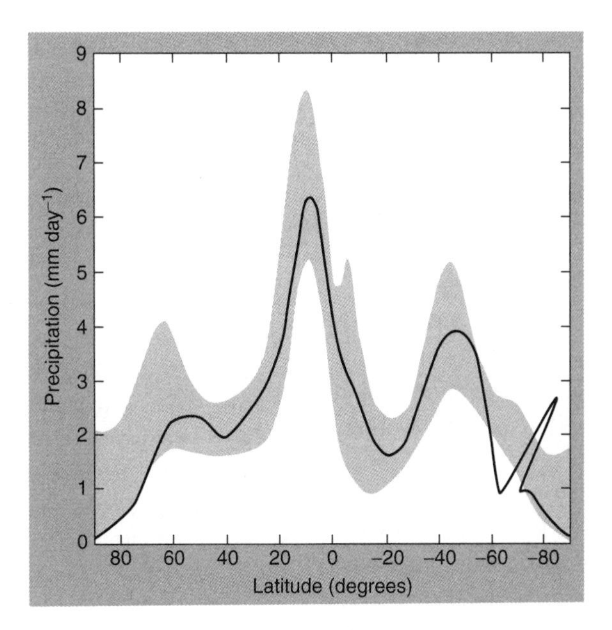

Fig. 4.4 Solid line is the observed global mean summer (June–August) precipitation (mm day^{-1}) by latitude. The shaded area encloses the range of predictions of 11 different global climate models (From Gates WL, Henderson-Sellers A, Boer GJ, Folland CK, Kitoh A, et al. 1996. In: Houghton JT, Filho LGM, Callender BA, Harris N, Kattenberg A, Maskell K, eds *Climate Change 1995: The Science of Climate Change*. Intergovernmental Panel on Climate Change, World Meterological Organization and United Nations Environment Program. Cambridge: Cambridge University Press, p. 241. Reproduced by permission of Intergovernmental Panel on Climate Change).

release of methane, a greenhouse gas, to the atmosphere. This would contribute to additional warming. Uncertainties and feedbacks affecting climate include the following factors.

1 Variations in solar activity

Some researchers suggest that climate is influenced more by variations in solar output than by greenhouse gases. For example, from 1645 to 1715, sunspot (solar flare) activity was minimal and the Earth experienced a "Little Ice Age." However, several studies show that although it may have some influence, global mean temperature changes, recently and in the future, are due less to solar variability and more to greenhouse gases (Hansen and Lacis 1990, Thomson 1995).

2 Changes in the hydrosphere and cryosphere

Several interactions of the climate system with water and ice produce important feedbacks. First, water vapor is a greenhouse gas. In response to ocean warming and increased evaporation, concentrations of atmospheric water vapor increase. This process is a classical positive feedback loop, that is, warming produces a gas that causes even more warming. The positive contribution of increasing atmospheric water vapor to global warming is significant (Manabe and Wetherald 1967).

Also, clouds are a very important factor in determining the amount of heat retained by the Earth and many of the differences in GCM predictions have to do with their treatment of clouds. As the ocean warms and evaporation increases, cloudiness increases. The role of clouds in greenhouse warming is the subject of intense research. Low cumulus clouds reflect solar energy and have a net cooling effect, whereas high stratus clouds trap solar energy and contribute to additional warming (Figure 4.5). Model simulations suggest that overall clouds may amplify warming by a factor of 1.3 to 1.8 (Cess et al. 1989). Increased cloudiness should increase nighttime and wintertime minimum temperatures and lead to a decrease in both daily and seasonal temperature ranges. Data for the United States since 1900 confirm just such a trend (see Chapter 3).

Finally, snow cover and glaciers are white and highly reflective (have a high solar albedo). As they melt, the albedo decreases and a greater quantity of heat is absorbed by the darker earth or vegetation surface. Snow cover in the Northern Hemisphere has declined by about 10% over the past 20 years. As a result, spring warming has been significantly enhanced during the twentieth century (Groisman et al. 1994).

3 Chemical interactions and the oxidative state of the atmosphere

Greenhouse gas emissions may affect climate indirectly through chemical interactions taking place in the atmosphere (Box 4.2). These interactions are complex and some are not well understood. Nevertheless, studies suggest that they can have important consequences for climate change (Fuglestvedt et al. 1996).

The stratospheric ozone layer protects the Earth from excessive ultraviolet radiation (UV). Stratospheric ozone depletion leads to increases in ground-level UV. Formation of tropospheric ozone (O_3) (an air pollutant and greenhouse gas, Box 10.1) is enhanced by UV.

4 Sulfate aerosols and dust

Several natural sources have a cooling effect on the planet. Volcanic activity and human-induced increases in deforestation and desertification lead to increased dust levels in the atmosphere. The resulting reduction in

(a)

(b)

Fig. 4.5 (a) Cumulus clouds reflect solar energy back to space and have a cooling effect; (b) cirrus clouds allow solar energy through, but trap heat and have a warming effect. However, how global warming will affect clouds and, in turn, the overall radiation balance of the Earth remains one of the largest uncertainties in global climate modeling.

sunlight reaching the Earth's surface has a cooling effect on climate.

Sulfur comes from both natural and anthropogenic sources. Natural sulfur compounds are emitted by some plankton species in the ocean and serve as cloud condensation nuclei (Charlson et al. 1987). Combustion of fossil fuel, especially high-sulfur coal, releases sulfur to the atmosphere. Such emissions are expected to increase atmospheric sulfur by 160 to 270% between 1994 and 2040. Inhalation of sulfur particles contributes to human respiratory illness and is a significant air pollutant causing acid rain and damage to ecosystems.

These detrimental effects of atmospheric sulfur are offset to some degree by sulfur's role in decreasing greenhouse warming. Small

Box 4.2 Feedbacks and uncertainties in atmospheric chemistry

Stratospheric ozone depletion increases the formation of hydroxyl radicals (OH^-) in the atmosphere.

$$O_3 + \text{UV light} \longrightarrow O_2 + O$$
$$O + H^+ \longrightarrow OH^-$$

Reaction with hydroxyl radicals in turn is a primary sink for many gases including CO, CH_4, and CFCs, and nonmethane hydrocarbons (NMHC), all of which decrease the oxidation state of the atmosphere by neutralizing hydroxyl radicals. For example, methane is oxidized to CO_2 (which has only 1/20th of the per molecule warming potential as CH_4):

$$CH_4 + 6OH^- \longleftrightarrow CO + 5H_2O$$
$$CO + OH^- \longleftrightarrow CO_2 + H^+$$

Counterbalancing this removal of OH^- could be an increase in OH^- from increasing amounts of water vapor in the atmosphere. The net result of these and other chemical interactions are not well understood.

sulfur particles or aerosols in the atmosphere reflect solar energy. In the Northern Hemisphere, the negative radiative forcing (cooling) due to sulfate aerosols is about equal to the positive radiative forcing (warming) due to anthropogenic greenhouse gases (Charlson et al. 1992, Charlson and Wigley 1994). Thus, the current negative forcing from anthropogenic sulfur emissions substantially offsets global greenhouse warming, especially in the Northern Hemisphere (Taylor and Penner 1994, Figure 4.6). Ironically, reductions in sulfur emission to improve public health and environmental quality will probably exacerbate global warming.

5 The CO_2 fertilization effect

Plants, through photosynthesis, take up CO_2 and convert it to organic carbon. Studies, primarily in greenhouses, indicate that increased atmospheric CO_2 increases plant growth. If true on a larger scale, this process could act as a negative feedback on warming, since increased plant biomass would mean more carbon stored organically rather than as atmospheric CO_2. Since the 1960s, the amplitude of the seasonal atmospheric CO_2 variation, that is, low in summer as plants take up CO_2 and high in winter as plants die, has been increasing. This suggests that plant biomass might indeed be increasing and having a greater effect on seasonal increases and decreases in atmospheric CO_2.

6 Changes in forest carbon

Forest harvesting releases huge quantities of stored carbon into the atmosphere. Globally, forests and their soils contain about 2,000 billion tonnes (Gt) of carbon. The carbon storage of the Amazon forest alone is equal to about 30 years of current fossil-fuel combustion. Trees are converted to lumber and paper – products that store carbon for much shorter time periods than forests. The

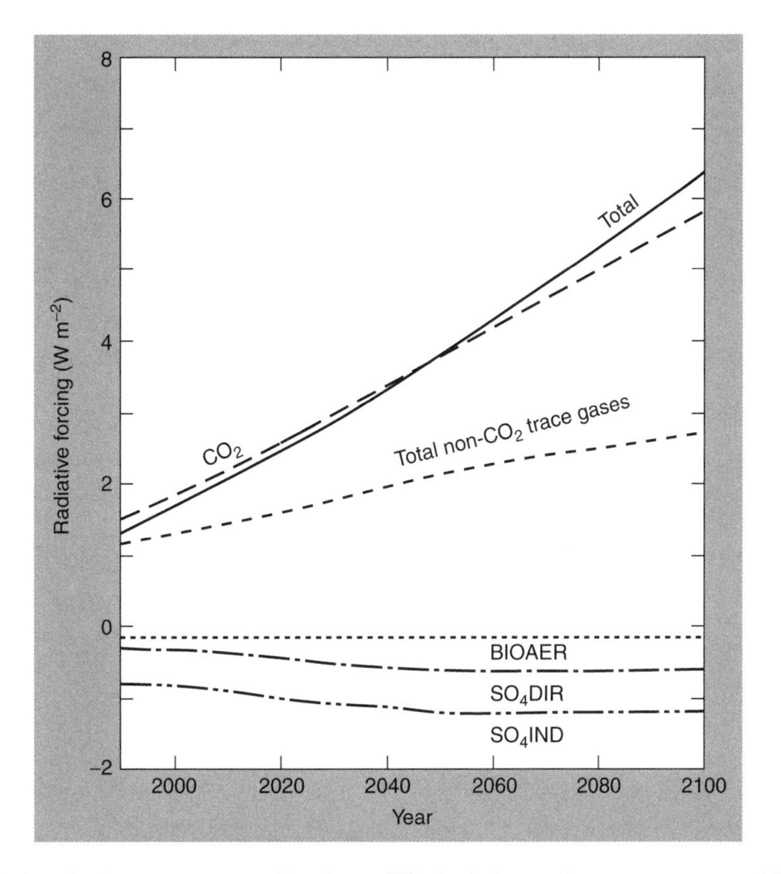

Fig. 4.6 Radiative forcing components. Total non-CO_2 includes methane, water vapor, N_2O, tropospheric ozone, and halocarbons. Negative forcing is from SO_4 aerosols and biomass burning (From Kattenberg A, Giorgi F, Grassl H, Meehl GA, Mitchell JFB, et al. 1996. Climate models – projections of future climate. In: Houghton JT, Filho LGM, Callender BA, Harris N, Kattenberg A, Maskell K, eds *Climate Change 1995: The Science of Climate Change*. Intergovernmental Panel on Climate Change, World Meterological Organization and United Nations Environment Program. Cambridge: Cambridge University Press, p. 321. Reproduced by permission of Intergovernmental Panel on Climate Change).

destruction of forests, largely for agriculture, contributed an estimated one-third of the CO_2 emitted to the atmosphere during the last century. In 1990, deforestation in the low latitudes emitted 1.6 Gt of carbon per year, whereas forest expansion and growth in mid- and high-latitude forests sequestered 0.7 Gt of carbon, for a net flux to the atmosphere of 0.9 Gt of carbon per year (Dixon et al. 1994).

Global warming could cause latitudinal and elevational shifts in forest habitat (Chapter 6). Forests may not be able to migrate fast enough. Any large-scale forest demise could release considerable CO_2 to the atmosphere – a positive feedback to warming. Also, warming could increase respiration rates in living plants and lead to increased release of CO_2. The correlation between CO_2 and temperature for the Little Ice Age suggests that a 2 °C warming might result in a total release of 80 Gt of carbon over decades, but the range of possibilities is large (Woodwell et al. 1998).

In contrast, some studies suggest a negative feedback on warming from changes in terrestrial carbon. For example, a doubled CO_2 climate could be an average 5.4 °C warmer and 17.5 mm day^{-1} wetter. Such change would lead to an increase in plant growth, a 75% increase in the area of tropical rain forests (discounting harvesting), and a reduction in desert and semidesert areas by 60% and 20%, respectively. The overall effect would be a removal of 235 Gt from the atmosphere and its storage as organic carbon in the terrestrial biosphere (Prentice and Fung 1990). However, much, if not all, of the cooling accomplished by expanding forests biomass (and carbon storage) could be offset because forests reflect less solar energy than lighter cleared areas or grasslands (Betts 2000).

7 Changes in soil organic matter

Carbon released by soil respiration accounts for about 10% of the carbon in the atmospheric pool, and increased temperatures and respiration rates could create a strong positive feedback – releasing additional CO_2 to the atmosphere. Soils are the largest terrestrial reservoir of carbon, holding almost three times more carbon than vegetation (Chapter 1). About two-thirds of carbon in forest ecosystems is contained in the soil and in peat deposits. Greenhouse warming should increase the metabolic rate of microorganisms that oxidize organic matter in the soil. A greenhouse temperature increase of 0.03 °C year^{-1} (a likely scenario) from 1990 to 2050 would release 61 Gt of carbon as CO_2 from soil organic matter – that is, about 19% of the total CO_2 released from the combustion of fossil fuel during the same period (Jenkinson et al. 1991). However, studies in the tall grass prairie of North America suggest that soil respiration acclimatizes somewhat to increased temperature, thus weakening the strength of the positive feedback (Luo et al. 2001).

As soils warm, farmlands, wetlands, tundra, and peat could all become sources of potential positive warming feedbacks. Higher temperature will promote the conversion of soil nitrogen to atmospheric nitrous oxide (N_2O) (a greenhouse gas). Soil warming will also increase CO_2 and CH_4 loss from wetlands, which contain about 15% of the global soil carbon. If the high-latitude permafrost melts, microbial breakdown of soil organic matter will release additional carbon to the atmosphere. In fact, some evidence suggests that the Arctic tundra, in response to warming, is already changing from a net CO_2 sink to a source (Oechel et al. 1993). Also, across northern latitudes an estimated 450 billion tonnes of carbon is tied up in peat – a carbon-rich soil deposit. Rising temperatures may already be responsible for triggering the release of carbon from peat bogs to the atmosphere, which in some areas increased 65% in 12 years (Freeman et al. 2001).

8 Ocean feedbacks

The ocean plays an important role in global climate (see Chapter 1). Some of the additional heat from greenhouse warming is absorbed by the surface layer of the ocean and transported by mixing to the deep ocean – a negative feedback that acts to substantially slow global warming. On the other hand, the water solubility of CO_2 and CH_4 decreases by approximately 1 to 2% per 1 °C increase in ocean temperature. Therefore, as the oceans warm, CO_2 will move from the large ocean reservoir into the atmosphere, a positive feedback that will further enhance the greenhouse effect.

Reservoirs of solid carbon hydrates represent a significant potential source of greenhouse gases. Given the proper temperature and pressure, carbon dioxide and methane form icelike crystalline solids called *clathrates*, or

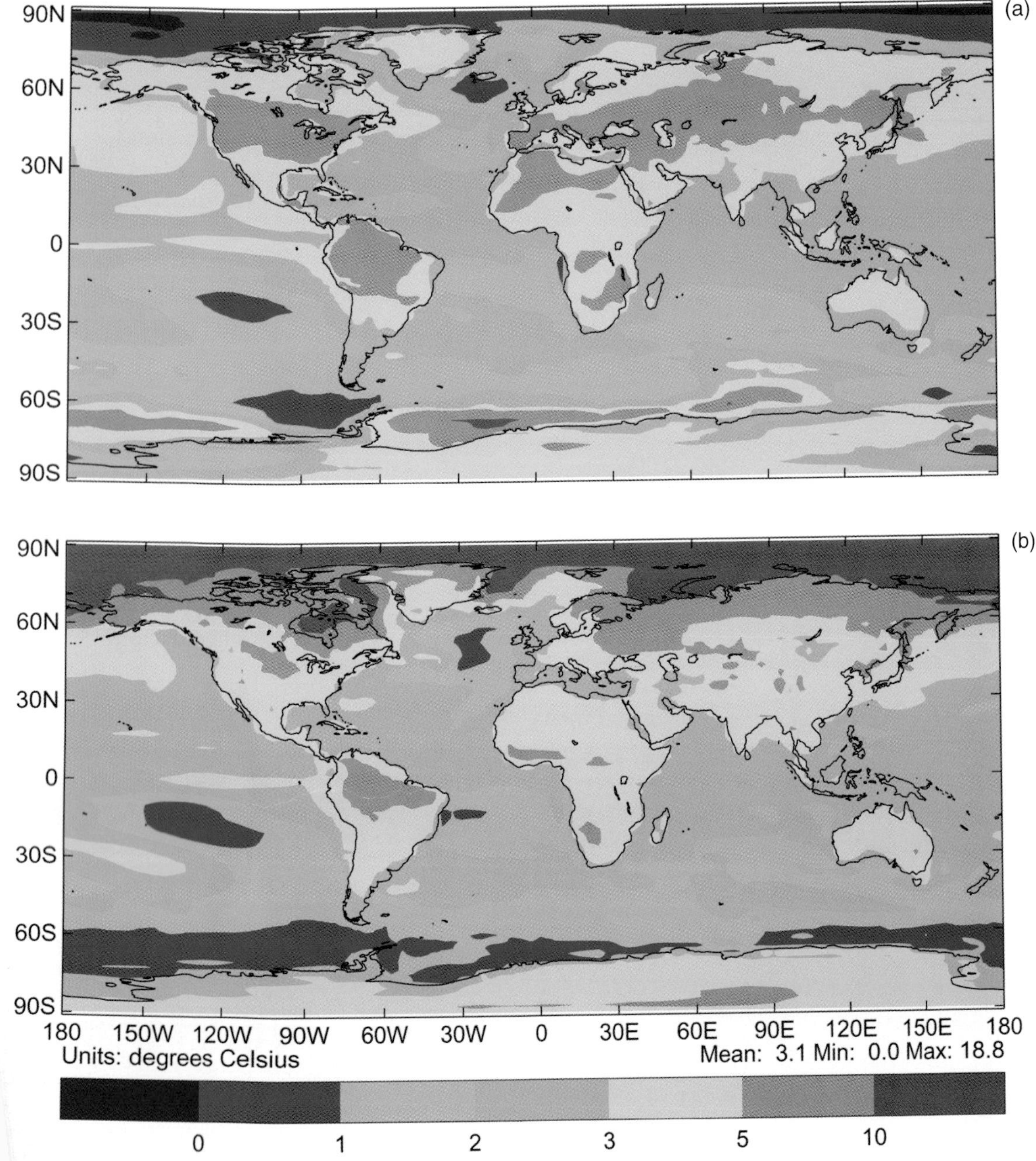

ate 3 Change in average surface air temperature from 1960 to 1990 to 2070 to 2100 based on IPCC scenario IS92a ıd the HadCM3 global climate model. (a) June–July–August; (b) December–January–February. From Hadley Centre)02, Hadley Centre for Climate Prediction and Research, Meteorological Office, UK *http://www.meto.gov.uk/research/ dleycentre/index.html.* © Crown Copyright. Reproduced with permission. (See Chapter 4).

gas hydrates. These forms of carbon occur in some shallow sea sediments and tundra. Warming could melt these clathrates, releasing large quantities of methane or carbon dioxide into the water column and from there into the atmosphere, increasing global warming (Wilde and Quinby-Hunt 1997).

Studies using a coupled ocean-atmosphere model suggest that increased rainfall from greenhouse warming will result in surface freshening (decreased density) and stratification of seawater over a large area of the Southern Ocean. The decrease in downward mixing and transport of heat and carbon to the deep ocean could substantially decrease the oceanic uptake of CO_2 over the next few decades (Sarmiento et al. 1998).

9 Overall climate carbon cycle feedbacks

At least one study linking a carbon cycle model to an AOGCM suggests that global warming will reduce both terrestrial and oceanic uptake of CO_2. Net ecosystem productivity (carbon storage) will be strongly reduced in the subtropics by increases in soil aridity. At the same time, three factors will reduce oceanic uptake of carbon, especially at high latitudes: (1) decreased solubility of CO_2 at higher temperatures, (2) increased density stratification with warm waters lying on the surface that reduce vertical mixing and transport of CO_2 to the deep ocean, and (3) changes in the biogeochemical cycle of CO_2. The gain in atmospheric CO_2 from these feedbacks is 10% with a doubled and 20% with a quadrupled CO_2 atmosphere. This translates into a 15% higher mean global temperature increase than would occur in the absence of these feedbacks (Friedlingstein et al. 2001).

10 The human dimension

Perhaps the greatest sources of uncertainty in predicting future climate change arise from the human dimension (Chapters 9 to 12). Future greenhouse gas emissions depend on human population and economic growth, and per capita emissions. New technologies might directly reduce or sequester greenhouse gas emissions from fossil-fuel combustion. Continued increases in energy efficiency and use of alternative (nonfossil fuel) energy could reduce emission growth rates. Mitigation and adaptation could, in a variety of ways, offset the negative impacts of climate change. Continuing attempts at international agreements could fail or could succeed in greatly reducing greenhouse gas emissions.

Scenario-Based Climate Predictions

Despite the complexity of the Earth-climate system and the uncertainties involved in modeling, our ability to predict the climatic effects of greenhouse gas emissions continues to improve. New models consider the effects of sulfate aerosol cooling, of changes in the Earth's albedo due to melting ice, and the changes in atmospheric trace gas concentrations other than CO_2. The ability to "predict" changes that already occurred during the twentieth century or earlier has led to improved confidence in these models. Nevertheless, models often differ significantly in their predicted outputs. The Intergovernmental Panel on Climate Change (IPCC), to encompass the range of uncertainty in predicted changes, collected results from numerous models incorporating transient increases in greenhouse gases. Most studies assume a 1% per year increase in atmospheric greenhouse gas concentrations until CO_2 doubling, tripling, or quadrupling. In addition, the IPCC developed a series of emission scenarios (SRES) to encompass possible ranges of future human population and economic growth that influence fossil-fuel consumption (Box 4.3).

Box 4.3 Emission scenarios (SRES)

In earlier studies, the IPCC used a set of scenarios called IS92 for a range of future economic and population assumptions. For example, IS92a, often called the *business as usual scenario*, assumes continuation of current rates of population and economic growth. Following 1996, the IPCC developed a set of 40 new scenarios covering a wide range of possible future demographic, economic, and technological forces that influence future greenhouse gas and aerosol emissions. In the Special Report on Emission Scenarios (SRES), each scenario represents a specific quantification of four primary "storylines." However, for illustrative purposes, six Illustrative Marker Scenarios are most often cited (A1B, A1T, A1FI, A2, B1, and B2). Detailed definitions and references to these scenarios can be found in IPCC (2001).

A1. The A1 storyline and scenario family describes a future world of very rapid economic growth, global population that peaks in midcentury and declines thereafter, and the rapid introduction of new and more efficient technologies. Major underlying themes are convergence among regions, infrastructure capacity building, and increased cultural and social interactions, with substantial reduction in regional differences in per capita income. The A1 scenario family develops into three groups that describe alternative directions of technological change in the energy system. The three A1 groups are distinguished by their technological emphasis: fossil-intensive (A1FI), nonfossil energy sources (A1T), or a balance across all sources (A1B) (where balance is defined as not relying too heavily on one particular energy source, on the assumption that similar improvement rates apply to all energy supply and end-use technologies).

A2. The A2 storyline and scenario family describe a very heterogeneous world. The underlying theme is self-reliance and preservation of local identities. Fertility patterns across regions converge very slowly, which results in continuously increasing populations. Economic development is primarily region-oriented, and *per capita* economic growth and technological change are more fragmented and slower than in other storylines.

B1. The B1 storyline and scenario family describes a convergent world with the same global population that peaks in midcentury and declines thereafter, as in the A1 storyline, but with rapid change in economic structures toward a service and information economy, with reduction in material intensity and the introduction of clean and resource-efficient technologies. The emphasis is on global solutions to economic, social, and environmental sustainability, including improved equity, but without additional climate initiatives.

B2. The B2 storyline and scenario family describe a world in which the emphasis is on local solutions to economic, social, and environmental sustainability. It is a world with continuously increasing global population, at a rate lower than A2, intermediate levels of economic development, and less rapid and more diverse technological change than in the B1 and A1 storylines. While the scenario is also oriented toward environmental protection and social equity, it focuses on local and regional levels.

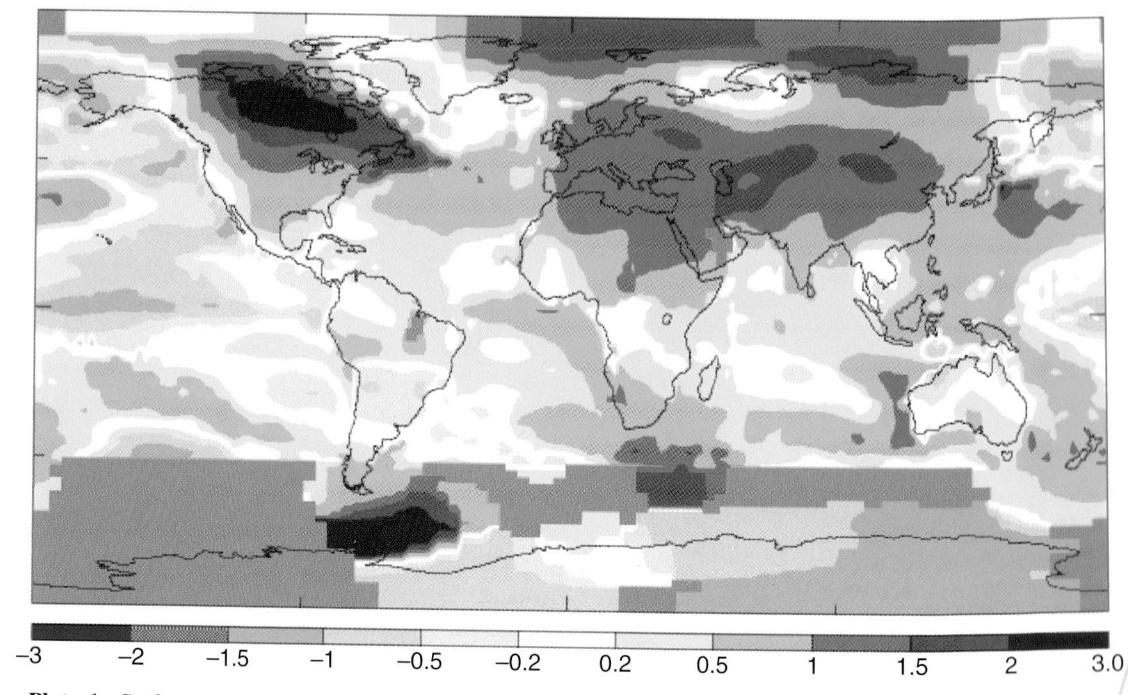

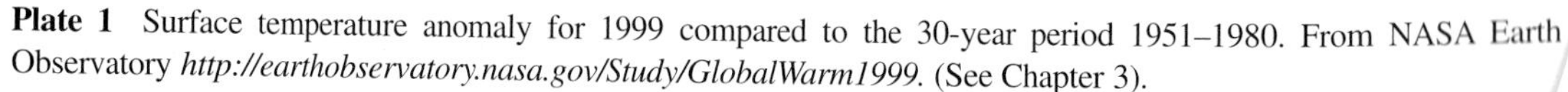

Plate 1 Surface temperature anomaly for 1999 compared to the 30-year period 1951–1980. From NASA Earth Observatory *http://earthobservatory.nasa.gov/Study/GlobalWarm1999*. (See Chapter 3).

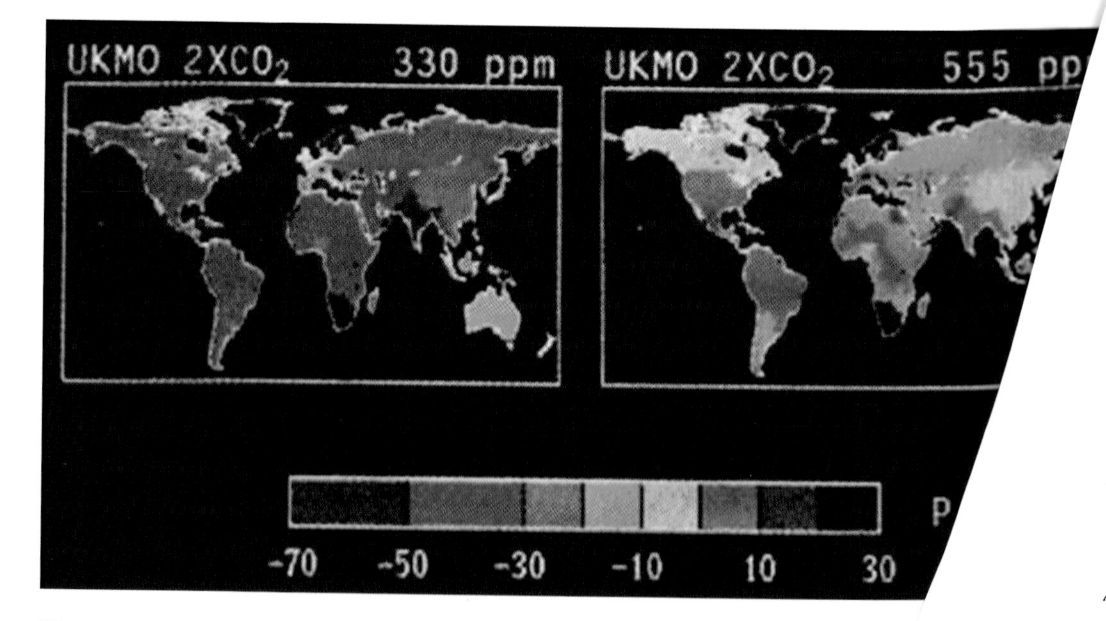

Plate 2 Predicted change in national grain yield (wheat, rice, coarse grains, and protein feed) ir of atmospheric CO_2. Includes direct CO_2 fertilization effect. Level 1 (left) adaptation signifies n agricultural systems; level 2 (right) adaptation signifies major changes. Based on UK Mete Climate Model. Reproduced with permission from Rosenzweig C and Parry M 1994, Potentia on world food supply. *Nature* **367**: 133–138. Copyright © 1994 Macmillan Magazines Limi

gas hydrates. These forms of carbon occur in some shallow sea sediments and tundra. Warming could melt these clathrates, releasing large quantities of methane or carbon dioxide into the water column and from there into the atmosphere, increasing global warming (Wilde and Quinby-Hunt 1997).

Studies using a coupled ocean-atmosphere model suggest that increased rainfall from greenhouse warming will result in surface freshening (decreased density) and stratification of seawater over a large area of the Southern Ocean. The decrease in downward mixing and transport of heat and carbon to the deep ocean could substantially decrease the oceanic uptake of CO_2 over the next few decades (Sarmiento et al. 1998).

9 Overall climate carbon cycle feedbacks

At least one study linking a carbon cycle model to an AOGCM suggests that global warming will reduce both terrestrial and oceanic uptake of CO_2. Net ecosystem productivity (carbon storage) will be strongly reduced in the subtropics by increases in soil aridity. At the same time, three factors will reduce oceanic uptake of carbon, especially at high latitudes: (1) decreased solubility of CO_2 at higher temperatures, (2) increased density stratification with warm waters lying on the surface that reduce vertical mixing and transport of CO_2 to the deep ocean, and (3) changes in the biogeochemical cycle of CO_2. The gain in atmospheric CO_2 from these feedbacks is 10% with a doubled and 20% with a quadrupled CO_2 atmosphere. This translates into a 15% higher mean global temperature increase than would occur in the absence of these feedbacks (Friedlingstein et al. 2001).

10 The human dimension

Perhaps the greatest sources of uncertainty in predicting future climate change arise from the human dimension (Chapters 9 to 12). Future greenhouse gas emissions depend on human population and economic growth, and per capita emissions. New technologies might directly reduce or sequester greenhouse gas emissions from fossil-fuel combustion. Continued increases in energy efficiency and use of alternative (nonfossil fuel) energy could reduce emission growth rates. Mitigation and adaptation could, in a variety of ways, offset the negative impacts of climate change. Continuing attempts at international agreements could fail or could succeed in greatly reducing greenhouse gas emissions.

Scenario-Based Climate Predictions

Despite the complexity of the Earth-climate system and the uncertainties involved in modeling, our ability to predict the climatic effects of greenhouse gas emissions continues to improve. New models consider the effects of sulfate aerosol cooling, of changes in the Earth's albedo due to melting ice, and the changes in atmospheric trace gas concentrations other than CO_2. The ability to "predict" changes that already occurred during the twentieth century or earlier has led to improved confidence in these models. Nevertheless, models often differ significantly in their predicted outputs. The Intergovernmental Panel on Climate Change (IPCC), to encompass the range of uncertainty in predicted changes, collected results from numerous models incorporating transient increases in greenhouse gases. Most studies assume a 1% per year increase in atmospheric greenhouse gas concentrations until CO_2 doubling, tripling, or quadrupling. In addition, the IPCC developed a series of emission scenarios (SRES) to encompass possible ranges of future human population and economic growth that influence fossil-fuel consumption (Box 4.3).

Box 4.3 Emission scenarios (SRES)

In earlier studies, the IPCC used a set of scenarios called IS92 for a range of future economic and population assumptions. For example, IS92a, often called the *business as usual scenario*, assumes continuation of current rates of population and economic growth. Following 1996, the IPCC developed a set of 40 new scenarios covering a wide range of possible future demographic, economic, and technological forces that influence future greenhouse gas and aerosol emissions. In the Special Report on Emission Scenarios (SRES), each scenario represents a specific quantification of four primary "storylines." However, for illustrative purposes, six Illustrative Marker Scenarios are most often cited (A1B, A1T, A1FI, A2, B1, and B2). Detailed definitions and references to these scenarios can be found in IPCC (2001).

A1. The A1 storyline and scenario family describes a future world of very rapid economic growth, global population that peaks in midcentury and declines thereafter, and the rapid introduction of new and more efficient technologies. Major underlying themes are convergence among regions, infrastructure capacity building, and increased cultural and social interactions, with substantial reduction in regional differences in per capita income. The A1 scenario family develops into three groups that describe alternative directions of technological change in the energy system. The three A1 groups are distinguished by their technological emphasis: fossil-intensive (A1FI), nonfossil energy sources (A1T), or a balance across all sources (A1B) (where balance is defined as not relying too heavily on one particular energy source, on the assumption that similar improvement rates apply to all energy supply and end-use technologies).

A2. The A2 storyline and scenario family describe a very heterogeneous world. The underlying theme is self-reliance and preservation of local identities. Fertility patterns across regions converge very slowly, which results in continuously increasing populations. Economic development is primarily region-oriented, and *per capita* economic growth and technological change are more fragmented and slower than in other storylines.

B1. The B1 storyline and scenario family describes a convergent world with the same global population that peaks in midcentury and declines thereafter, as in the A1 storyline, but with rapid change in economic structures toward a service and information economy, with reduction in material intensity and the introduction of clean and resource-efficient technologies. The emphasis is on global solutions to economic, social, and environmental sustainability, including improved equity, but without additional climate initiatives.

B2. The B2 storyline and scenario family describe a world in which the emphasis is on local solutions to economic, social, and environmental sustainability. It is a world with continuously increasing global population, at a rate lower than A2, intermediate levels of economic development, and less rapid and more diverse technological change than in the B1 and A1 storylines. While the scenario is also oriented toward environmental protection and social equity, it focuses on local and regional levels.

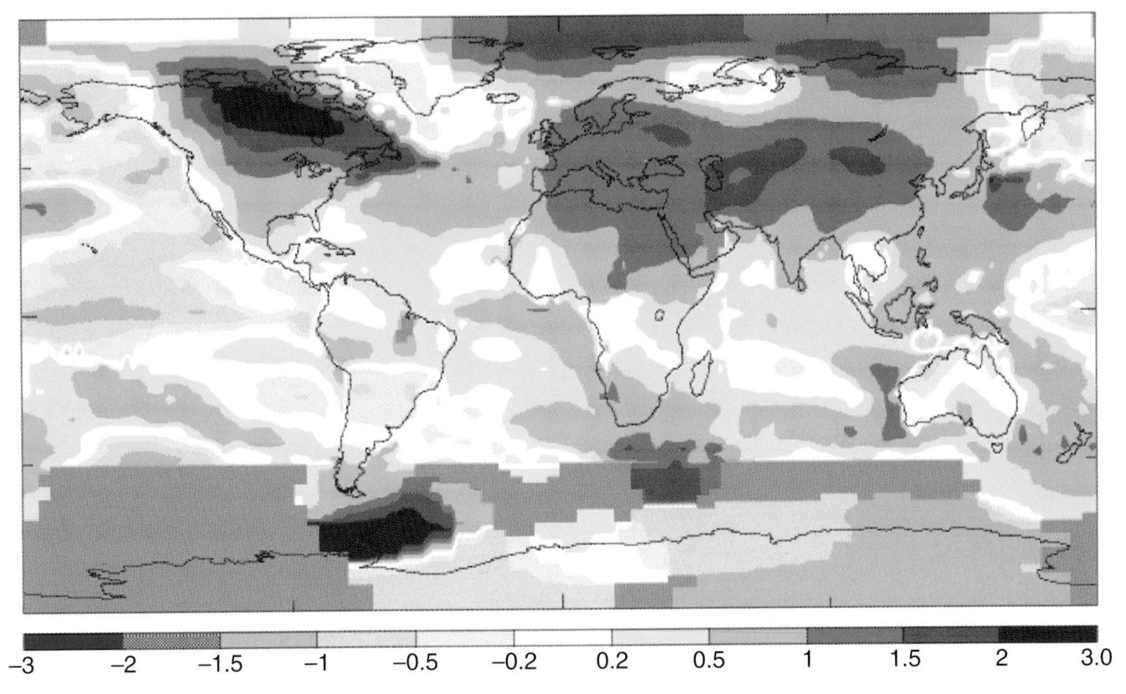

Plate 1 Surface temperature anomaly for 1999 compared to the 30-year period 1951–1980. From NASA Earth Observatory *http://earthobservatory.nasa.gov/Study/GlobalWarm1999*. (See Chapter 3).

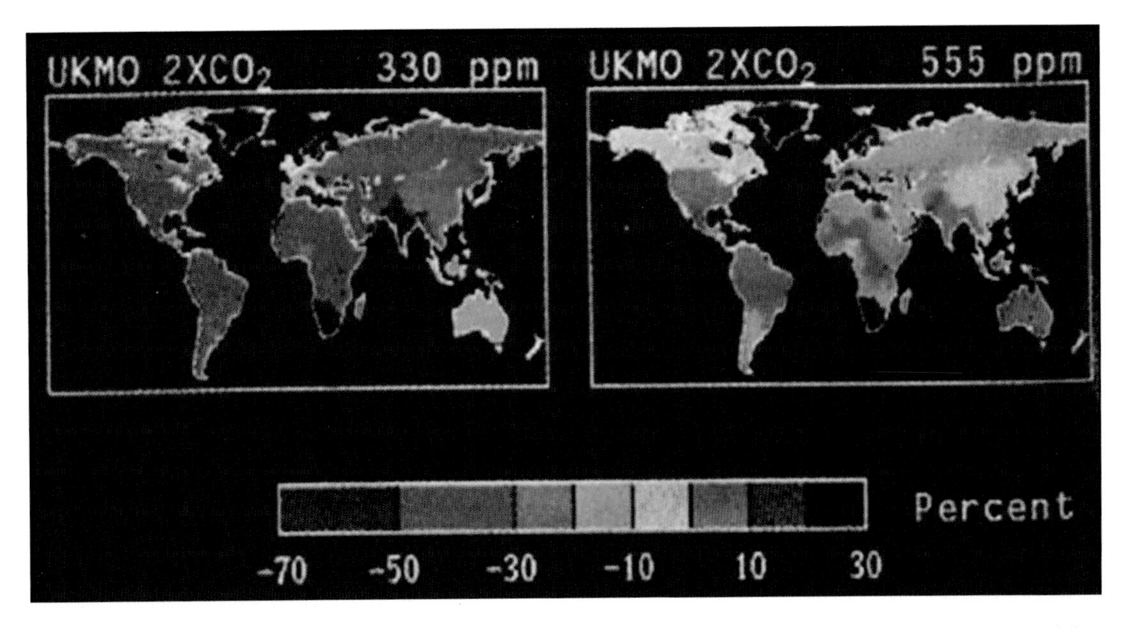

Plate 2 Predicted change in national grain yield (wheat, rice, coarse grains, and protein feed) in response to a doubling of atmospheric CO_2. Includes direct CO_2 fertilization effect. Level 1 (left) adaptation signifies minor changes to existing agricultural systems; level 2 (right) adaptation signifies major changes. Based on UK Meteorological Office Global Climate Model. Reproduced with permission from Rosenzweig C and Parry M 1994, Potential impact of climate change on world food supply. *Nature* **367**: 133–138. Copyright © 1994 Macmillan Magazines Limited. (See Chapter 7).

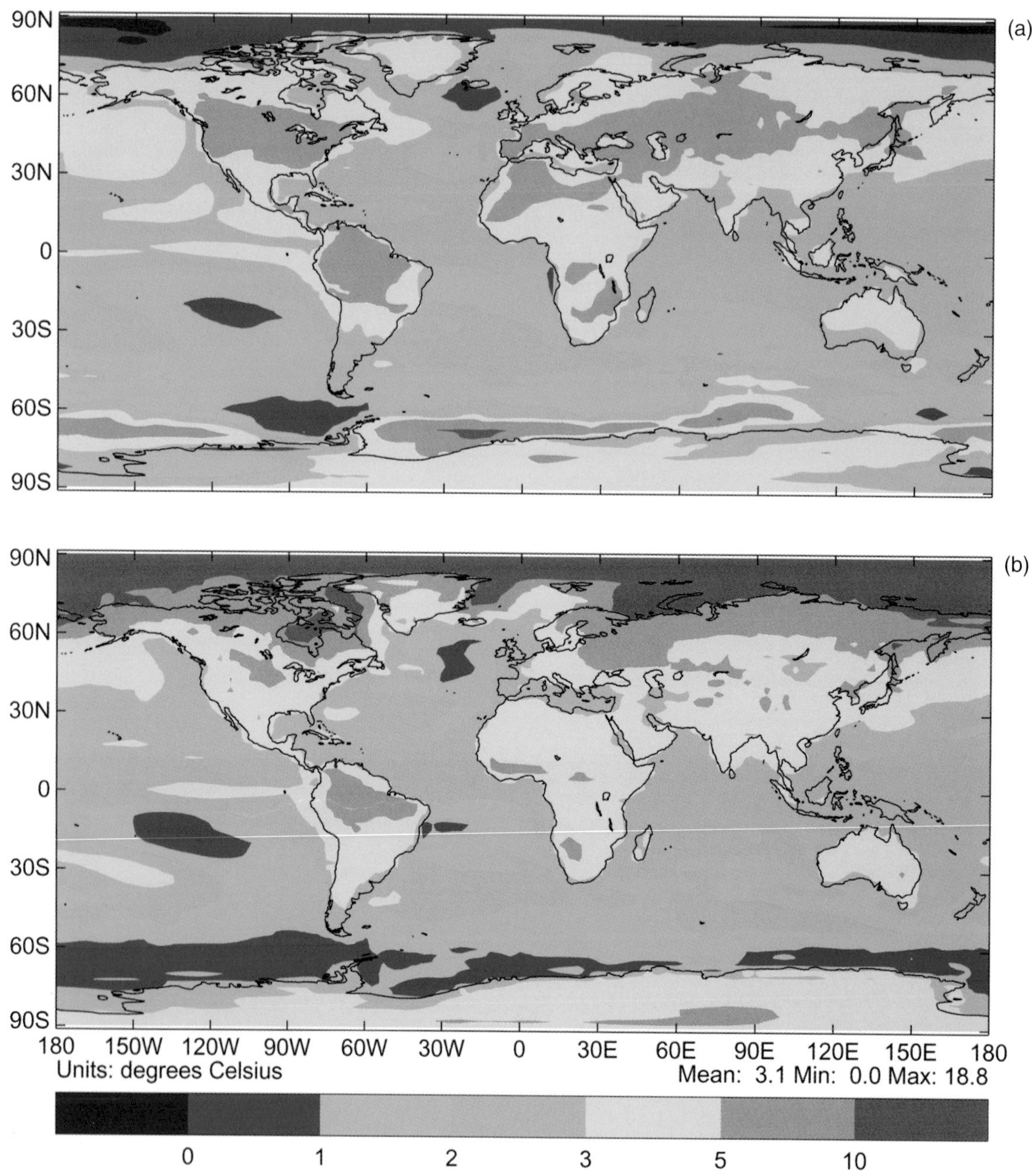

Plate 3 Change in average surface air temperature from 1960 to 1990 to 2070 to 2100 based on IPCC scenario IS92a and the HadCM3 global climate model. (a) June–July–August; (b) December–January–February. From Hadley Centre 2002, Hadley Centre for Climate Prediction and Research, Meteorological Office, UK *http://www.meto.gov.uk/research/hadleycentre/index.html.* © Crown Copyright. Reproduced with permission. (See Chapter 4).

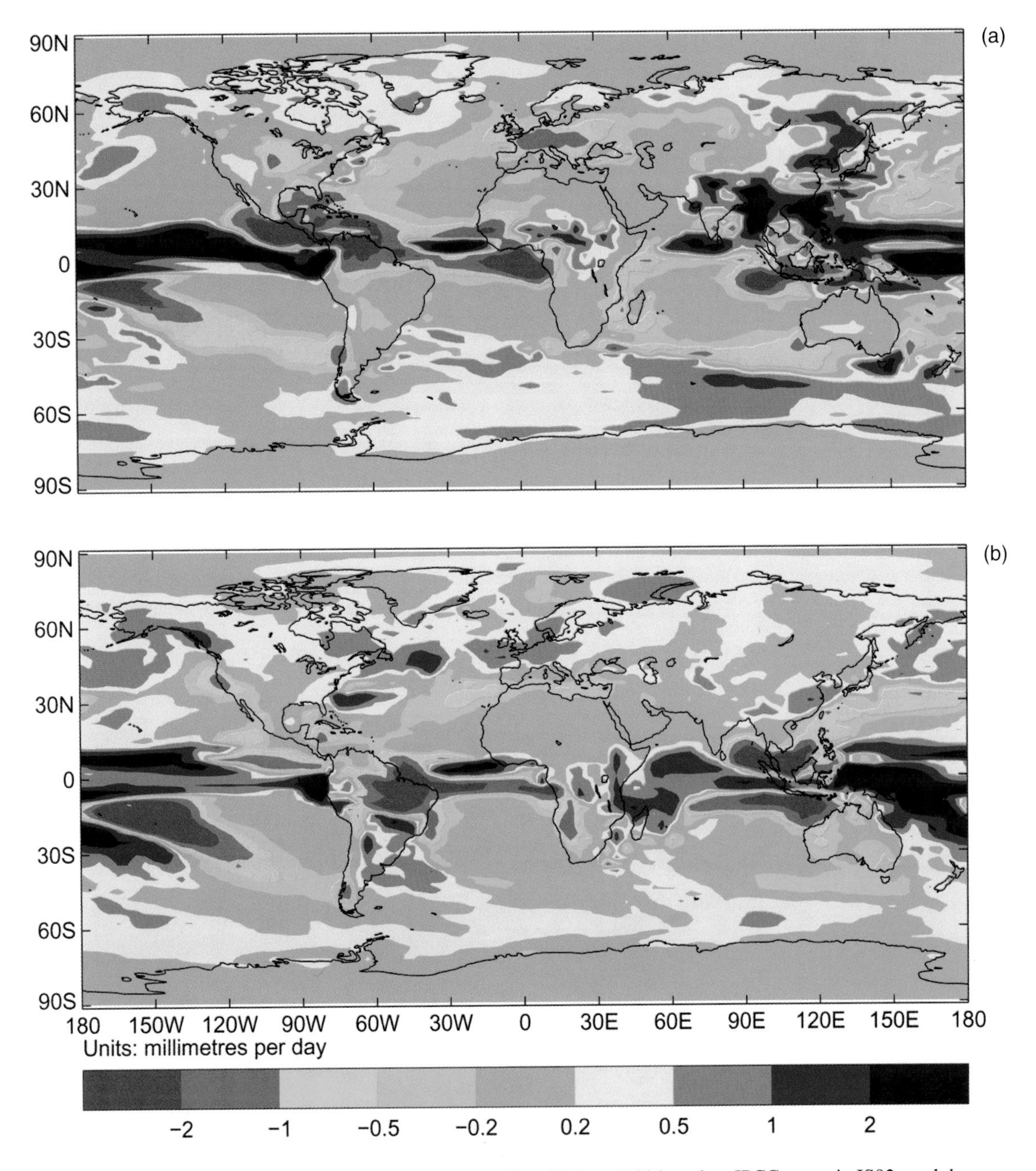

Plate 4 Change in average precipitation from 1960 to 1990 to 2070 to 2100 based on IPCC scenario IS92a and the HadCM3 global climate model. (a) June–July–August; (b) December–January–February. From Hadley Centre 2002, Hadley Centre for Climate Prediction and Research, Meteorological Office, UK *http://www.meto.gov.uk/research/hadleycentre/index.html.* © Crown Copyright. Reproduced with permission. (See Chapter 4).

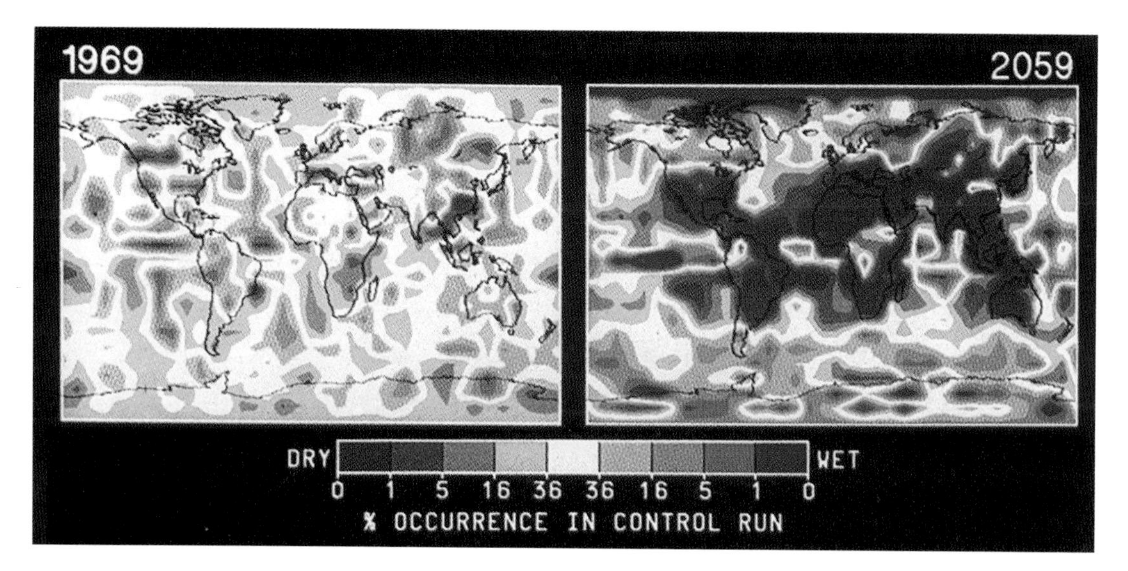

Plate 5 The occurrence of drought for the period June through August. The color scale is set by the frequency of occurrence of drought in a 100-year control run using a 1958 atmospheric composition. From Rind D, Goldberg R, Hansen J, Rosenweig C and Ruedy R, Potential evapotranspiration and the likelihood of future drought. *Journal of Geophys. Res.* (atmospheres) **95**: p. 9994. Copyright © 1990 American Geophysical Union. Reproduced with permission from the American Geophysical Union. (See Chapter 5).

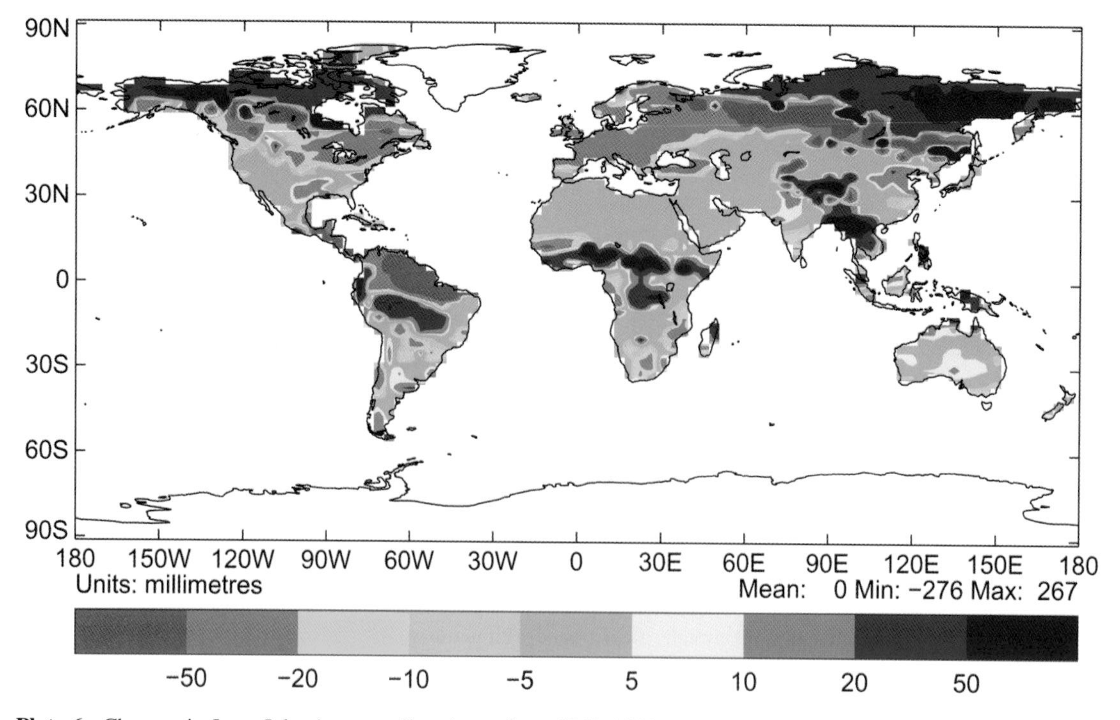

Plate 6 Changes in June–July–August soil moisture from 1960–1990 to 2070–2100 from the Hadley Centre Model HadCM3 using a IS92a "business as usual" scenario. Hadley Centre for Climate Prediction and Research, Meteorological Office, UK. © Crown Copyright. Reproduced with permission. (See Chapter 5).

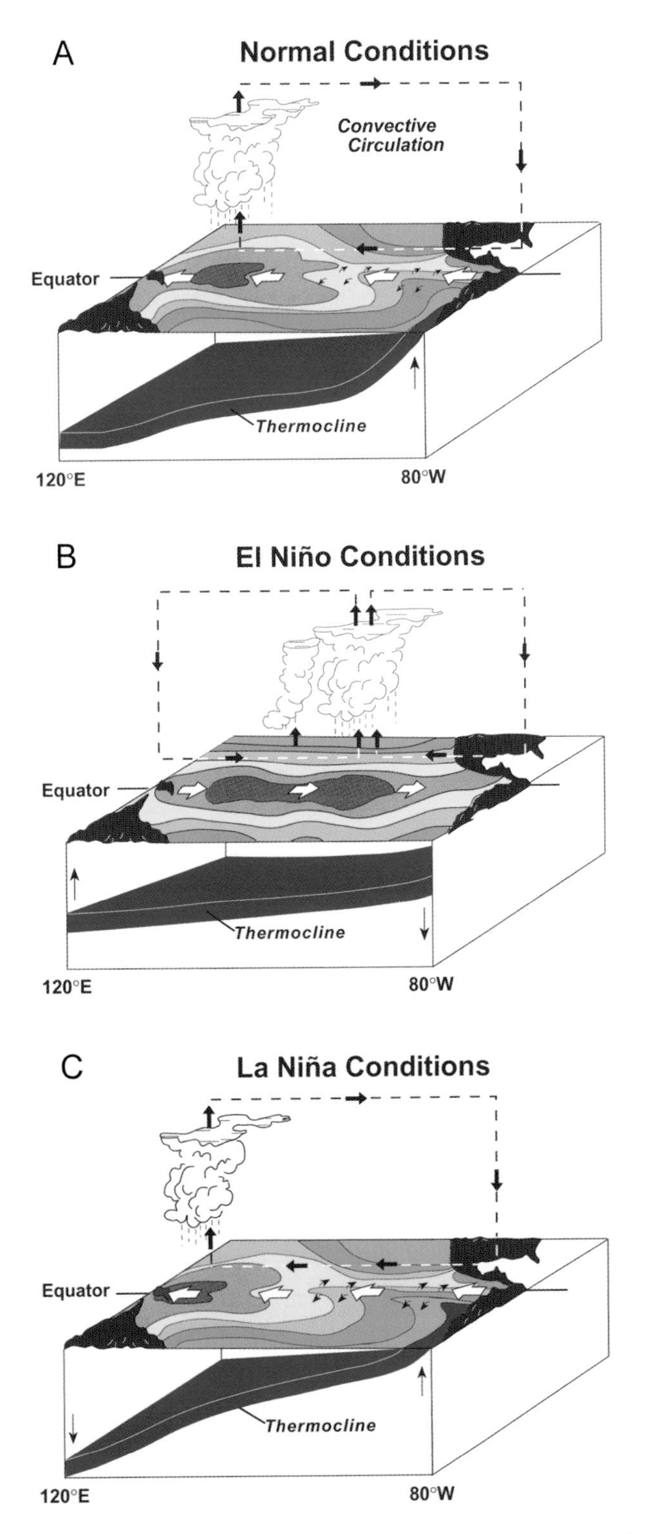

Plate 7 The El Niño Southern Oscillation (ENSO). Reproduced with permission from NOAA/PMEL/TAO Project Office, Dr MJ McPhaden, Director. (See Chapter 8).

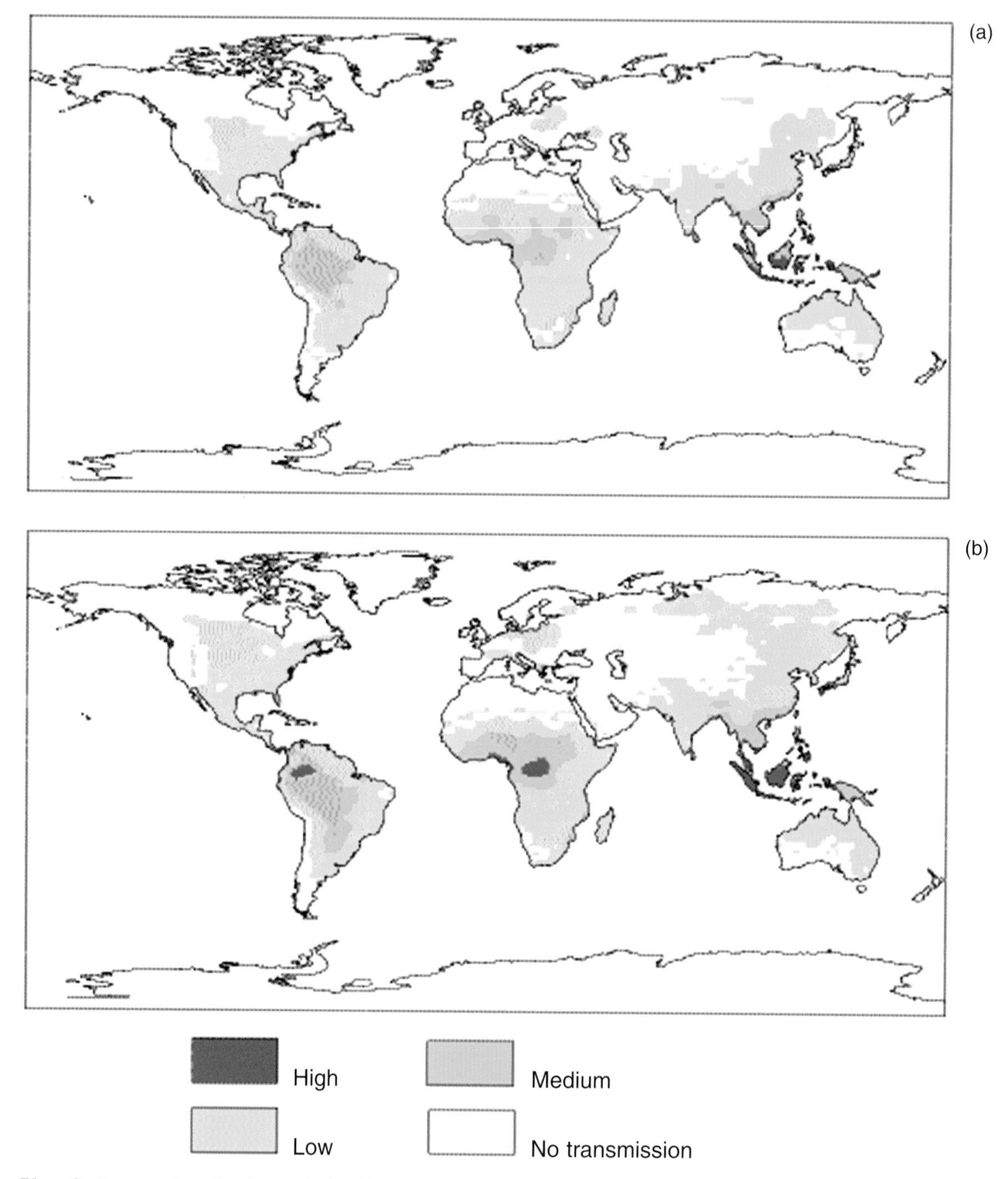

Plate 8 Increase in risk of malaria by 2050s from the expanding habitat of the mosquito vector *Plasmodium falciparum.* (a) 1961–1990 baseline conditions; (b) under the HadCM3 model climate change projection. Adapted from McMichael AJ and Haines A 1997, Global climate change: the potential effects on health. *British Medical Journal* **315**: 805–809. (See Chapter 10).

Greenhouse gases and aerosols

Concentrations of greenhouse gases will continue to increase during this century under virtually all scenarios. Atmospheric CO_2 will reach double preindustrial levels well before 2100. In a typical forecast, based on a number of different models and assuming an SRES A1B population-economic scenario, by 2100 atmospheric CO_2 will increase to more than 700 ppm (parts per million) and CH_4, after peaking about the year 2050 at 2,400 ppb (parts per billion), will decrease somewhat (Figure 4.7). Overall, a variety of model approaches and scenarios predict CO_2 concentrations of 540 to 970 ppm by 2100, compared to the 250-ppm concentration in the year 1750. However, uncertainties, especially about the feedback from the terrestrial biosphere, expand the total range of possibilities to 490 to 1,260 ppm (IPCC 2001). Predicted changes in other greenhouse gases and aerosols vary widely.

Temperature

Predicted temperatures from numerous transient models, incorporating several greenhouse gases as well as the effects of water vapor and sulfate aerosols, and based on 35 different SRES predict a global average warming of 1.4 to 5.8 °C for the period 1990 to 2100 (IPCC 2001). However, on the basis of a more selected "ensemble" of climate models, the global average temperature increase is most likely to range between 2.0 and 4.5 °C (Figure 4.8). Temperatures in winter and at higher latitudes may increase to more than twice the global average (Plate 3).

Precipitation

Average global precipitation will increase by >10%, but change differs both seasonally and regionally. Model experiments at the UK Hadley Centre assume a midrange economic growth and "business as usual" emission scenario in which CO_2 more than doubles over the course of this century without

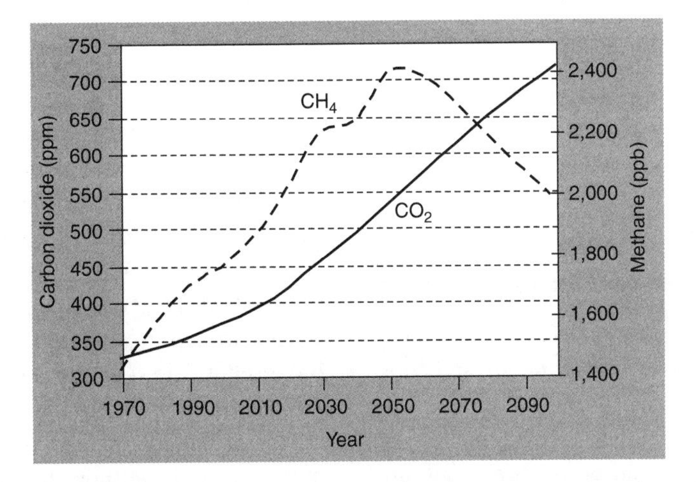

Fig. 4.7 Recent and future atmospheric abundances of carbon dioxide and methane. Projections based on a single example scenario SRES A1B (Box 4.3) (Based on data from IPCC 2001. Houghton JT, Ding Y, Griggs DJ, Noguer M, van der Linden PJ, Dai X, et al. eds *Climate Change 2001: The Scientific Basis*. Intergovernmental Panel on Climate Change, Working Group 1. Cambridge: Cambridge University Press, Appendix II. Reproduced by permission of Intergovernmental Panel on Climate Change).

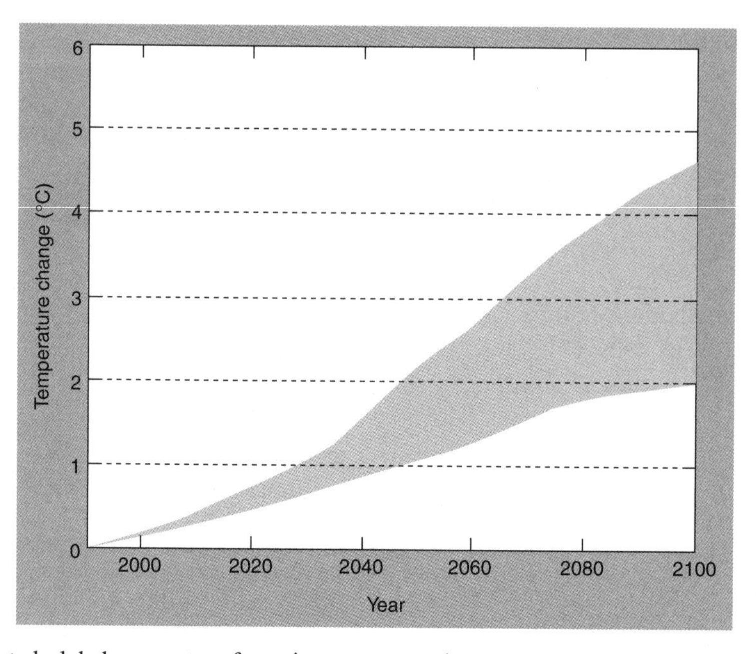

Fig. 4.8 Predicted global average surface air temperature increase to 2100. The shaded area represents the range of outputs for the full set of 35 SRES scenarios based on the mean results of seven different AOGCMs for a doubling of CO_2. The range of the global mean temperature increase from 1990 to 2100 is 2.0 to 4.5 °C (Adapted from Cubasch U and Meehl GA 2001. Projections of future climate change. In: Houghton JT, Ding Y, Griggs DJ, Noguer M, van der Linden PJ, Dai X et al., eds *Climate Change 2001: The Scientific Basis*. Intergovernmental Panel on Climate Change, Working Group 1. Cambridge: Cambridge University Press, pp. 525–582. Reproduced by permission of Intergovernmental Panel on Climate Change).

measures to reduce emissions. In that case, many areas between 5 and 25° latitude, and midcontinental areas elsewhere, will become dryer (Plate 4).

Regional Climates and Extreme Events

Regional climate models (RCMs) are being developed to improve spatial detail and look at local and regional change. Course resolution AOGCMs simulate ocean and atmosphere general circulation features and are used for predicting global change. However, changes at finer scales can be different in magnitude or direction from the larger-scale AOGCMs. Topography, land-use patterns, and the surface hydrologic cycle strongly affect climate at the regional to local scale. These models reveal a number of differences between regions. For example, compared to the global mean, warming will be greater over land areas, especially at high latitudes in winter, while it will be less in June–August in South Asia and Southern South America. European summer temperatures will increase by about 1.5 to 4 °C by 2080 (Figure 4.9).

Precipitation will increase over northern midlatitude regions in winter and over northern high-latitude regions and Antarctica in both winter and summer. In December–January–February, rainfall will increase in tropical Africa and decrease in Central America. Precipitation will decrease over Australia in winter and over the

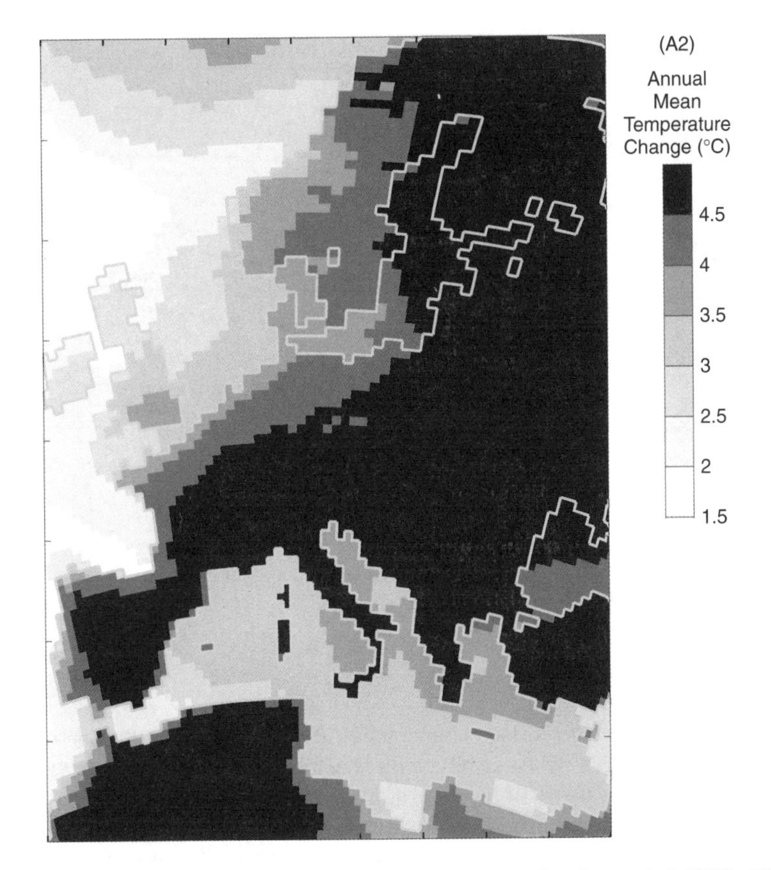

Fig. 4.9 Predicted annual mean temperature increase for Europe for the period 2070–2099 compared to the 1961–1990 measured mean. Based on the UK Meteorological Office Hadley Centre model HadRM3 and the average of three IPCC SRES A2 scenarios (Courtesy D. Viner, LINK Project, University of East Anglia).

Mediterranean region in summer (Giorgi and Hewitson 2001).

Models predict, in addition to warmer average temperatures, a greater frequency of extremely warm days and a lower frequency of extremely cold days. Extremes of temperature and precipitation that now occur on average every 20 years will probably occur more frequently leading, in some areas, to increased "urban heat waves" or flooding. There will be a general drying trend of the midcontinent areas during summer and an increased chance of drought. The Indian monsoon variability will increase, thus increasing the chances of extreme dry and wet monsoon seasons (Meehl et al. 2000).

The Persistence of a Warmer Earth

The climate may take a long time and ecosystems even longer (Chapters 5 to 8) to heal from the wounds inflicted by human-induced climate change. The Earth and the oceanic heat sink respond slowly to insult. Models suggest that the human-induced global warming may continue for centuries. The uncertainty of predicted climate change increases

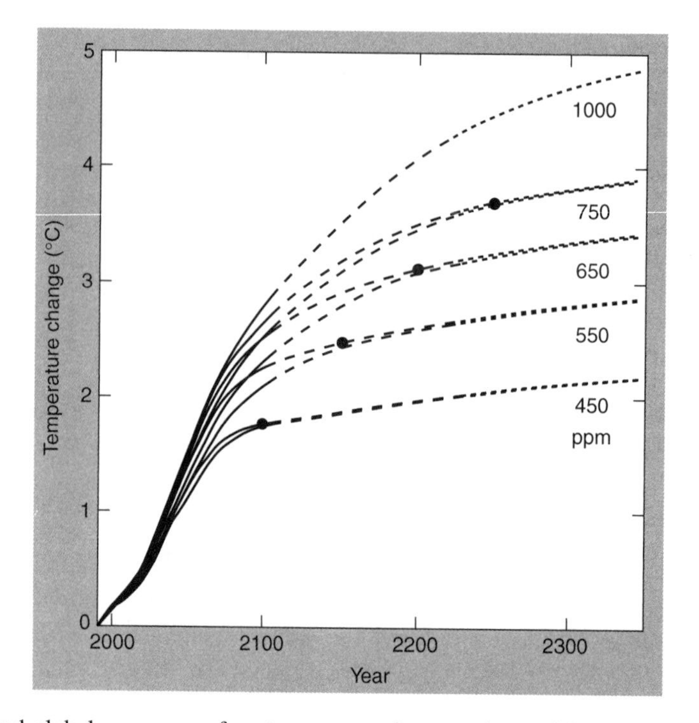

Fig. 4.10 Predicted global average surface temperature increase beyond 2100. The black dots represent the time and concentration of CO_2 when stabilization is achieved. Projections assume that emissions of CO_2 and non-CO_2 greenhouse gases will increase in accordance with the A1B scenario (Box 4.3) out to 2100. In 2100 sulfur dioxide emissions will stabilize. After 2100 the emissions of non-CO_2 gases will remain constant (From Cubasch U and Meehl GA 2001. Projections of future climate change. In: Houghton JT, Ding Y, Griggs DJ, Noguer M, van der Linden PJ, Dai X et al., eds *Climate Change 2001: The Scientific Basis*. Intergovernmental Panel on Climate Change, Working Group 1. Cambridge: Cambridge University Press, pp. 525–582. Reproduced by permission of Intergovernmental Panel on Climate Change).

as we project beyond the twenty-first century. However, various scenarios can be examined to illustrate the range of possibilities. Long-term temperature trends will depend on when emissions are reduced enough to stabilize the atmospheric concentrations of greenhouse gases. The longer it takes to stabilize atmospheric CO_2, the greater will be its concentration and the resultant warming potential. Temperatures will continue to increase after CO_2 stabilization owing to the inertia of the climate system, which will require several centuries to come into equilibrium with a particular level of radiative forcing (Figure 4.10).

Even if all emissions of greenhouse gases decline linearly to zero from 2100 to 2200, the Earth's climate will probably remain altered for centuries to come (IPCC 2001). These changes will have serious effects on the Earth's ecosystems that support human civilization. If we continue our current pattern of fossil-fuel consumption, the concentration of atmospheric CO_2 could quadruple over the next several centuries. This could lead to a 7 °C increase in global average temperature over the next 500 years and result in a climate not experienced on Earth since the early tertiary or late

Cretaceous period – over 140,000 years ago (Manabe 1998).

Summary

The science of climate prediction has improved immensely during the past few decades. New and more refined models, incorporating many of the known feedbacks, indicate that the climate is already warming in response to anthropogenic emissions of greenhouse gases. Furthermore, changes in temperature and precipitation patterns will continue and accelerate during this century and possibly for centuries to come. Several important uncertainties in the current numerical models frustrate predictions of climate. Chief among these is the prediction of human behavior. How will human population growth, economic change, and improvements in technology alter our consumption of fossil fuel? Despite the uncertainties, policy decisions to deal with fossil-fuel emissions must be made (see Chapter 12). We could decide to reduce emissions immediately as an insurance policy against future damage. On the other hand, we could wait until predictions are even more certain, delaying immediate costs of emissions reductions, but risking the possibility of environmental catastrophe. Our choice will be critical in determining the future health of the Earth's ecosystems and the human population.

References

Arrhenius S 1896 On the influence of carbonic acid in the air upon the temperature of the ground. *Philosophical Magazine and Journal of Science. Series 5* **41**(251): 237–276.

Baskin Y 1993 Ecologists put some life into models of a changing world. *Science* **259**: 1694–1696.

Betts RA 2000 Offset of the potential carbon sink from boreal forestation by decreases in surface albedo. *Nature* **408**: 187–190.

Cess RD, Potter GL, Blanchet JP, Boer GJ, Ghan SJ, Kiehl JT, et al. 1989 Interpretation of cloud-climate feedback as produced by 14 atmospheric circulation models. *Science* **245**: 513–516.

Charlson RJ, Lovelock JE, Andreae MO and Warren SG 1987 Oceanic phytoplankton, atmospheric sulphur, cloud albedo and climate. *Nature* **326**(6114): 655–661.

Charlson RJ, Schwartz SE, Hales JM, Cess RD, Coakley Jr JA, Hansen JE, et al. 1992 Climate forcing by anthropogenic aerosols. *Science* **255**: 423–430.

Charlson RJ and Wigley TML 1994 Sulfate aerosol and climatic change. *Scientific American* **February**: 48–57.

Cubasch U and Meehl GA 2001 Projections of future climate change. In: Houghton JT, Ding Y, Griggs DJ, Noguer M, van der Linden PJ, Dai X, et al. eds *Climate Change 2001: The Scientific Basis*. Intergovernmental Panel on Climate Change, Working Group 1. Cambridge: Cambridge University Press, pp. 525–582.

Dixon RK, Brown S, Houghton RA, Solomon AM, Trexler MC and Wisniewski J 1994 Carbon pools and flux of global forest ecosystems. *Science* **263**: 185–190.

Freeman C, Evans CD, Monteith DT, Reynolds B and Fenner N 2001 Export of organic carbon from peat soils. *Nature* **412**: 785.

Friedlingstein P, Fairhead L, LeTreut H, Monfray P and Orr J 2001 Positive feedback between future climate change and the carbon cycle. *Geophysical Research Letters* **28**(8): 1543–1546.

Fuglestvedt JS, Isaksen ISA and Wang WC 1996 Estimates of indirect global warming potentials for CH_4, CO and NO_X. *Climatic Change* **34**: 405–437.

Gates WL, Henderson-Sellers A, Boer GJ, Folland CK, Kitoh A, et al. 1996 In: Houghton JT, Filho LGM, Callender BA, Harris N, Kattenberg A, Maskell K, eds *Climate Change 1995: The Science of Climate Change*. Intergovernmental Panel on Climate Change, World Meterological Organization and United Nations Environment Program. Cambridge: Cambridge University Press, pp. 228–284.

Giorgi F and Hewitson B 2001 Regional climate information – evaluation and projections. In: Houghton JT, Ding Y, Griggs DJ, Noguer M, van der Linden PJ, Dai X, et al. eds *Climate Change 2001: The Scientific Basis*. Intergovernmental

Panel on Climate Change, Working Group 1. Cambridge: Cambridge University Press, pp. 583–638.

Groisman PY, Karl TR and Knight RW 1994 Observed impact of snow cover on the heat balance and the rise of spring continental temperatures. *Science* **263**: 198–200.

Hadley Centre 2002 *Hadley Centre for Climate Prediction and Research*, UK: Meteorological Office, *http://www.meto.gov.uk/research/hadleycentre/index.html*.

Hansen JE and Lacis AA 1990 Sun and dust versus the greenhouse gases: an assessment of their relative roles in global climate change. *Nature* **346**: 713–719.

Hulme M and Carter TR 2000 The changing climate of Europe. In: Parry ML, ed. *Assessment of Potential Effects and Adaptations for Climate Change in Europe: The Europe ACACIA Project*. Norwich, UK: The Jackson Environment Institute, UEA, pp. 47–84.

IPCC 1996 *Climate Change 1995: The Science of Climate Change*. Intergovernmental Panel on Climate Change, World Meteorological Organization and United Nations Environment Program. Cambridge: Cambridge University Press, p. 37.

IPCC 2001 Houghton JT, Ding Y, Griggs DJ, Noguer M, van der Linden PJ, Dai X, et al. eds *Climate Change 2001: The Scientific Basis*. Intergovernmental Panel on Climate Change, Working Group 1. Cambridge: Cambridge University Press.

Jenkinson DS, Adams DE and Wild A 1991 Model estimates of CO_2 emissions from soil in response to global warming. *Nature* **361**: 304–306.

Kattenberg A, Giorgi F, Grassl H, Meehl GA, Mitchell JFB, et al. 1996 Climate models – projections of future climate. In: Houghton JT, Filho LGM, Callender BA, Harris N, Kattenberg A, Maskell K, eds *Climate Change 1995: The Science of Climate Change*. Intergovernmental Panel on Climate Change, World Meterological Organization and United Nations Environment Program. Cambridge: Cambridge University Press, pp. 285–357.

Kerr RA 1993 Ocean-in-a-machine starts looking like the real thing. *Science* **260**: 32,33.

Luo Y, Wan S, Hui D and Wallace LL 2001 Accli matization of soil respiration to warming in a tall grass prairie. *Nature* **413**: 622–624.

Manabe S 1998 Study of global warming by GFDL climate models. *Ambio* **27**(3): 182–186.

Manabe S and Wetherald RT 1967 Thermal equilibrium of the atmosphere with a given distribution of relative humidity. *Journal of the Atmospheric Sciences* **24**: 241–259.

Meehl GA, Zwiers F, Evans J, Knutson T, Mearns L and Whetton P 2000 Trends in extreme weather and climate events: issues related to modeling extremes in projections of future climatic change. *Bulletin of the American Meteorological Society* **81**(3): 427–436.

Oechel WC, Hastings SJ, Vourlitis G, Jenkins M, Riechers G and Grulke N 1993 Recent evidence of Arctic tundra ecosystems from a net carbon dioxide sink to a source. *Science* **361**: 520–523.

Prentice K and Fung IY 1990 The sensitivity of carbon storage to climate change. *Nature* **346**: 48–51.

Sarmiento JL, Hughes TMC, Stouffer RJ and Manabe S 1998 Simulated response of the ocean carbon cycle to anthropogenic warming. *Nature* **393**: 245–249.

Schlesinger ME and Mitchell JFB 1987 Climate model simulations of the equilibrium climatic response to increased carbon dioxide. *Reviews of Geophysics* **25**: 760–798.

Taylor KE and Penner JE 1994 Response of the climate system to atmospheric aerosols and greenhouse gases. *Nature* **369**: 734–737.

Thomson DJ 1995 The seasons, global temperature, and precession. *Science* **268**: 59–69.

Viner D 2002 *Elements of a Global Climate Model*. Norwich, UK: Climatic Research Unit, University of East Anglia, *http://www.cru. uea.ac.uk/cru/info/modelcc/*.

Wilde P and Quinby-Hunt MS 1997 Methane clathrate outgassing and anoxic expansion in southeast Asian deeps due to global warming. *Environmental Monitoring and Assessment* **44**: 149–153.

Woodwell GM, MacKensie FT, Houghton RA, Apps M, Gorham E and Davidson E 1998 Biotic feedbacks in the warming of the Earth. *Climatic Change* **40**: 495–518.

SECTION II

Ecological Effects of Climate Change

Chapter 5

Effects on Freshwater Systems

"A greenhouse warming is certain to have major impacts on both water availability and water quality."

Kenneth Frederick and Peter Gleick

Introduction

Climate-induced changes in precipitation, surface runoff, and soil moisture will probably have profound impacts on natural systems and human populations. Life on land, in streams, and in lakes depends on the availability of freshwater. Globally, precipitation averages about 86 cm (34 inches) per year, and ranges from 25 to 254 cm (10 to 100 inches) per year over most of the world. These differences in precipitation patterns, along with temperature, largely determine the geographic distribution of major terrestrial ecosystems (biomes) from deserts to rain forests (Chapter 6). Surface and groundwater sources supply water to humans for domestic use, agricultural irrigation, industry, transportation, recreation, waste disposal, and hydroelectric power generation. Human settlements have been, and continue to be, linked closely to the availability of freshwater. In fact, many historians believe successful early agricultural civilizations such as those in Africa (Egypt) or Central America (Mayan) developed out of the social organization necessary for large-scale water-management projects. Today, about 1.7 billion people or one-third of the world's population live in areas of water scarcity, that is, where they use more than 20% of their renewable water supply. This number is projected to grow, depending on population growth rates, to about 5 billion by 2025.

Because of the basic importance of water to living systems, changes in water availability represent one of the most serious potential consequences of greenhouse warming. Global warming could create new challenges for areas already facing water supply problems. Warming will accelerate oceanic evaporation and increase overall global average precipitation. On the basis of the SRES A2 scenario (see Chapter 4), the 30-year average precipitation for the period 2071 to 2100 will be 3.9% (range = 1.3 to 6.8%) greater

Climate Change: Causes, Effects, and Solutions John T. Hardy
 ISBNs: 0-470-85018-3 (HB); 0-470-85019-1 (PB)

than the period 1961 to 1990 (Cubasch and Meehl 2001). However, regional and seasonal changes are very important and could differ greatly from the global mean. Moist air will penetrate to higher latitudes and result in large increases in precipitation, soil moisture, and runoff there, except in summer. At the same time, precipitation will decrease at lower latitudes between about 5 and 30°. Climate change will alter water availability, water demand, and water quality.

Surface and Groundwater

The study of the movement and fate of water constitutes the science of hydrology. Water can move from the atmosphere to the surface (precipitation), accumulate for some time as snow or ice, evaporate, penetrate the soil to aquifers, or be taken up by plants and transpired through leaves back to the atmosphere (Figure 5.1).

A basic water balance model is

$$Q = P - E$$

where Q = runoff, P = precipitation, and E = evaporation and other losses.

Huge quantities of water are taken up from the soil by plants. Much of this is lost through pores in the leaves (stomata) back to the atmosphere, a process known as transpiration. Transpiration loss is influenced by the concentration of CO_2 in the atmosphere. As CO_2 concentration increases, it lowers the pH within the stomata cells, leading to a closing of the pore openings. This "stomatal resistance" slows water loss. Some researchers, on the basis of greenhouse and field studies, estimate that a doubling of atmospheric CO_2 and reduction in transpirational water loss from plants will result in more soil-water saturation and increases in water surface runoff by as much as 60 to 85%.

Potential evapotranspiration (PE) is *the amount of water that could evaporate and transpire from a landscape fully covered by a homogenous stand of vegetation without any shortage of soil moisture within the rooting zone*. Evapotranspiration rates differ according to the type of vegetation cover, for example, grassland, pine forest, and so on

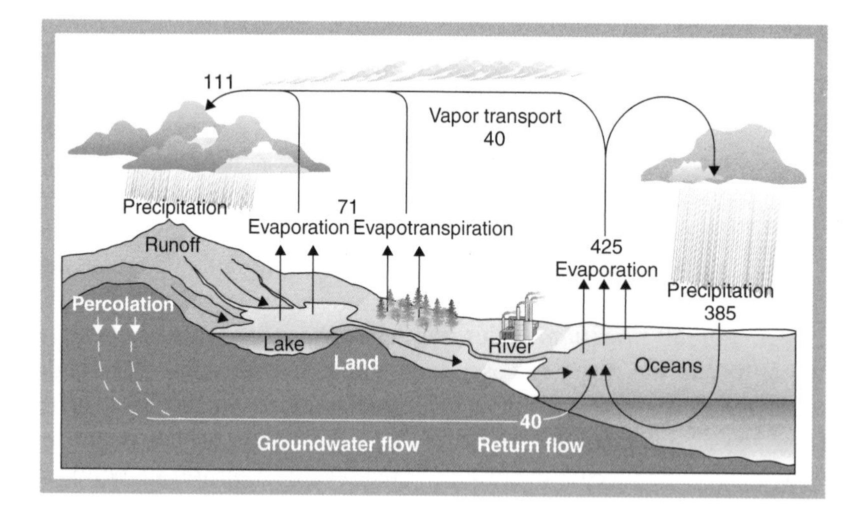

Fig. 5.1 The global water budget. Flows in thousands of km^3 per year (Iken The Global Water Budget, p. 82. From Mauritis la Rivière JW 1989. Threats to the world's water. *Scientific American* **261**(3): 80–94).

across a landscape. The resulting movement of water from soil to atmosphere is the actual evapotranspiration (AE). Enhanced atmospheric CO_2 levels will probably suppress both PE and AE, and for some vegetation covers it will significantly change the relationship between the two (Lockwood 1999).

Model studies suggest that in many regions the combination of increased temperature and evaporation together with decreased precipitation will lead to severe water shortages. Even small increases in temperature, when coupled with changing precipitation, can lead to rather large changes in surface runoff. With a doubling of CO_2, predicted regional changes in precipitation are on the order of plus or minus 20%, while changes in runoff and soil moisture are on the order of plus or minus 50% (Schneider et al. 1990). For example, in an area that experiences a 1 to 2 °C warming and a 10% decrease in precipitation, available surface water runoff will decrease by 40 to 70% (Waggoner 1991). The US Geological Survey estimates that in the United States, a 2 °C temperature increase and a 10% precipitation decrease would result in an average decrease in runoff of 35%.

Variability of the hydrologic cycle is often more important than long-term averages. The supply of available water is never uniform over time. Studies often include statistical analyses to determine averages and variabilities in precipitation, soil moisture, reservoir or groundwater recharge, or the frequency of floods and droughts. Climate change may lead to more frequent extremes. Crops could be devastated by water shortage (drought) or water excess (floods) (Chapter 7).

Surface runoff is an important source of water and humans increasingly rely on reservoirs or other storage systems to store surface water. Water reservoirs must be designed to store excess in times of ample runoff in order to meet demands in times of shortage. Reservoirs are generally planned according to certain guidelines, that is, for the sake of economy they are not designed to supply water under all circumstances, but only most of the time. Thus, in the rare instance of a 100-year drought, a reservoir may not hold enough water to meet the demand.

The frequency of water shortages can be described as a series of storage–yield curves that allow one to predict the percentage of years in which a given reservoir will be inadequate to meet a certain yield (water use) (Figure 5.2). For example, in the United Kingdom, in a year when stream flows are good, that is, in the upper 90th percentile of typical flows, a reservoir designed to be inadequate only 5% of the years would need a volume of storage equal to slightly more than 1% of the annual runoff. If it were acceptable for the reservoir to be inadequate 40% of the time, it would only need to hold about 0.3% of the annual runoff. However, if stream flow decreases to the 60th percentile of typical flows, then the reservoir will need to have a storage equal to about 15% of the annual average runoff to have only 5% of the years be inadequate to meet the needs. Greenhouse warming may change hydrological extremes more than average conditions. Therefore, in areas where stream flows decrease, major expansions in water storage facilities will be needed.

In Europe, a doubling of CO_2 would probably result in decreases in runoff in the Mediterranean region (Giorgi and Hewitson 2001). In fact, this predicted pattern might already be taking place. Lake Iliki, situated northeast of Athens, Greece, suffered nearly a decade of drought in the 1990s, which left the lake level low and threatened the water supply of four million people. In Northern Europe, an

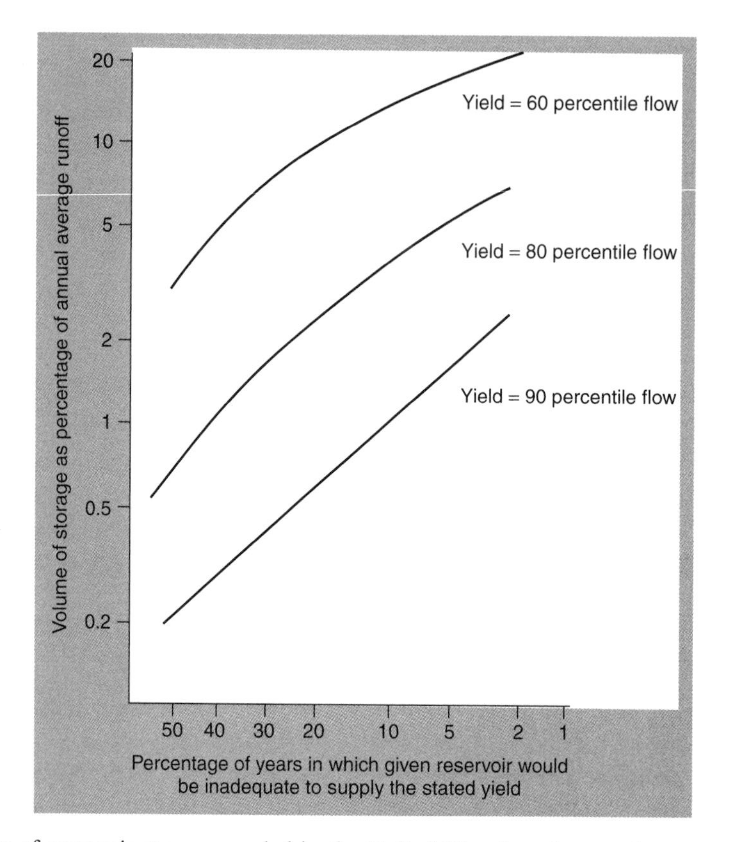

Fig. 5.2 Volume of reservoir storage needed in the United Kingdom for an adequate supply in years with different stream flows (Beran M 1986. The water resource impact of future climate change and variability. In: Titus JG, ed. *Effects of Changes in Stratospheric Ozone and Global Climate: Vol. 1.* US. EPA, pp. 299–328).

analysis of 19 climate model projections suggests that extreme winter rainfall events will become five times more likely within the next 50 to 100 years (Palmer and Räisänen 2002).

In Britain, climate models predict significant regional changes in precipitation and potential evaporation by 2050. Average annual surface runoff will increase in the north. However, annual potential evaporation in Southern Britain will increase by 30% and average annual runoff will decrease by 20% (Figure 5.3). Summer runoff in Britain will decrease by up to 50% in the south and east, and groundwater recharge will decrease (Arnell 1998). Increasing stream temperatures in Britain would shrink the optimum thermal habitat for some fish and lower stream flows. Reduced flow would, in turn, allow the buildup of pollutants. Even as precipitation and runoff in the south decrease, the demand for water will substantially increase. For example, with climate change, the demand for spray irrigation could increase 115% between 1991 and 2021 (Herrington 1996). However, winter precipitation and flooding in most areas of Britain will increase. By 2050, extreme floods with historical frequencies of 20 years would occur, on average, every five years. These changes in hydrology in Britain may have serious implications for everything from

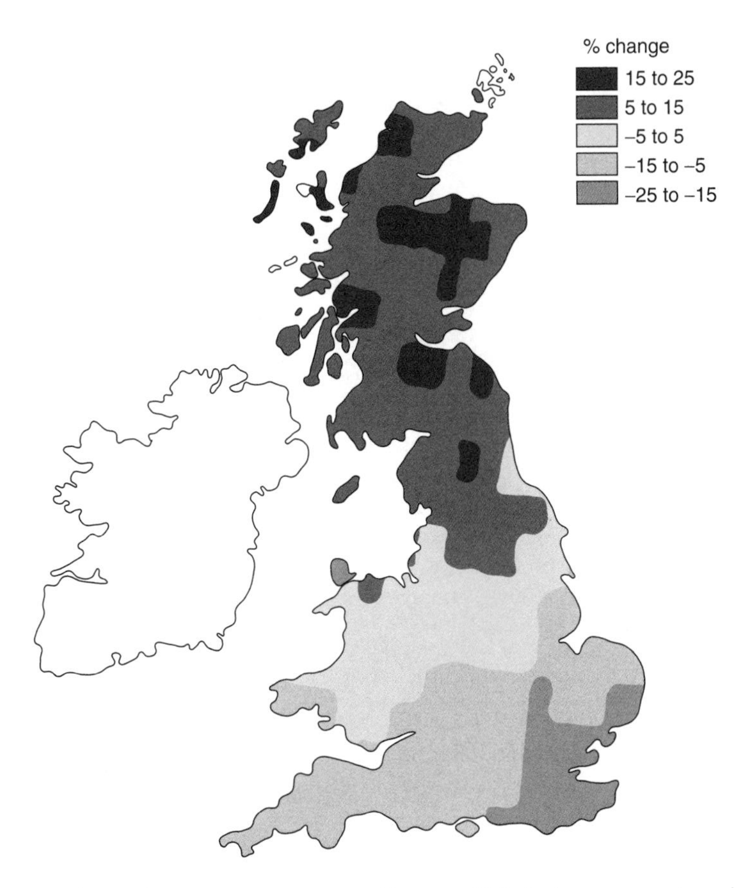

Fig. 5.3 Percentage change in average annual runoff across Britain in response to a doubled atmospheric CO_2 warming (Arnell NW 1998. Climate change and water resources in Britain. *Climatic Change* **39**: 83–110, original copyright notice with kind permission of Kluwer Academic Publishers).

the reliability of domestic water supplies to navigation, aquatic ecosystem health, recreation, and power generation.

Climate change in many countries will add to existing water shortages. The IPCC (1996) examined potential changes in per capita water availability between 1990 and 2050 in 21 countries using four climate-change scenarios. Water demand will increase even without climate change, and countries with high population growth rates (e.g. Kenya and Madagascar) will experience sharp declines in per capita water availability (Figure 5.4). Africa is particularly vulnerable to factors that affect water supply. Per capita water availability has diminished by 75% during the past 50 years. This developing scarcity results largely from rapid population growth. However, in regions such as sub-Saharan West Africa, river flows have declined in the past 20 years and climate change may accelerate this decline (McCarthy et al. 2001).

Predictions of regional changes in the water budget remain uncertain. For example, for the United States, the Hadley Centre, UK and the Canadian climate models both predict significant increases in temperature and PE by 2100. However, the Hadley

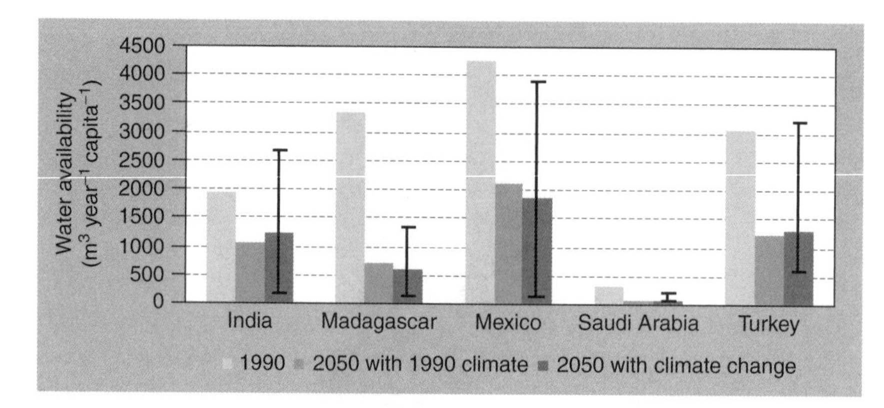

Fig. 5.4 Per capita water availability in 1990 and in 2050. Vertical bars represent the predicted range of availability based on three transient global climate models for the year 2050 (Based on data in Zdzislaw K 1996. Water resources management. In: Watson RT, Zinyowera MC and Moss RH, eds *Climate Change 1995: Impacts, Adaptations and Mitigation of Climate Change: Scientific-Technical Analyses*. Intergovernmental Panel on Climate Change. Cambridge: Cambridge University Press, pp. 469–486. Reproduced by permission of Intergovernmental Panel on Climate Change).

model predicts increased flooding, while the Canadian model predicts water scarcity for much of the country (Frederick and Gleick 1999). Numerous studies, however, suggest a variety of mostly negative impacts on water resources in North America (Box 5.1).

Groundwater is also an important source for human use. For example, the Ogallala

Box 5.1 Water resources and climate change in North America (From Zdzislaw K 1996. Water resources management. In: Watson RT, Zinyowera MC and Moss RH, eds *Climate Change 1995: Impacts, Adaptations and Mitigation of Climate Change: Scientific-Technical Analyses*. Intergovernmental Panel on Climate Change. Cambridge: Cambridge University Press, p. 58)

I. *Alaska, Yukon, and Coast British Columbia.*
Lightly settled/water-abundant region; potential ecological, hydropower, and flood impacts:

- Increased spring flood risks
- Glacial retreat/disappearance in south, advanced in north; changing flows impacts stream ecology
- Increased stress on salmon and other fish species
- Flooding of coastal wetlands
- Changes in estuary salinity and ecology

II. *Pacific Coast States (USA).*
Large and rapidly growing population; water abundance decreases north to south; intensive irrigated agriculture; massive water-control infrastructure; heavy reliance on hydropower; endangered species issues; and increasing competition for water:

- More winter rainfall and less snowfall – earlier seasonal peak in runoff, increased fall and winter flooding, decreased summer water supply
- Possible increases in annual runoff in Sierra Nevada and Cascades
- Possible summer salinity increase in San Francisco Bay, Sacramento, and San Joaquin Delta
- Changes in lake and stream ecology – warm water species benefiting; damage to cold-water species (e.g. trout and salmon)

III. *Rocky Mountains (USA and Canada).*
Lightly populated in north, rapid population growth in south; irrigated agriculture, recreation, and urban expansion increasingly competing for water; headwaters area for other regions:

- Rise in snow line in winter–spring, possible increases in snowfall, earlier snowmelt, more frequent rain on snow – changes in seasonal stream flow, possible reductions in summer stream flow, reduced soil moisture

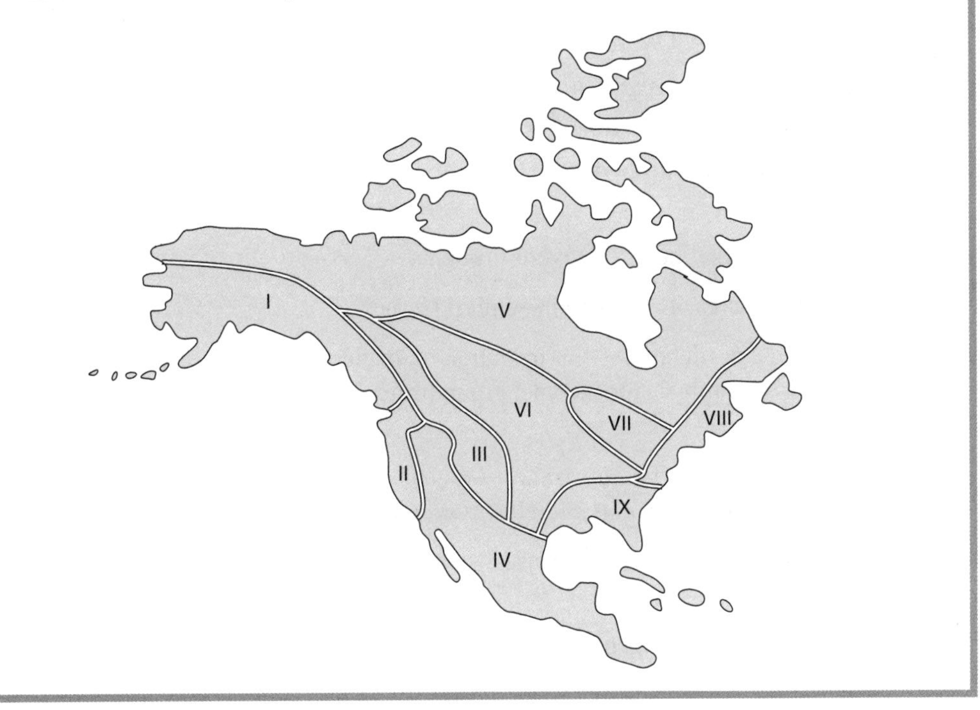

- Stream temperature changes affecting species composition; increased isolation of cold-water stream fish

IV. *Southwest.*
Rapid population growth; dependence on limited groundwater and surface water supplies; water quality concerns in border region; endangered species concerns; vulnerability to flash flooding:

- Possible changes in snowpacks and runoff
- Possible declines in groundwater recharge – reduced water supplies
- Increased water temperatures – further stress on aquatic species
- Increased frequency of intense precipitation events – increased risk of flash floods

V. *Sub-Arctic and Arctic.*
Sparse population (many dependent on natural systems); winter ice cover important feature of hydrologic cycle:

- Thinner ice cover, one- to three-month increase in ice-free season, increased extent of open water
- Increased lake-level variability, possible complete drying of some delta lakes
- Changes in aquatic ecology and species distribution as a result of warmer temperatures and longer growing season

VI. *Midwest USA and Canadian Prairies.*
Agricultural heartland – mostly rain-fed, with some areas relying heavily on irrigation:

- Annual stream flow decreasing/increasing; possible large declines in summer stream flow
- Increasing likelihood of severe droughts
- Possible increasing aridity in semiarid zones
- Increases or decreases in irrigation demand and water availability – uncertain impacts on farm-sector income, groundwater levels, stream flows, and water quality

VII. *Great Lakes.*
Heavily populated and industrialized region; variations in lake levels/flows now affect hydropower, shipping, and shoreline structures:

- Possible precipitation increases coupled with reduced runoff and lake-level declines
- Reduced hydropower production; reduced channel depths for shipping

- Decreases in lake ice extent – some years without ice cover
- Changes in phytoplankton/zooplankton biomass, northward migration of fish species, possible extinctions of cold-water species

VIII. *Northeast USA and Eastern Canada.*
Large, mostly urban population – generally adequate water supplies; large number of small dams but limited total reservoir capacity; heavily populated floodplains:

- Decreased snow cover amount and duration
- Possible large reductions in stream flow
- Accelerated coastal erosion, saline intrusion into coastal aquifers
- Changes in magnitude, timing of ice freeze-up/breakup, with impacts on spring flooding
- Possible elimination of bog ecosystems
- Shifts in fish species distributions, migration patterns

IX. *Southeast, Gulf, and Mid-Atlantic USA.*
Increasing population – especially in coastal areas; water quality/nonpoint source pollution problems; stress on aquatic ecosystems:

- Heavily populated coastal floodplains at risk to flooding from extreme precipitation events, hurricanes
- Possible lower base flows, large peak flows, longer droughts
- Possible precipitation increase – possible increases or decreases in runoff/river discharge, increased flow variability
- Major expansion of northern Gulf of Mexico hypoxic zone possible – other impacts on coastal systems related to changes in precipitation/nonpoint source pollutant loading
- Changes in estuary systems and wetland extent, biotic processes, species distribution.

Aquifer underlying the US Great Plains provides water for about 20% of the irrigated land in the United States. It is fully utilized and is currently being depleted because the 20 km^3 of water withdrawn annually exceeds the long-term recharge rate. Researchers at the Pacific Northwest National Laboratory and the US Department of Agriculture used three global climate models and evaluated the effects of three different future atmospheric CO_2 concentrations. Outputs from the climate models, such as predicted precipitation, were linked to a hydrologic model that estimates parameters such as surface runoff and groundwater recharge. Recharge of the aquifer will be reduced under all three severities of climate change. For example, assuming an atmospheric CO_2 concentration of 560 ppmv and a global mean temperature increase of 2.5 °C, recharge decreases by 18 to 45% over the two major drainage basins feeding the aquifer. Climate change, forced by global warming, will make mining the aquifer's water even less sustainable than it is now (Rosenberg et al. 1999).

Drought and Soil Moisture

Greenhouse warming, in some areas, will lead to higher surface air temperatures, greater evapotranspiration, lower soil moisture, and increasingly frequent droughts. A "drought index," based on the atmospheric supply of moisture minus the atmospheric demand for moisture, can be used together with climate model outputs of precipitation and potential evapotranspiration to evaluate the probable occurrence of droughts. Summer droughts will become increasingly frequent at low latitudes and less frequent at high-latitude areas such as Canada and Siberia (Plate 5).

Soil moisture is often the most important factor controlling plant growth and agricultural production. Globally, models predict significant soil moisture changes with decreases in some regions and increases in others (Plate 6). Large reductions in summer soil moisture will occur in midcontinental regions of mid to high latitudes, for example, the North American Great Plains, Western Europe, Northern Canada, and Siberia (Manabe et al. 1981). For example, researchers at the University of Delaware (Mather and Feddema 1986) examined the effects of a doubling of CO_2 on soil moisture at 12 selected regions of the globe. They applied two global climate models to the climatic water budget on a course grid of 8 × 10 or 4 × 5 degrees latitude by longitude and calculated the soil moisture index (I_m):

$$I_m = 100\,[(P/PE) - 1]$$

where P = precipitation, PE = potential evapotranspiration (based on field capacity of the soil and air temperatures).

Their results for the region of North America covering Southeastern Texas and Northern Mexico indicate that, under changed conditions, the months of May through September will experience no more than 5 cm of precipitation and that peak summer evapotranspiration will increase by 27%. Of the 12 global regions examined, they predict an increase in PE in all regions. Because of low temperatures, the warming would have little influence on evapotranspiration in winter at higher latitudes. Precipitation would increase overall, but the additional water will be less than the loss due to evapotranspiration. Therefore, the annual water deficit would increase in all regions except south central Canada (NOAA model) and the Ukraine (GISS model). In 7 of 12 areas, both models agree on a shift to drier conditions, most markedly in the upper Midwest of the United States, the Texas–Mexico area, and Northeast Brazil. Overall, they conclude that water demand will increase more than precipitation and most regions will experience an increase in annual water deficit, a decrease in annual water surplus, and a decrease in summer soil moisture storage. They predict changes in vegetation to more drought-tolerant species in about two-thirds of the 12 regions studied, the exceptions being the Pacific Northwest, the Ukraine, and West Central Africa.

Finally, greenhouse warming will lead to greater heating in the eastern tropical Pacific than in the western tropical Pacific, that is, an intensification of the El Niño pattern. This will increase the intensity of future droughts in the Australasian region (Meehl 1997).

Lake and Stream Biota

Continued warming will alter the thermal structure of lakes and the impacts on lake biota are likely to be largely negative. Individual aquatic species, including fish, have an optimum and range of temperatures for growth and reproduction (their thermal habitat). Heat from global warming will be transferred to streams directly from the air and indirectly from warmer groundwater, shrinking the available thermal habitat for aquatic species. Climate warming will affect each species, as well as the prey on which they feed, differently. Thus, warming would lead to changes in the species composition, stability, and food web dynamics of aquatic ecosystems (Beisner et al. 1997). Direct effects of climatic change on lakes include changes in temperature, water level, ice-free period, and concentrations of dissolved gases. Elevated temperatures decrease oxygen solubility and increase the rates of microbial oxygen demand, both leading to a decrease in dissolved oxygen available for fish and other animals. Changing precipitation and evaporation patterns are likely to alter the chemistry of lakes, especially in mid-continental semiarid regions.

Lakes throughout the world will be affected by climate change. For example, climate change is already warming lake waters in Canada. Shallow closed-basin saline lakes in semiarid central areas of Canada on the Northern Great Plains are especially sensitive to changes in precipitation and evaporation that can alter lake salinities. On the basis of paleoecological data and projections of climate change for the region, the salinity of these lakes will probably increase, leading to a shift in phytoplankton species and an altered food web (Evans and Prepas 1996).

Application of climate model projections, based on a doubling of CO_2, shows that in temperate lakes (Minnesota, USA), ice formation will be delayed by about 20 days and the ice-covered period shortened by up to 58 days. Compared to baseline data (1961–1979), winter water temperatures will change little, but summer temperatures are projected to increase by 3 to 4 °C. Because of the warmer less dense surface water, summer water column stratification will also increase (Stefan et al. 1998).

Projected temperature increases of 3 to 6 °C over the next 50 years in the Canadian and the Alaskan Arctic are about double that of the global average. Studies at the Alaska Long-Term Ecological Research Site indicate that climate change will probably alter the entire food web structure of Arctic Lakes (McDonald et al. 1996). Like most fish, the sustained yield of lake trout decreases rapidly as the area of its preferred thermal habitat shrinks (Figure 5.5). Using a bioenergetics model, researchers subjected lake trout, a keystone predator in many of these lakes, to a simulated 3 °C increase in upper water column temperatures during July. They found that, because of the increase in metabolism at higher temperatures, trout would need to consume eight times more food. Toolik Lake, used in the study, has warmed over a recent 16-year period by 3 °C and experienced a decrease in primary productivity that supplies food for fish. If warming continues, lake trout are liable to perish from lack of food.

In streams, invertebrates serve as a food source for many higher organisms including fish. A 2 to 3.5 °C increase in stream temperature can decrease the population density, reduce the size, and alter the sex ratio of insects and other invertebrates (Hogg and Williams 1996).

Cool groundwater discharge, in many areas, maintains cold-water habitats for fish such as salmon and trout. The temperature of groundwater is often warmer than the air in winter, but cooler than the air in summer, and is expected to increase with global warming. Air temperature fluctuations are reflected in seasonal ground temperature fluctuations down to a depth called the *neutral zone*. In

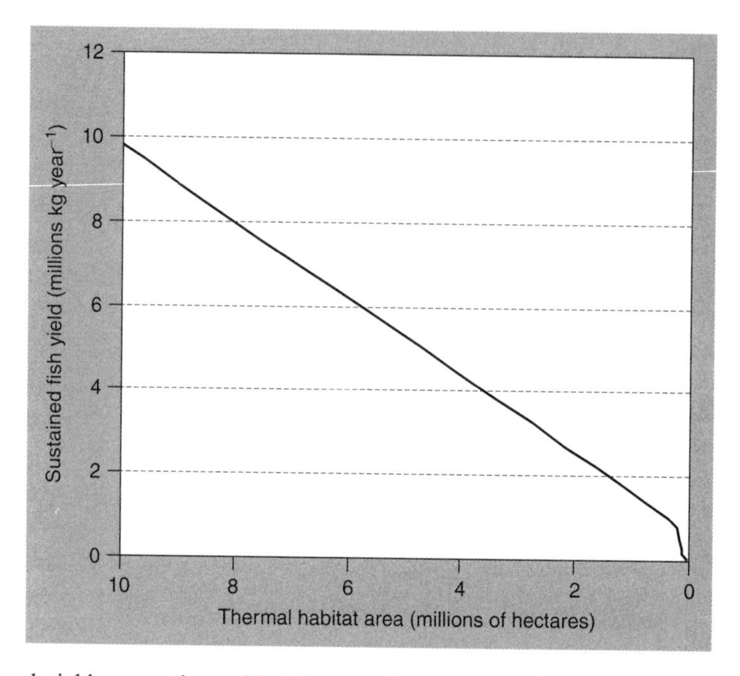

Fig. 5.5 Sustained yield versus thermal habitat area (hectares per 10 days over the summer) for a set of 21 large north temperate lakes (Christie GC and Regier HA 1988. Relationship of fish production to a measure of thermal habitat niche. *Canadian Journal of Fisheries and Aquatic Sciences* **45**: 301–314).

the United States, for example, groundwater temperatures at 10- to 20-m depth are about 1.1 to 1.7 °C greater than the mean annual air temperature. In mountainous areas, both air and groundwater temperatures generally decrease with increasing altitude. For a temperate region such as Switzerland the average decrease in temperature with altitude is about 4.1 °C per 1,000 m elevation. Also, shading by vegetation cover reduces the mean and variance of groundwater temperature (Meisner et al. 1988).

Several species of trout and salmon prefer temperatures of 12 to 18 °C, and the size of the population depends on the amount of area available within this near-optimal range, that is, the thermal habitat. An increase in summer temperature of the base flow of 4 to 5 °C in low elevation streams will shrink the available thermal habitat (area of optimum temperature) of trout. One might hypothesize that higher stream temperatures could increase the growth rates of trout at higher elevations (currently colder) in streams and that this would offset the negative effects of higher summer temperatures. However, simulations of three different feeding rates and food abundances at increased temperatures indicate that a 15 to 20% increase in food consumption would be necessary to maintain growth rates with even a 2 °C increase in temperature (Ries and Perry 1995). Climate change in Southern Ontario, Canada, will have a severe impact on populations of brook trout. Air and groundwater stream source temperatures will increase by more than 4 °C. This will shift the thermal habitat maximum temperature upstream to cooler higher elevations and shrink the available trout habitat by 30 to 42% (Meisner 1990).

In the Yakima Basin watershed in the northwestern United States, a 2 °C rise in temperature (conditions present 6,000 to 8,000 BP) would reduce numbers of Chinook, and possibly other species of salmon by 50% (Chatters et al. 1991).

Human Infrastructure

Climate-change alterations of some lakes and streams will negatively impact their utilization for human activities such as navigation, irrigation, power generation, and waste disposal. The North American Great Lakes represent a case study of what could happen to other lakes around the world in response to climate change (Marchand et al. 1988). The Great Lakes store 20% of the world's fresh surface water and about 12% of the American and 27% of the Canadian population live within the Great Lakes Basin. As a navigable waterway, the Great Lakes serve the cargo shipping needs of 17 US States or Canadian Provinces.

Climate change will have significant impacts on lake ecology, shipping, and the economics of the Great Lakes region. A doubling of CO_2 will reduce runoff in streams feeding the Lakes by 15 to 21% and increase average temperatures by 5 °C in winter and 3 °C in summer. As evaporation will exceed runoff, water levels in the Lakes will fall by one half meter. Benefits to commerce could result from decreases in seasonal ice cover that will expand the shipping season to 11 months of the year. On the other hand, lower lake levels will prevent some ships from entering ports unless the harbors are dredged to deeper depths, or more water is diverted into the lakes. As a "best guess," shipping costs are likely to increase 30% by 2035, with much of the increase attributable to climate change. Overall, climate change is likely to have serious socioeconomic impacts on the Great Lakes (Marchand 1988).

Climate-change effects on lakes and streams could impact several aspects of human infrastructure. In areas where stream flows and lake levels decline, hydroelectric energy production is likely to suffer. Stream flows entering lakes or estuaries often serve to flush out introduced pollutants from the receiving water. Decreasing flows could impair this ecosystem service. Finally, rising sea level (Chapter 8), in some coastal communities, will result in saltwater intrusion into groundwater, thus compromising its quality.

Wetlands

Rising temperatures and changes in surface water could damage wetland communities. Wetlands are permanently or temporarily submerged systems with vegetation adapted to water-saturated soil conditions. Coastal and freshwater wetlands are critical habitats for many species of plants, waterfowl, fish, and other animals. Seasonal and annual changes in water level regulate vegetation growth, which in turn strongly influences the biota of the wetland. Each different wetland community occupies a preferred position along the water-depth gradient and shoreline slope (Figure 5.6). Urbanization, recreational development, and conversion to agricultural land all threaten wetlands, because man-made structures impede the migration of wetland biota and inhibit their adjustment to changing water levels. In the Great Lakes, climate models suggest that key wetlands will be at risk from declining water levels (Mortsch and Quinn 1996).

The Cryosphere

Much of the world's available water remains frozen in the polar caps and alpine glaciers. The effects of global warming on this ice volume could be complex. In most temperate areas, warming will decrease the percentage

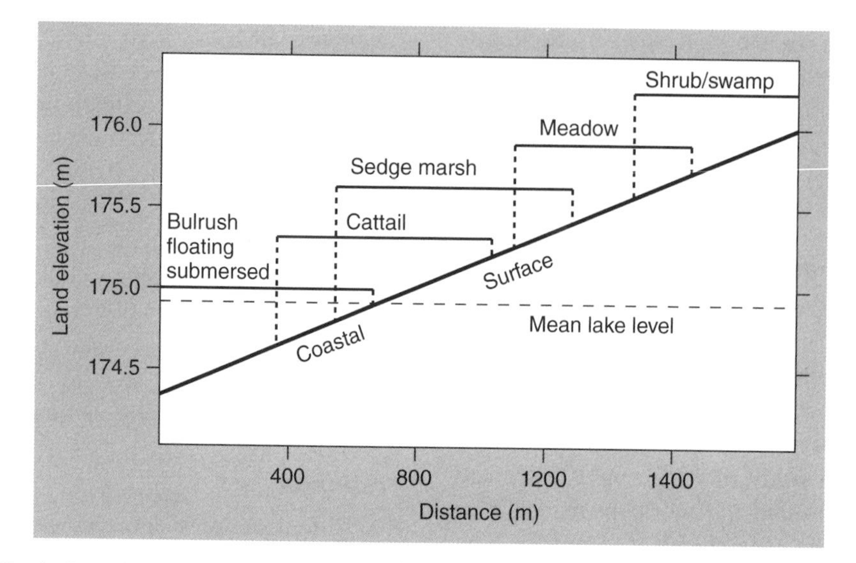

Fig. 5.6 Typical vertical shoreline distribution of lake vegetation (Herdendorf CE, Hartley SM and Barnes MD, eds 1981. *Fish and Wildlife Resources of the Great Lakes Coastal Wetlands Within the United States: Vol. 1.* U.S. Fish and Wildlife Service, Washington, DC, p. 289).

of annual precipitation that comes as snow and increase the length of the frost-free season. Satellite data reveal that between 1970 and 1990, the overall Northern Hemisphere winter snow cover decreased by about 10% (Folland et al. 1990).

Many alpine glaciers in the European Alps and North America have lost 30 to 40% of their glacial area and about half of their total volume between the 1850s and 1980s – an extremely rapid rate (Haeberli and Beniston 1998). These trends, including earlier seasonal melting of lake ice and alpine snowpacks and shrinking alpine glacial volumes (Figure 5.7) will continue and perhaps accelerate. Erosion following the disappearance of snow and permafrost could lead to increased sedimentation of alpine streams and degradation of fish habitat. With a 4 °C average global warming, about one-third to one-half of the mountain glacier mass of the world would disappear during this century.

Even with global warming, temperatures of polar regions, at least in the interior away from the warmer ocean, will remain below freezing for most of the year. Thus, predicted increases in precipitation for high latitudes could increase ice volumes in the interior Arctic or Antarctic. However, researchers suggest that in Greenland at least, snow accumulation is influenced more by specific weather patterns than by general changes in atmospheric water vapor and precipitation. Storms tend to move toward Greenland during major warming periods and away from Greenland during cooling periods. Thus, a simple warming is unlikely to lead to large polar increases in snow accumulation and "it may be prudent to plan for somewhat larger future sea-level rises than that of the IPCC best estimate case" (Kapsner et al. 1995). The British Antarctic survey reported that by 1998 Antarctic warming was occurring five times faster than the global average and large sections of the Larsen Ice Shelf, the size of the state of Rhode

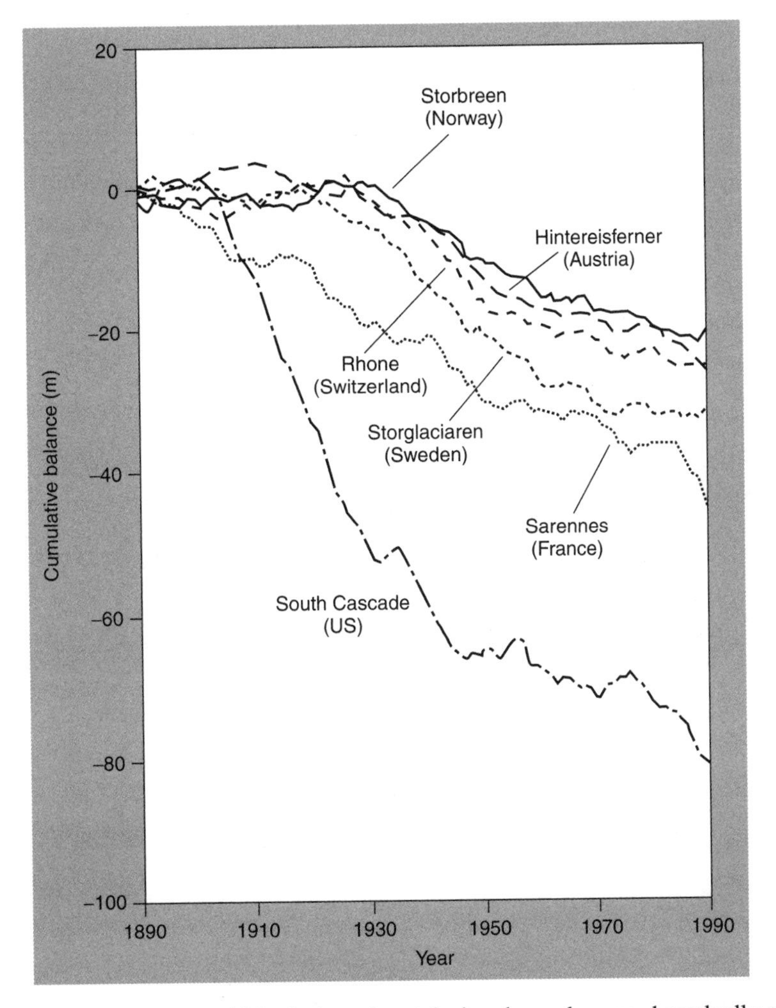

Fig. 5.7 The mass of European and North American glaciers have decreased markedly since 1890 (in meters of water equivalent) (From Warrick RA, Le Provost C, Meier MF, Oberlemans J and Woodworth PL 1996. Changes in sea level. In: Houghton JT, Filho LGM, Callender BA, Harris N, Kattenberg A and Maskell K, eds *Climate Change 1995: The Science of Climate Change*. Intergovernmental Panel on Climate Change, World Meteorological Organization and United Nations Environment Program. Cambridge: Cambridge University Press, p. 371. Reproduced by permission of Intergovernmental Panel on Climate Change).

Island, were disintegrating (Figure 5.8). Continued warming of the Antarctic coastal area could threaten the stability of the entire Ross Ice Shelf – an area the size of France.

As the Earth warms, precipitation in alpine areas will come more as rain and less as snow. Skiing, important as a tourist industry in many alpine areas, could suffer serious economic impacts. In the Western United States (and in many other areas), climate change will reduce the amount of water stored in the winter alpine snowpack. The alpine area covered by snow in the Pacific Northwest could decrease by nearly half by 2055 (Pelley 1999).

Fig. 5.8 Iceberg broken off from the Antarctic shelf (From NOAA photolibrary *http://www.photolib.noaa.gov*).

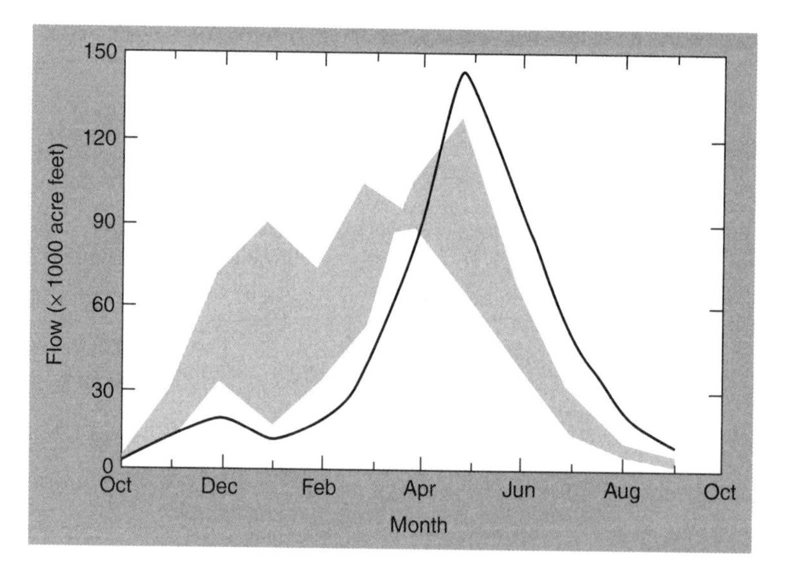

Fig. 5.9 Mean monthly stream flow for the Merced Watershed in Central California. Baseline (solid dark line). Shaded area represents range of predictions from three global climate models for a doubling of atmospheric CO_2 (Smith JB and Tirpak DA, eds 1990. *The Potential Effects of Global Climate Change on the United States* New York: Hemisphere Publishing, p. 81. Reproduced by permission of Routledge, Inc., part of The Taylor & Francis Group).

In the Cascade Mountains of the Pacific Northwest and the Sierra Nevada of California, rather than a gradual release and runoff from the snowpack lasting through the summer, rapid runoff could lead to flooding. With a 3 °C annual average temperature increase, about one-third of the present spring snowmelt will be shifted into increased winter runoff. For example, in the Merced watershed of central California, peak runoff into streams normally occurs in late April or early May. With a doubling of atmospheric CO_2, peak flow will occur in March and decline prior to the peak irrigation demand in the summer. Reservoirs would be depleted by midsummer. This could pose a serious threat to water supplies during the summer months (Gleick 1987) (Figure 5.9).

Climate not only affects the cryosphere but the cryosphere also influences climate (Chapter 4). Sea ice affects ocean temperatures and circulation. Also, snow and ice have a high reflectivity (albedo); thus as ice cover decreases, the darker Earth absorbs more solar energy, becoming warmer and further melting the nearby ice (Clark 1999).

Managing Water

Climate change, coupled with human population growth and increasing water demand, will create new challenges for managing water for beneficial human uses. Managing water for human use has always been a challenging and expensive activity. Humans will need to adapt to increased flooding in some areas and increased drought in others. Climate change will probably exacerbate conflicts over water resources (Figure 5.10). Management for sustainable water supplies will require new policy approaches that plan ahead to accommodate climate change (Box 5.2).]

Box 5.2 Summary of recommendations for Water Managers from the American Water Works Association Public Advisory Committee (Adapted from Journal of the American Water Works Association **89**(11): 107–110, by permission. Copyright 1997, American Water Works Association)

- While water-management systems are often flexible, adaptation to new hydrologic conditions may come at substantial economic costs. Water agencies should begin now to reexamine engineering design assumptions, operation rules, system optimization, and contingency planning for existing and planned water-management systems under a wider range of climatic conditions than traditionally used.
- Water agencies and providers should explore the vulnerability of both structural and nonstructural water systems to plausible future climate changes, not just past climatic variability.
- Governments at all levels should reevaluate legal, technical, and economic approaches to managing water resources in the light of possible climate changes.
- Water agencies should cooperate with leading scientific organizations to facilitate the exchange of information on the state-of-the-art thinking about climatic change and impacts on water resources.
- The timely flow of information from the scientific global change community to the public and water-management community would be valuable. Such lines of communication need to be developed and expanded.

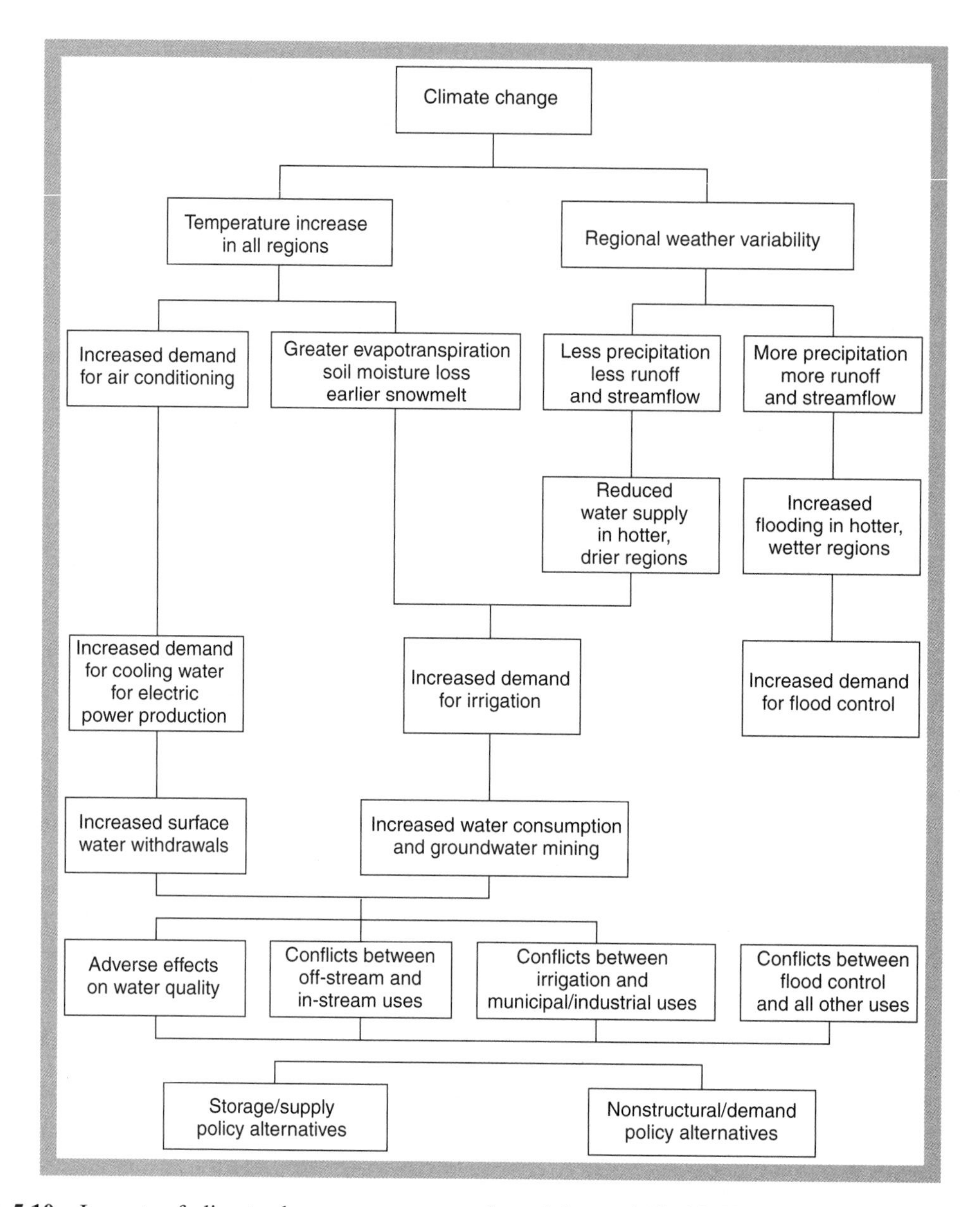

Fig. 5.10 Impacts of climate change on water supply and demand (Smith JB and Tirpak DA, eds 1990. *The Potential Effects of Global Climate Change on the United States*. New York: Hemisphere Publishing, p. 297. Reproduced by permission of Routledge, Inc., part of The Taylor & Francis Group).

Although climate change will affect many elements of the hydrologic cycle, human population growth and economic development over the next few decades will probably outweigh climate change in terms of per capita water availability (Vörösmarty et al. 2000). In the next 30 years, accessible water runoff should increase by about 10%, but during the same period world population will increase by about 33%. Unless the efficiency of water use can be dramatically increased, per capita freshwater availability will decline further. Recognizing the growing importance of water resources and their management, UNESCO established the World Water Assessment Programme. The objectives of the Programme are to develop the tools and skills needed to achieve a better understanding of basic processes, management practices, and policies that will help improve the supply and quality of global freshwater resources (UNESCO 2002).

Summary

The US EPA summarized the results of numerous studies on the effects of climate change on water resources of the United States (Smith and Tirpak, 1990, p. 281). Their findings, for the most part, still hold not only for the United States but also for the world. Climatic change during this century will probably lead to the following:

Global and regional changes in precipitation and evaporation:

- increasing precipitation at higher latitudes leading to increased winter/spring runoff and flooding in some areas;
- decreasing precipitation and increasing drought frequencies at lower latitudes;
- increased summer-time evaporation and decreasing surface flow and soil moisture at mid to high latitudes;
- decreasing lake levels in some areas;
- changes in wetland communities;
- decreasing per capita water availability, particularly in low-latitude countries with high population growth rates.

Higher temperatures in lakes, streams, and groundwater sources:

- decreases in dissolved oxygen;
- changes in freshwater invertebrate and fish species populations.

Continued and possibly accelerated shrinking of snow and ice cover:

- earlier spring melting of ice cover in some lakes and polar seas;
- increased exposure of darker land masses and decreased reflectivity positively contributing to further warming;
- earlier runoff of spring melt water from alpine areas.

Altogether, the changes imposed on freshwater systems by climate change translate into major regional impacts. In subsequent chapters we consider how these changes in freshwater, along with other changes, will influence natural ecosystems, agriculture, and human settlements and infrastructure.

References

AWWA 1997 Climate change and water resources. Committee report of the American Water Works Association Public Advisory Forum. *Journal of the American Water Works Association* **89**(11): 107–110.

Arnell NW 1998 Climate change and water resources in Britain. *Climatic Change* **39**: 83–110.

Beran M 1986 The water resource impact of future climate change and variability. In: Titus JG, ed. *Effects of Changes in Stratospheric Ozone and Global Climate: Vol. 1.* US. EPA, pp. 299–328.

Beisner BE, McCauley E and Wrona FJ 1997 The influence of temperature and food chain length on plankton predator-prey dynamics. *Canadian Journal of Fisheries and Aquatic Science* **54**: 586–595.

Chatters JC, Neitzel DA, Scott MJ and Shankle SA 1991 Potential impacts of global climate change on Pacific Northwest Chinook salmon (Oncorhynchus tshawytschia): an exploratory case study. *The Northwest Environmental Journal* **7**: 71–92.

Christie GC and Regier HA 1988 Relationship of fish production to a measure of thermal habitat niche. *Canadian Journal of Fisheries and Aquatic Sciences* **45**: 301–314.

Clark PU, Alley RB and Pollard D 1999 Northern Hemisphere ice-sheet influences on global climate change. *Science* **286**(5442): 1104–1111.

Cubasch U and Meehl GA Projections of future climate change. In: Houghton JT, Ding Y, Griggs DJ, Noguer M, van der Linden PJ, Dai X, et al., eds *Climate Change 2001: The Scientific Basis Intergovernmental Panel on Climate Change, Working Group 1*. Cambridge: Cambridge University Press, pp. 525–582.

Evans C and Prepas EE 1996 Potential effects of climate change on ion chemistry and phytoplankton communities in prairie saline lakes. *Limnology and Oceanography* **41**(5): 1063–1066.

Frederick KD and Gleick PH, Pew Center on Global Climate Change, 1999 *Water and Global Climate Change: Potential Impacts on US Water Resources*, p. 48.

Folland CK, Karl TR and Vinnikov KYO 1990 Observed climate variations and change. In: Houghton JT, Jenkins GJ, and Ephraums JJ, eds *Climate Change: the IPCC Scientific Assessment*. Cambridge: Cambridge University Press, pp. 194–238.

Giorgi F and Hewitson B 2001 Regional climate information – evaluation and projections. In: Houghton JT, Ding Y, Griggs DJ, Noguer M, van der Linden PJ, Dai X, et al. eds. *Climate Change 2001: The Scientific Basis. Intergovernmental Panel on Climate Change, Working Group 1* Cambridge: Cambridge University Press, pp. 583–638.

Gleick PH 1987 The development and testing of a water balance model for climate impact assessment: modeling the Sacramento Basin. *Water Resources Research* **23**: 1049–1061.

Haeberli W and Beniston M 1998 Climate change and its impact on glaciers and permafrost in the Alps. *Ambio* **27**(4): 258–265.

Herdendorf CE, Hartley SM and Barnes MD, eds 1981 *Fish and Wildlife Resources of the Great Lakes Coastal Wetlands Within the United States*. Vol. 1. Washington, DC: U.S. Fish and Wildlife Service, p. 289.

Herrington P 1996 *Climate Change and the Demand for Water* London: HMSO.

Hogg ID and Williams DD 1996 Response of stream invertebrates to a global-warming thermal regime: an ecosystem-level manipulation. *Ecology* **77**(2): 395–407.

Kapsner WR, Alley RB, Shuman CA, Anandakrishnan S and Grootes PM 1995 Dominant influence of atmospheric circulation on snow accumulation in Greenland over the past 18,000 years. *Nature* **373**: 52–54.

Lockwood, JG 1999 Is potential evapotranspiration and its relationship with actual evapotranspiration sensitive to elevated atmospheric CO_2 levels? *Climatic Change* **41**: 193–212.

Manabe S, Wetherald RT and Stouffer RJ 1981 Summer dryness due to an increase of atmospheric CO_2 concentration. *Climatic Change* **3**: 347–385.

Marchand D, Sanderson M, Howe D and Alpaugh C 1988 Climatic change and great lakes levels the impact on shipping. *Climatic Change* **12**: 107–133.

Mather JR and Feddema J 1986 Hydrologic consequences of increases in trace gases and CO_2 in the atmosphere. In: Titus JG, ed. Effects of Changes in Stratospheric Ozone and Global Climate, Vol. 3: Climate Change. *Proceedings of the International Conference on Health and Environmental Effects of Ozone Modification and Climate Change*, October 1986. UNEP and US EPA, pp. 251–271.

Mauritis la Rivière JW 1989 Threats to the world's water. *Scientific American* **261**(3): 80–94.

McCarthy JJ, Canziani OF, Leary NA, Dokken DJ and White KS, eds *Climate Change 2001: Impacts, Adaptation and Vulnerability. Intergovernmental Panel on Climate Change, Working Group II.* Cambridge: Cambridge University Press, p. 45.

McDonald ME, Hershey AE and Miller MC 1996 Global warming impacts on lake trout in arctic lakes. *Limnology and Oceanography* **41**(5): 1102–1108.

Meehl GA 1997 Pacific region climate change. *Ocean and Coastal Management* **37**(1): 137–147.

Meisner JD 1990 Potential loss of thermal habitat for brook trout, due to climatic warming, in two southern Ontario streams. *Trans. Amer. Fisheries Soc.* **119**: 282–291.

Meisner JD, Rosenfeld JS and Regier HA 1988 The role of groundwater in the impact of climate warming on stream salmonids. *Fisheries* **13**(3): 2–7.

Mortsch L and Quinn F 1966 Climate change scenarios for Great Lakes Basin ecosystem studies. *Limnology and Oceanography* **41**: 903–911.

Palmer T and Räisänen J 2002 Quantifying the risk of extreme seasonal precipitation events in a changing climate. *Nature* **415**: 512–514.

Pelley J 1999 Predicted summer water shortages attributed to climate change. *Environ. Sci. Technol.* **33**(15): 305A.

Ries RD and Perry SA 1995 Potential effects of global climate warming on brook trout growth and prey consumption in central Appalachian streams, USA. *Climate Research* **5**(3): 197–206.

Rind D, Goldberg R, Hansen J, Rosenweig C and Ruedy R 1990 Potential evapotranspiration and the likelihood of future drought. *Journal of Geophys. Res.* **95**: 9983–10004.

Rosenberg NJ, Epstein DJ, Wang D, Vail L, Srinivasan R and Arnold JG 1999 Possible impacts of global warming on the hydrology of the Ogallala aquifer region. *Climatic Change* **42**: 677–692.

Schneider SH, Gleick PH and Mearns LO 1990 Prospects for climate change. In: Waggoner PE, ed. *Climate Change and U.S. Water Resources* New York: John Wiley and Sons, pp. 41–73.

Smith JB and Tirpak DA, eds 1990 *The Potential Effects of Global Climate Change on the United States* New York: Hemisphere Publishing.

Stefan HG, Fang X and Hondzo M 1998 Simulated climate change effects on year-round water temperatures in temperate zone lakes. *Climatic Change* **40**: 547–576.

UNESCO 2002 *World Water Assessment Programme.* Available from: *http://www.unesco.org/water/wwap/.*

Vörösmarty CJ, Green P, Salisbury J and Lammers RB 2000 Global water resources: vulnerability from climate change and population growth. *Science* **289**: 284–288.

Waggoner PE 1991 U.S. water resources versus an announced but uncertain climate change. *Science* **251**: 1002.

Warrick RA, Le Provost C, Meier MF, Oberlemans J and Woodworth PL 1996. Changes in sea level. In: Houghton JT, Filho LGM, Callender BA, Harris N, Kattenberg A and Maskell K, eds *Climate Change 1995: The Science of Climate Change*. Intergovernmental Panel on Climate Change, World Meteorological Organization and United Nations Environment Program. Cambridge: Cambridge University Press, p. 371.

Zdzislaw K 1996 Water resources management. In: Watson RT, Zinyowera MC and Moss RH, eds *Climate Change 1995: Impacts, Adaptions and Mitigation of Climate Change: Scientific-Technical Analyses*. Intergovernmental Panel on Climate Change. Cambridge: Cambridge University Press, pp. 469–486.

Chapter 6

Effects on Terrestrial Ecosystems

"As I did stand my watch upon the hill,
I look'd toward Birnam, and anon, methought,
The wood began to move."

William Shakespeare, Macbeth

Introduction

Climate, primarily temperature and precipitation, determine the geographic distribution of major terrestrial ecosystems (biomes) from deserts to rain forests. Local and regional differences in soil types, watershed conditions, and slope angle (sun exposure) influence the success of different plants. However, seasonal patterns of rainfall and temperature largely dictate the type of plant associations that dominate an area – associations we call tundra, desert, grassland, rainforest, and so on. (Figure 6.1). Each plant association has an optimum "climate space," that is, a specific combination of temperature and precipitation conditions in which it best thrives.

Terrestrial ecosystems are an integral part of the global carbon cycle. Grasslands and forests sequester atmospheric carbon (CO_2) through photosynthesis and store it temporarily as organic carbon. Below ground, organic carbon is decomposed by microorganisms and released back into the atmosphere. Both these processes are influenced by temperature and could be altered by global warming.

Forests and other terrestrial biomes provide habitats for a diversity of plants and animals. If the forest is damaged or removed, habitat loss can endanger the survival of the associated living organisms. Climate change can directly affect many plants and animals by altering the growing season or temperature patterns that trigger life cycle changes.

In some countries, forests occupy a major portion of the total land area (e.g. 33% in the United States) and serve many important functions. They influence the availability and runoff of water (see Chapter 5) and provide sites for recreation and for the harvesting

Climate Change: Causes, Effects, and Solutions John T. Hardy
 ISBNs: 0-470-85018-3 (HB); 0-470-85019-1 (PB)

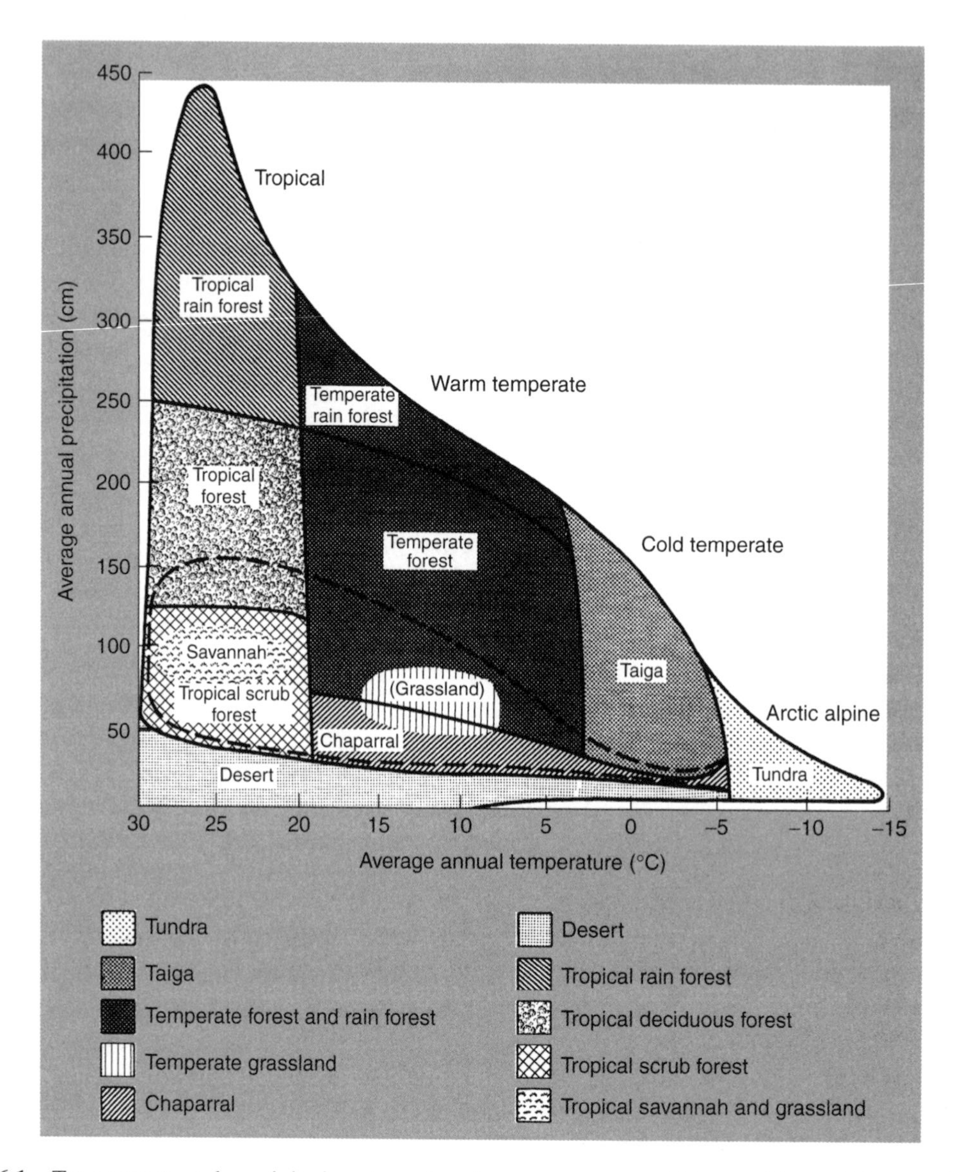

Fig. 6.1 Temperature and precipitation determine the major terrestrial biomes (From Stiling P 1996. *Ecology: Theories and Applications*. Upper Saddle River NJ: Prentice Hall, p. 403).

of timber for lumber, wood pulp (paper), and firewood fuel. The total commercial value of forest products can be large (e.g. $290 billion in the United States in 1999) (Howard 1999).

Finally, climate change presents a challenge to managers – both those who regulate timber harvesting and those charged with protecting and conserving terrestrial ecosystems. Here we examine how terrestrial systems have changed in the past, how they have changed recently, and how they probably will change in the future in response to anthropogenic greenhouse warming.

Geographic Shifts in Terrestrial Habitats

Past migrations

Historically, the spatial distribution of major vegetation types throughout the world has changed markedly in response to climate change (Chapter 3). For example, in the Northern Hemisphere during the height of the last glaciation (18,000 years ago), spruce forests were restricted to a small area south of the North American Great Lakes and oak trees to small pockets in the eastern Mediterranean. As the climate warmed during the last 18,000 years, spruce forests moved northward, occupied their present area in Northern Europe, Russia, and Canada, and became virtually extinct in the United States. At the same time, oak forests expanded in the Southeastern United States and Western and Southern Europe (Figure 6.2).

Such changes in vegetation distribution have occurred slowly over thousands of years. However, recent trends suggest a more rapid geographic shift in species distributions. According to historical written records and recent field studies of tree distribution, the percentage of red spruce making up forests in New Hampshire decreased from 40% in 1830 to 6% in 1987 (Figure 6.3). The decrease is not believed to result from land clearing or from pollutant stress but from a 2.2 °C increase in the average summer temperature during the same period.

Temperate vegetation

The climate space (optimum temperature and precipitation patterns) defining the current distribution of vegetation types can be measured. Global climate models (Chapter 4) can predict future geographic shifts in defined climate spaces. Thus, the possible geographic distribution of available future habitat for a vegetation type can be mapped. For example, studies suggest that large-scale changes in the distribution of major vegetation types in the United States will take place by the end of this century in response to anthropogenic climate change. Arid lands (desert) in the Southwestern United States will shrink as precipitation increases. Savanna/shrub/woodland systems will replace grasslands in parts of the Great Plains. In the Eastern United States under more moderate scenarios, forests would expand, but, under more severe climate scenarios, decreased moisture and catastrophic fires in the southeast would trigger a rapid conversion from broadleaf forest to savanna (Figure 6.4).

In response to a doubled atmospheric CO_2 warming, temperatures within the range of several major US timber trees in the Great Lakes region will increase by 7 to 10 °C. This will exceed the climatic tolerance of any of the species under any moisture regime. Beech, hemlock, and yellow birch will all be reduced in abundance in the Great Lakes region and their optimum habitat space will shift northwest into Canada. Because of the longevity of such trees, changes in distribution will lag in time behind the actual climate space shift, but all these tree species would become extinct in the Great Lakes area. Extinction will result from both the failure of young seedlings to establish and from mortality of adult trees. Some species may not survive anywhere in their present range except Nova Scotia. Hardwood logging would be eliminated as an economic resource. Salvage logging of the increased number of dead trees in the southern part of the current range would provide a short-term economic gain (Zabinski and Davis 1989).

A doubled CO_2 atmosphere (likely within the next 50 years) will also affect the distribution of sugar maple in Eastern North

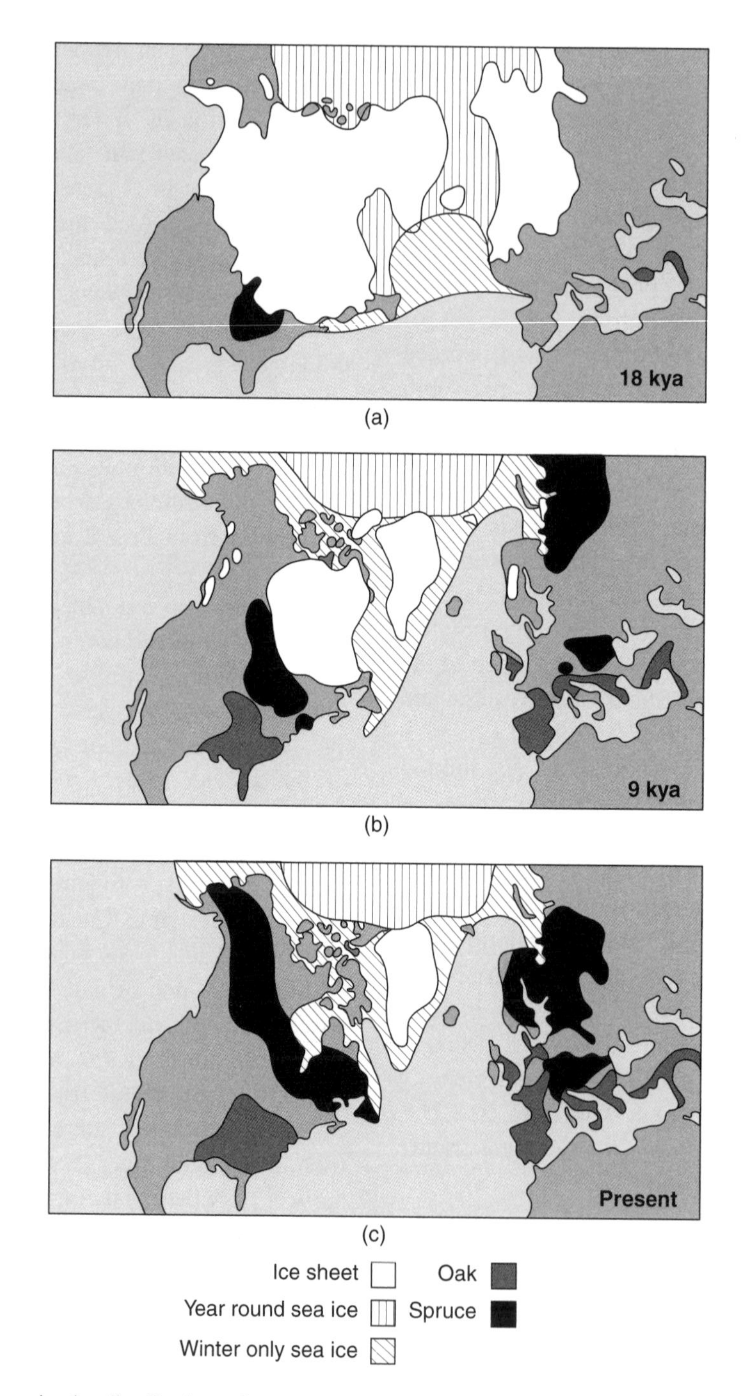

Fig. 6.2 Changes in the distribution of spruce and oak forests in the Northern Hemisphere since the last glacial period 18,000 years ago (18 kya) (Reprinted with permission from COHMAP 1988. Climatic changes of the last 18,000 years: observations and model simulations. *Science* **241**: 1043–1052. Copyright (1988) American Association for the Advancement of Science).

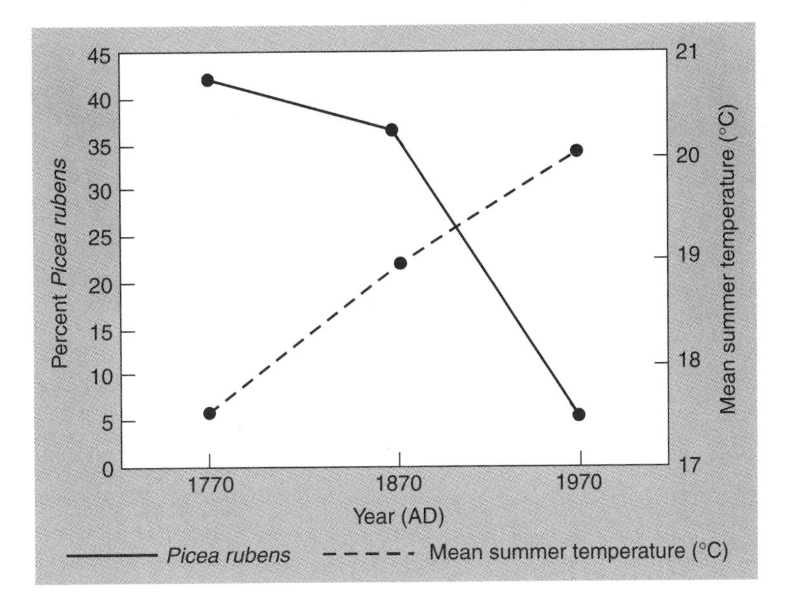

Fig. 6.3 Decrease in percent of red spruce (*Picea rubens*) in old growth stands in New England in response to increasing temperature since 1770 (Based on data from Hamburg SP and Coghill CV 1988. Historical decline in red spruce populations and climatic warming. *Nature* **331**: 428–431. Copyright (1988) Macmillan Magazines Limited).

America. Increased temperature, combined with a decrease in soil moisture, will shift sugar maple habitat northeast. The species will completely disappear from much of its current range and survive only in a much-reduced area to the northeast (Figure 6.5).

In the wet coastal mountains of California and Oregon, Douglas fir will shrink in the lowlands and be replaced by more drought-tolerant western pine species. In the Sierra Nevada mountains of California and the Cascade mountains of Oregon and Washington, a 2.5 to 5 °C warming will shift the current species composition. The postwarming species composition on the west slope will more closely match that of the currently less dense east slope forests, reducing biomass to about 60% of present levels (Franklin et al. 1991). Current east slope forests will gradually shift to a drier juniper and sagebrush system (Figure 6.6). Overall, in the Western United States, climate shifts will favor more drought-tolerant species such as pine at the expense of other species. The frequency of forest fires could increase, reducing the total forest area.

However, considering shifting climate spaces alone may be misleading when it comes to predicting future plant distribution. Microcosm research suggests that dispersal mechanisms and species interactions are important and must be included for accurate predictions of biotic change in relation to climate (Davis et al. 1998). During past glacial–interglacial climatic changes, tree species populations expanded into favorable regions at rates averaging 10 to 40 km per century. Geographic shifts in climate space resulting from anthropogenic climate change will be much more rapid. Thus, rates of seed dispersal and colonization will be important limiting factors in forest survival in this century.

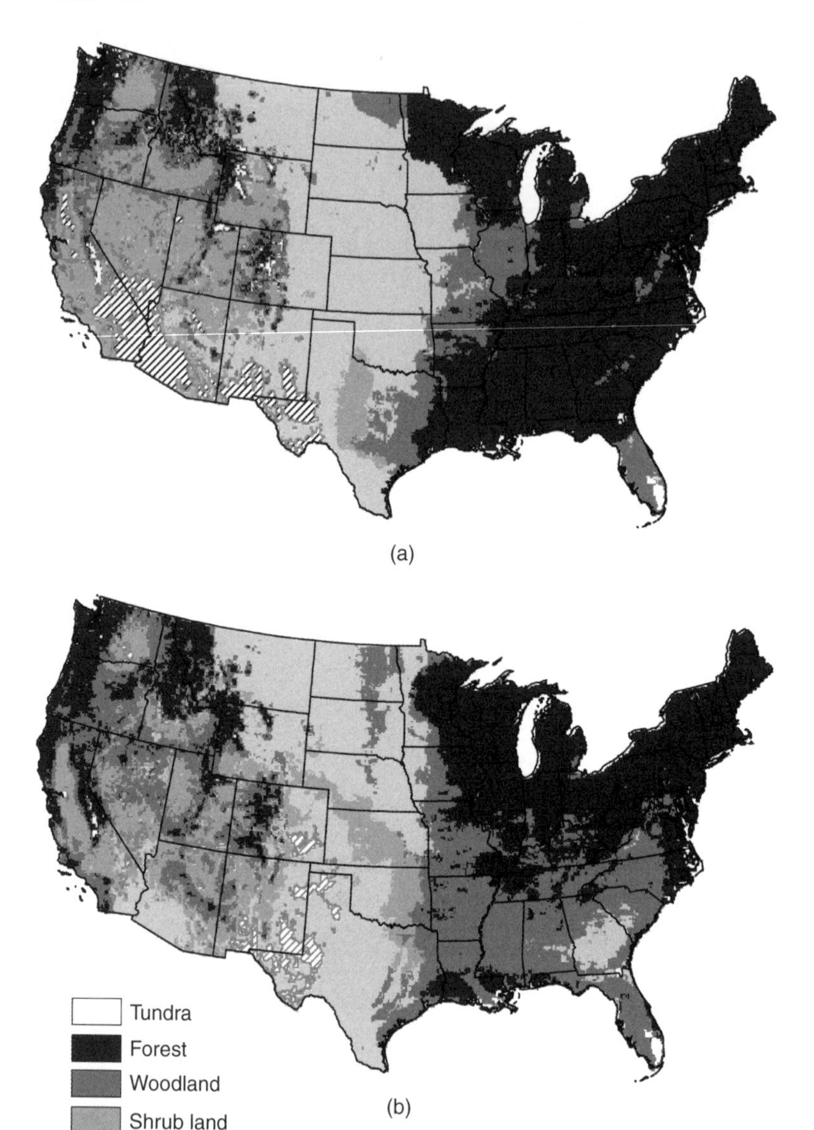

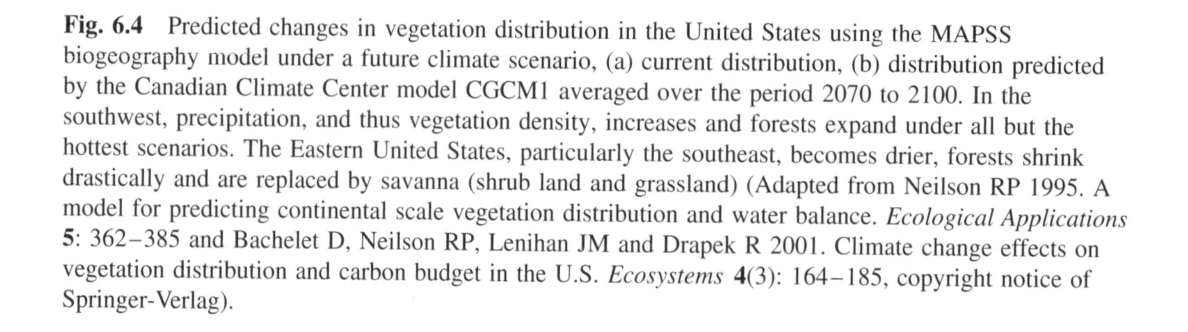

Fig. 6.4 Predicted changes in vegetation distribution in the United States using the MAPSS biogeography model under a future climate scenario, (a) current distribution, (b) distribution predicted by the Canadian Climate Center model CGCM1 averaged over the period 2070 to 2100. In the southwest, precipitation, and thus vegetation density, increases and forests expand under all but the hottest scenarios. The Eastern United States, particularly the southeast, becomes drier, forests shrink drastically and are replaced by savanna (shrub land and grassland) (Adapted from Neilson RP 1995. A model for predicting continental scale vegetation distribution and water balance. *Ecological Applications* **5**: 362–385 and Bachelet D, Neilson RP, Lenihan JM and Drapek R 2001. Climate change effects on vegetation distribution and carbon budget in the U.S. *Ecosystems* **4**(3): 164–185, copyright notice of Springer-Verlag).

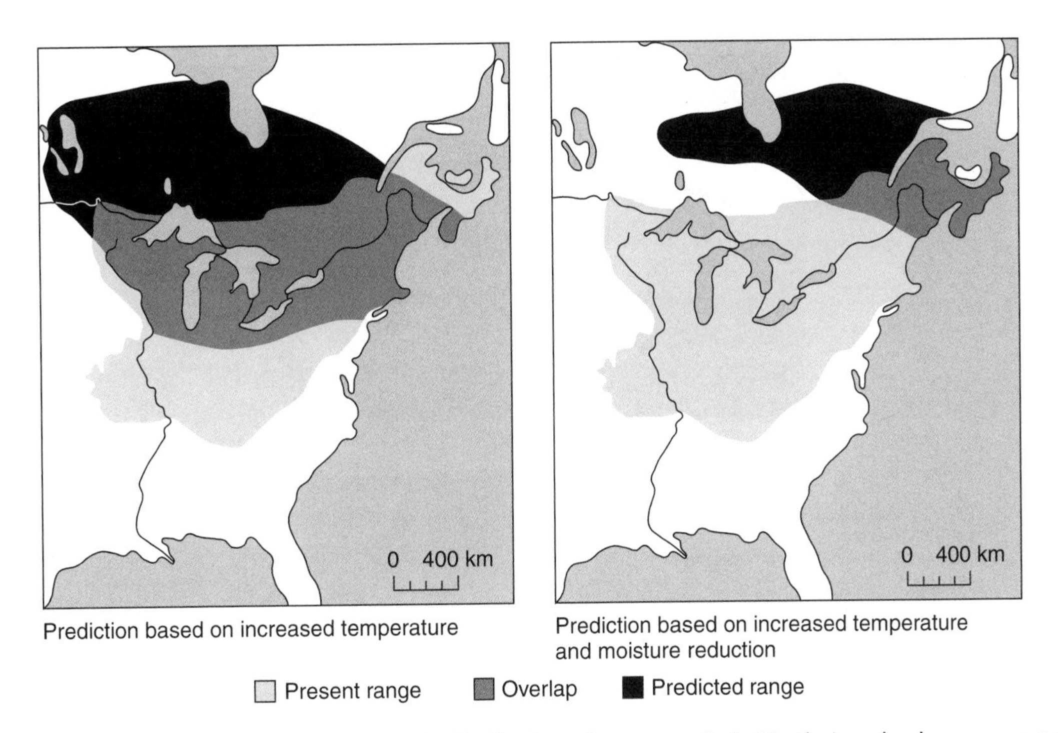

Fig. 6.5 Predicted shift in the geographic distribution of sugar maple in North America in response to a doubling of atmospheric carbon dioxide (Adapted from Davis MB and Zabinski C 1992. Changes in geographical range resulting from greenhouse warming: effects on biodiversity of forests. In: Peters RL and Lovejoy TE, eds *Global Warming and Biological Diversity*. New Haven: Yale University Press, pp. 297–308).

Boreal and alpine vegetation

Future climate changes will probably be greater at high latitudes (Chapter 4). Paleoclimatological records show past dramatic shifts in the distribution of vegetation at high latitudes. Northern Canada and Alaska are already experiencing rapid warming and reductions in ice cover (Chapter 3). A warmer and wetter climate at high latitudes will shift the vegetation away from cold-loving tundra plants toward more temperate forest species (Starfield and Chapin 1996). Tundra, taiga, and temperate forest systems will all migrate poleward (Monserud et al. 1993). For example, the distribution and area cover of nine different vegetation types in Siberia will be "almost completely changed" by a doubled CO_2 climate (Figure 6.7).

As the climate warms, alpine plant species may be forced to occupy higher mountain elevations. Studies in the European Alps suggest that such a trend may already be under way with the number of species of alpine plants increasing at higher elevations (Grabherr et al. 1994). Eventually, as their habitat shrinks to small mountaintops, plants could be forced to extinction.

Grassland and shrub land

Changes in precipitation and moisture will alter the composition and distribution of grasslands and shrub lands. Semiarid grass and shrub lands are under stress from human

(a)

(b)

(c)

Fig. 6.6 Models predict that climate change will force a shift toward more drought-tolerant vegetation in the Cascade mountains of Washington and Oregon (Pacific Northwest USA). On the wet western slope, the area now covered by western hemlock and Douglas fir (a) will shrink and be partially replaced by more drought-tolerant and less dense pine and oak now characteristic of the drier eastern slope (b). The eastern slope will become even drier and shift to Juniper savanna and sagebrush (c) (Courtesy of US Bureau of Land Management. Photos: (a) – D. Huntington, (b) – unknown, (c) – Mark Armstrong).

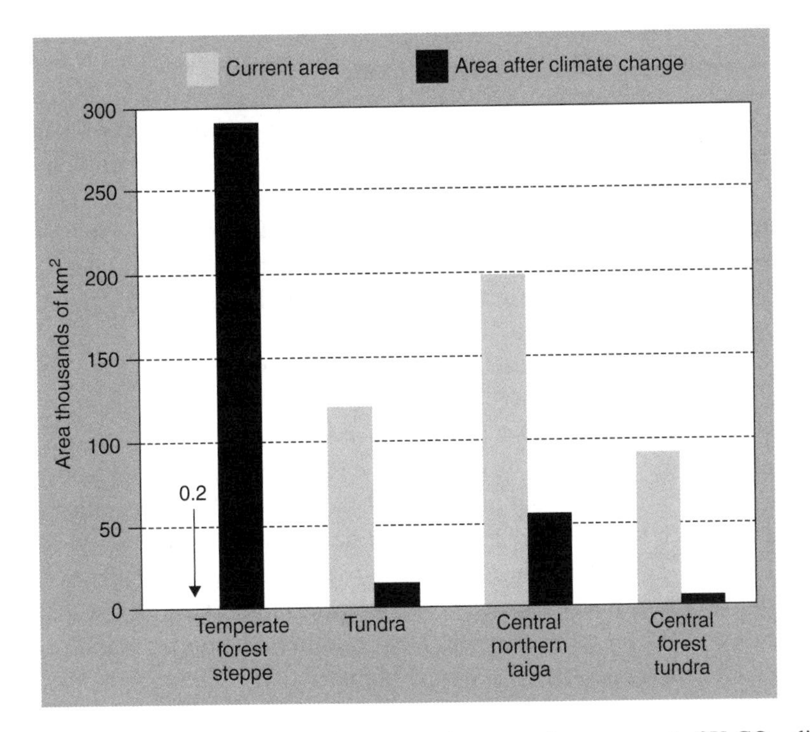

Fig. 6.7 Predicted changes in areas of Siberian vegetation types in response to 2X CO_2 climate change. Mean of four global climate models (Based on data of Tchebakova NM, Monserud RA, Leemans R and Nazimova DI 1995. Possible vegetation shifts in Siberia under climatic change. In: Pernetta, J, Leemans R, Elder D and Humphrey S, eds *The Impact of Climate Change on Ecosystems and Species: Terrestrial Ecosystems.* Gland, Switzerland IUCN, The World Conservation Union, pp. 67–82).

land-use practices, particularly animal grazing and row-crop agriculture. Decreased rainfall in some areas could, through a positive feedback loop, accelerate desertification, that is, the transformation of grasslands to shrub lands or desert. As grassland is replaced by shrub land, a greater percentage of the soil is exposed, and the temperature of the soil surface increases. Hot dry soils retard the accumulation of organic nitrogen – further inhibiting plant growth. Barren arid soils are then exposed to winds and transported into the atmosphere as dust. Dust over desert regions may act to trap infrared re-radiation and lead to warming, further exacerbating the problem (Schlesinger et al. 1990). Linked climate-change, plant-growth, and soil-water models predict shifts in the distribution of major types of North American prairie grasses over a 40-year period as the climate changes (Coffin and Lauenroth 1996).

Vegetation–Climate Interactions

Climate not only affects vegetation but the presence, absence, or type of vegetation can affect climate. Forests contain twice as much carbon as the atmosphere and metabolize more than 14% of atmospheric carbon each year. Forests and climate interact, and a disturbance in either can affect the other. Carbon released into the atmosphere from tropical forest harvesting totals 1.1 to 3.6 PgC year^{-1} (petagrams of carbon per year) (Houghton 1991). The Brazilian Amazon forest, host to

Box 6.1 How vegetation changes affect climate

Deforestation will probably have serious long-term irreversible effects on the climate of the Amazon Basin. Assuming that current trends in human habitat alteration continue, the tropical Amazon rainforest will be completely replaced by pastureland in a few decades. Researchers applied a climate model to describe the temperature, wind, and humidity of the Amazon at 18 elevations between the ground surface and 30-km altitude. The horizontal resolution was 1.8 by 2.8 km (Shulka et al. 1990). Within each grid, the vegetation was defined as 1 of 12 different vegetation types. They first ran the model under the present conditions of vegetation distribution and then, with the forest converted to pasture. Compared to the forest system, the pasture has a much greater stomatal resistance, that is, less capacity for transporting water from the soil to the atmosphere through plants and their leaves. A shallower, sparser root system results in a greatly reduced soil-water storage capacity. The surface and soil temperatures will be 1 to 3 °C warmer after deforestation. Less evapotranspiration leads to a reduction in atmospheric moisture, cloud formation, and precipitation over the entire Amazon region. The overall disruption of the hydrologic cycle will have negative impacts on plant–animal relationships. Once removed, the Amazon forest will be unable to reestablish itself. Deforestation of tropical rainforests elsewhere will probably have similar effects on regional climates.

half the world's species, and a storehouse of an estimated 70 PgC, is being deforested at a rate of 25,000 to 50,000 km^2 per year – much of it being converted to pastureland. This conversion will alter the regional climate (Box 6.1).

When old-growth forests are harvested and replaced by young forests, carbon storage as biomass decreases substantially. Even when storage of carbon as lumber in wooded buildings is included, timber harvesting results in a net positive flux of carbon to the atmosphere. In the United States, harvest of old-growth timber in Western Washington and Oregon alone over the last 100 years has added 1.5 to 1.8 PgC to the atmosphere (Harmon et al. 1990). One modeling strategy uses satellite imagery to map forest harvest activity and link it to a carbon flux model. This approach suggests that a single 1.2 million hectare area of Western Oregon used for timber lands contributed 0.018 PgC to the atmosphere between 1972 and 1991 or about 1.13 million g C ha^{-1} $year^{-1}$ (Cohen et al. 1996).

Increasing atmospheric CO_2 generally increases photosynthetic rates in individual plants. However, this increased productivity does not necessarily benefit plants. When several species are grown together, increased competition and nutrient availability diminishes any benefit of enhanced atmospheric CO_2. Overall, the effects of a CO_2-enhanced atmosphere on communities of vegetation are complex and not well understood (Bazzaz and Fajer 1992).

Effects of Disturbances

Greenhouse warming will increase the frequency of disturbance events that impact mid-latitude temperate forests (Overpeck et al. 1990). Models suggest increases in "disturbance weather," that is, summer/autumn drought and thunderstorms. These changing weather patterns with increases in lightning

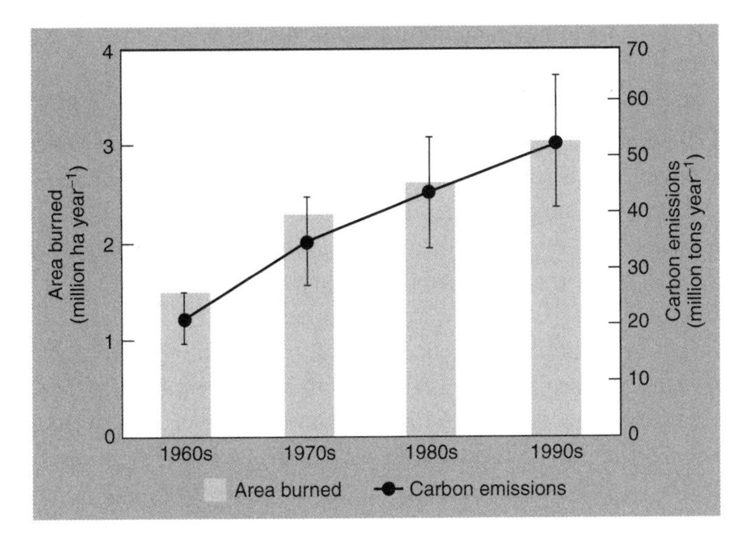

Fig. 6.8 In North American boreal forests, the average forest area that burned in 10-year periods doubled over 30 years. Carbon emissions into the atmosphere increased from 21 to 53 million tons per year (Senkowsky S 2001. A burning interest in boreal forests: researchers in Alaska link fires with climate change. *Bioscience* **51**(11): 916–921. Copyright, American Institute of Biological Sciences).

and wind, along with decreasing soil moisture, will lead to increases in forest fires. Wind damage from hurricanes and flooding from coastal seawater rise will also have negative impacts in some areas.

In response to climate change, the frequency and intensity of forest fires will probably increase in many regions of the world. In the tropics, fires in the Amazon will increase as a result of a longer dry season (Box 6.1). Also, in Borneo and Indonesia, strong El Niños, possibly the result of global warming, have already increased the incidence and extent of forest fires. In temperate and boreal regions of North America and Russia, a number of studies suggest climate change induced increases in forest fire seasonal severity, seasonal length, and areal extent (Stocks et al. 1998). Forest fires, through a positive feedback loop, could significantly affect climate. Warmer temperatures lead to more fires. These fires release greenhouse gases and contribute to additional warming. In fact, an increase in forest fires is already responsible for increasing releases of CO_2 into the atmosphere (Figure 6.8). Also, in boreal regions, soils and subsurface permafrost represent a large carbon reservoir. Fires remove the insulating vegetation cover that keeps the permafrost frozen. The subsequent microbial breakdown of the soil carbon can release large quantities of CO_2 into the atmosphere (Senkowsky 2001).

Loss of Biodiversity

Climate change threatens the very survival of some terrestrial species. Dominant forest tree species are the backbone of the actual forest ecosystem. Tree cover provides a habitat for numerous herbaceous plants, fungi, lichens, and small and large animals. Thus, loss of the tree species will affect virtually all the species that make up a complex forest ecosystem. The survival of species during climatic change will largely depend on their ability to migrate fast enough to keep up with their preferred climate space.

Studies using seven general circulation climate models (GCMs) and two biogeographic models suggest that very high (compared to fossil and historical records) migration rates will be necessary to keep pace with anthropogenic climate change. After the last glaciation, migration rates in North America were about 200 km or 20 km per century for spruce or beech trees, respectively. Projections suggest they will not be able to keep pace with an estimated 500-km migration needed during this century (Roberts 1988). Thus, climate change will radically increase species loss and reduce biodiversity, particularly in the higher latitudes of the Northern Hemisphere (Malcolm and Markham 2000).

Global warming has the potential during this century to significantly alter 35% of the world's existing terrestrial habitats (Figure 6.9). Warming will probably favor the most mobile species and eliminate the most sedentary ones. Populations at greatest risk are those that are rare and isolated species in fragmented habitats or those that are bounded by water bodies, human settlements, or agricultural areas. In northern countries, such as Russia, Sweden, and Finland, as well as in seven Canadian Provinces, half the existing terrestrial habitats are at risk. In Mexico, by 2055, the habitat area for many animal species will shrink significantly as a result of climate change. A predicted 2.4% of species will lose 90% of their range and be threatened with extinction. In the Chihuahuan desert, the habitat of about half of all species will probably disappear (Peterson et al. 2002). In the James and Hudson Bay areas of Canada, prolonged ice melt periods will delay the return of polar bears to their feeding areas. This, together with potential declines in seal populations, will put bears under nutritional stress (Stirling 1993). Rapid reductions in greenhouse gas emissions will be necessary to reduce the threat to global biodiversity.

Finally, temperature directly affects numerous functions of individual animals as well as the interactions between animal populations. Studies of plant and animal migrations

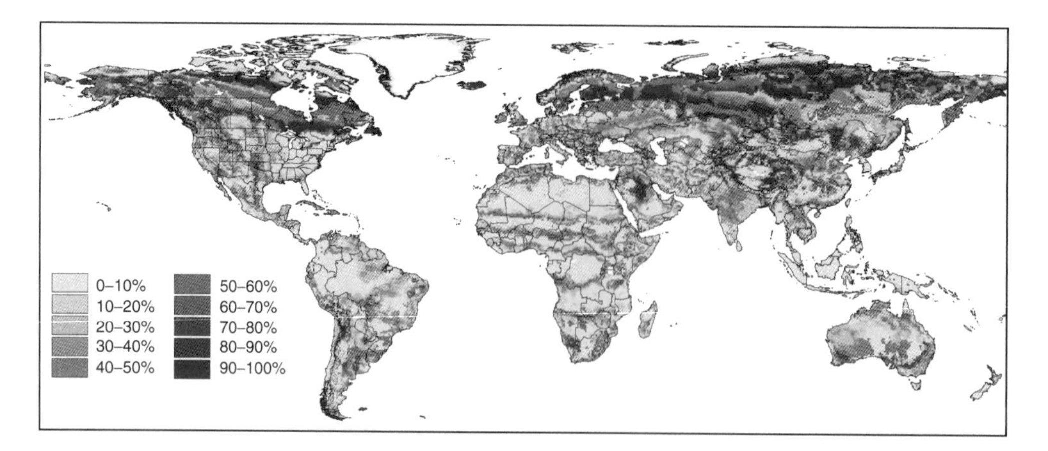

Fig. 6.9 Terrestrial biomes will probably change in response to a doubling of atmospheric CO_2. Shades indicate the percent of climate models that predict a change in the biome in that area (From Malcolm JR and Markham A 2000. *Global Warming and Terrestrial Biodiversity Decline*. Report of the World Wildlife Fund, Gland, Switzerland, p. 22).

during past climate changes suggest that habitats, along with their resident organisms, will not simply shift northward in response to warming. Rather, plant and animal communities will experience complex reorganizations. Invasive species with rapid reproductive cycles are likely to be favored in a changing environment. Weed species and pests could become more dominant in some areas (Malcolm and Markham 2000). Thus, changing climate may have many subtle and unforeseen effects on natural periodic events that occur in the life cycles of plants and animals (Box 6.2, Figure 6.10).

Box 6.2 Phenological changes

The natural periodic events (phenology) that occur as part of the life cycle of most organisms are closely linked to daily, seasonal, and long-term climate cycles. Temperature affects plants in many ways, including the length of the growing season, flowering time, and coordination with insect pollinators. The mating and migration times of many animals are closely linked to temperature. For example, in many reptiles, the male:female sex ratio of offspring is temperature-dependent. In the painted turtle, an increase of 2 °C in mean temperature may drastically increase this ratio and a climate change of 4 °C would effectively eliminate the production of male offspring and hence lead to species extinction (Janzen 1994). Like reptiles, the reproductive cycles of cold-blooded amphibians are also sensitive to climatic change. Thus, the migration to breeding ponds and time of spawning of two species of toad and frog in England has occurred earlier in recent years. During the period of study, from 1978 to 1994, the time for this reproductive behavior decreased 9 to 10 days per 1 °C increase in maximum temperature (Beebee 1995). Numerous studies suggest that phenological changes are already occurring in plants and animals in response to climate change (adapted from Peñuelas and Filella 2001):

- Mediterranean deciduous plants now leaf out 16 days earlier and fall 13 days later than 50 years ago.
- In Western Canada, trees (*Populus tremuloides*) bloom 26 days earlier than a century ago.
- In Europe and North America, biological spring occurs a week or more earlier.
- The growing season has increased 18 days in Eurasia and 12 days in North America over the past two decades.
- For many plants in the temperate zone, spring flowering is occurring earlier in the season.
- Insect larval development has accelerated, for example, 3 to 6 day advancement in the life cycle of aphids in the United Kingdom, over the past 25 years. Numbers of aphid eggs laid in Spitsbergen, Norway, increase greatly when small areas are artificially warmed (Young 1994).
- Butterflies in this decade metamorphose 11 days earlier in Northeast Spain than they did in 1952.

- Bird species surveyed in the United Kingdom shifted their egg laying to nine days earlier between 1971 and 1995.
- In New York State, frog calling occured an average 10 days earlier during 1990 to 1999 than in 1900 to 1912.
- In many areas of the world, dates for migratory birds to move have changed significantly.

Such phenological changes can have a number of detrimental consequences. One of the primary impacts may be a decoupling of species interactions, for example, between plants and their pollinators, or between birds or predators and their food supply. Phenological effects of climate change will probably be numerous, but at the same time, often subtle and difficult to detect. However, the changes described above, as well as many others, have occurred in response to a warming that is only 50% of what is expected for the twenty-first century.

Implications for Forest Management and Conservation Policy

Predicted changes raise numerous policy questions relevant to forest management. Long-term plans for forest management will need to take into account the likely effects of anthropogenic climate change. Changes in forest boundaries, brought about by climate change, will complicate land-use policies. If timber production is reduced by climate change, should governments open up currently protected parks and wilderness areas to ensure timber supplies?

It is unlikely that dispersal of most tree species will be able to keep up with the shifting climate space. Mitigation attempts may include artificial seed dispersal into climatically suitable regions. Transplantation of species into new areas may aid in preserving vegetation communities under the stress of shifting climate spaces. Such attempts, however, may be frustrated by the lack of suitable soil conditions in newer areas.

Should massive reforestation be undertaken now to help sequester CO_2 added to the atmosphere by fossil-fuel combustion? Estimates for the United States indicate that to keep pace with CO_2 emissions, reforestation efforts will need to be doubled or tripled at costs of hundreds of millions of dollars. It would take an estimated 100 years to reforest 40% of US forest lands. Such an effort would also require changes in the complex ownership patterns of forests. Application of new technologies of plant breeding, bioengineering, transplantations, fertilization, and irrigation could aid in mitigation. But who should pay the additional costs incurred for implementing new policies – landowners, forest users, consumers, or all taxpayers? Tropical forests are being harvested at a rapid pace. Any attempt at reversing this trend and using tropical forestation to sequester carbon will fail unless they address the economic, social, and political needs of the local people (Cairns and Meganck 1994).

There is even some doubt that massive reforestation will help mitigate greenhouse warming. The albedo (reflectivity) of forested land is usually much lower than agricultural or other types of land cover. Thus, converting high albedo land to forest may increase warming and offset much of the benefits of added carbon sequestration (Betts 2000).

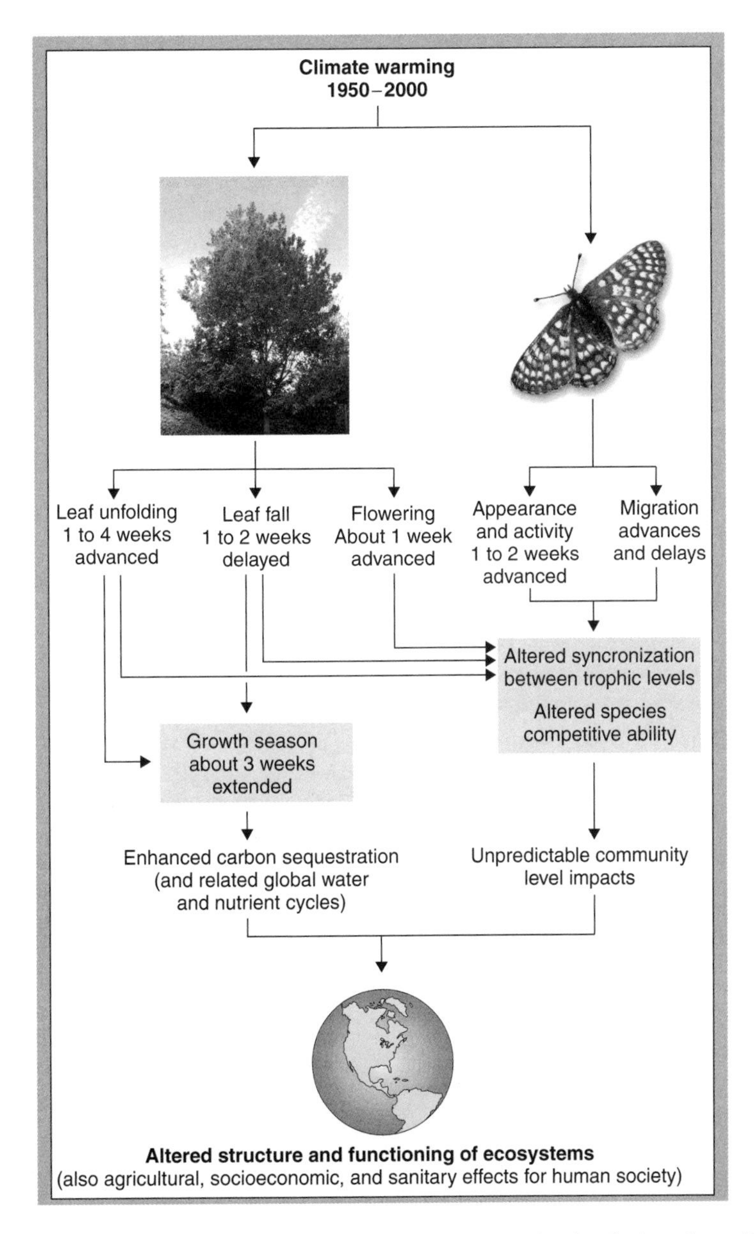

Fig. 6.10 Example of the effects of climate warming on plant and animal phenology (Reprinted with permission from Peñuelas J and Filella I 2001. Response to a warming world. *Science* **294**: 793–795. Copyright (2001) American Association for the Advancement of Science).

Our increasingly fragmented landscape poses special problems for the survival of species in an era of rapid climate change. Protected areas, such as parks and nature preserves, are especially at risk (Peters and Darling 1985). In most cases these are rather isolated areas, set aside for preservation amongst surrounding areas of urban growth or agricultural activity. As the climate changes, these nature reserves will be subject to unprecedented pressure. Only in areas with connecting corridors will species be able to migrate. Such a scenario suggests that new management and park design strategies may be necessary in order to preserve biodiversity.

Summary

The geographic distributions of major terrestrial ecosystems (desert, savanna, forest, etc.) are largely governed by patterns of temperature and precipitation. Such systems can migrate hundreds or thousands of kilometers over thousands of years in response to natural climate change, for example, during glacial–interglacial periods. However, such past migration rates are far too slow to keep pace with the rapid geographic shifts in regional climate expected from anthropogenic greenhouse warming. Some species may be able to migrate and keep pace with climate change, but many may not.

There can be little doubt that climate change during this century will significantly alter the distribution and abundance of terrestrial species. Additional research will undoubtedly provide even more accurate scenarios for the future. The interactions and feedbacks between vegetation and climate need to be more fully understood. As forests are weakened and stressed by climate change, will their susceptibility to insect and pathogen damage, air pollution, acid rain, and forest fires result in greatly increased mortality, and what will be the combined effects of all such stresses?

Changing climate appears to be responsible for many documented phenological (life cycle) changes in plants and animals over the past 50 years or more. Future climate change could further alter life cycle elements such as the timing of flowering in plants, metamorphosis in insects, or migration of animals.

Finally, if significant areas of forest are lost, what will this mean to the survival of the numerous animal species that inhabit the forests or streams that form part of the ecosystem? Predicted changes raise many questions regarding the best strategy for managing terrestrial ecosystems.

References

Bachelet D, Neilson RP, Lenihan JM and Drapek R 2001 Climate change effects on vegetation distribution and carbon budget in the U.S. *Ecosystems* **4**(3): 164–185.

Bazzaz FA and Fajer ED 1992 Plant life in a CO_2 rich world. *Scientific American* **January**: 68–74.

Beebee TJC 1995 Amphibian breeding climate. *Nature* **374**: 219, 220.

Betts RA 2000 Offset of the potential carbon sink from boreal forestation by decreases in surface albedo. *Nature* **408**: 187–190.

Cairns MA and Meganck RA 1994 Carbon sequestration, biological diversity, sustainable development: integrated forest management. *Environmental Management* **18**(1): 13.

Coffin DP and Lauenroth WK 1996 Transient responses of North-American grasslands to changes in climate. *Climatic Change* **34**: 269–278.

Cohen WB, Harmon ME, Wallin DO and Fiorella M 1996 Two decades of carbon flux from forests of the Pacific Northwest. *Bioscience* **46**(11): 836–844.

COHMAP 1988 Climatic changes of the last 18,000 years: observations and model simulations. *Science* **241**: 1043–1052.

Davis MB and Zabinski C 1992 Changes in geographical range resulting from greenhouse warming: effects on biodiversity of forests. In: Peters RL and Lovejoy TE, eds *Global Warming and Biological Diversity*. New Haven: Yale University Press, pp. 297–308.

Davis AJ, Jenkinson LS, Lawton JH, Shorrocks B and Wood S 1998 Making mistakes when predicting shifts in species range in response to global warming. *Nature* **391**: 783–786.

Franklin JF, Swanson FJ, Harmon ME, Perry DA, Spies TA, Dale VH, et al. 1991 Effects of global climate change on forests of northwestern North America. *The Northwest Environmental Journal* **7**: 233–254.

Grabherr G, Gottfried M and Pauli H 1994 Climate effects on mountain plants. *Nature* **369**: 448.

Hamburg SP and Coghill CV 1988 Historical decline in red spruce populations and climatic warming. *Nature* **331**: 428–431.

Harmon ME, Ferrell WK and Franklin JF 1990 Effects on carbon storage of conversion of old-growth forests to young forests. *Science* **247**: 699–701.

Houghton RA 1991 Tropical deforestation and atmospheric carbon dioxide. *Climatic Change* **19**: 99–118.

Howard JL 1999 *U.S. timber production, trade consumption, and price statistics 1965–1997*. General Technical Report FPL–GTR–116. Madison, WI: US Department of Agriculture, Forest Service, Forest Products Laboratory, p. 76.

Janzen FJ 1994 Climate change and temperature-dependent sex determination in reptiles. *Proceedings of the National Academy of Sciences* **91**: 7487–7490.

Malcolm JR and Markham A 2000 *Global Warming and Terrestrial Biodiversity Decline*. Report of the World Wildlife Fund, Gland, Switzerland, p. 22.

Monserud RA, Tchebakova NM and Leemans R 1993 Global vegetation change predicted by the modified Budyko model. *Climatic Change* **25**: 59–83.

Neilson RP 1995 A model for predicting continental scale vegetation distribution and water balance. *Ecological Applications* **5**: 362–385.

Overpeck JT, Rind D and Goldberg R 1990 Climate-induced changes in forest disturbance and vegetation. *Nature* **343**: 51–53.

Peñuelas J and Filella I 2001 Response to a warming world. *Science* **294**: 793–795.

Peters RL and Darling DS 1985 The greenhouse effect and nature reserves. *Bioscience* **35**(11): 707–717.

Peterson AT, Ortega-Huerta MA, Bartley J, Sánchez-Cordero V, Soberón J, Buddemeier RH, et al. 2002 Future projections for Mexican faunas under global climate change scenarios. *Nature* **416**: 626–629.

Roberts L 1988 Is there life after climate change? *Science* **242**: 1010–1012.

Schlesinger WH, Reynolds JF, Cunningham GL, Huenneke LF, Jarrell WM, Virginia RA, et al. 1990 Biological feedbacks in global desertification. *Science* **247**: 1043–1048.

Senkowsky S 2001 A burning interest in boreal forests: researchers in Alaska link fires with climate change. *Bioscience* **51**(11): 916–921.

Shulka J, Nobre C and Sellers P 1990 Amazon deforestation and climate change. *Science* **247**: 1322–1325.

Starfield A and Chapin III FS 1996 Model of transient changes in arctic and boreal vegetation in response to climate and land use change. *Ecological Applications* **6**(3): 842–864.

Stiling P 1996 *Ecology: Theories and Applications*. Upper Saddle River NJ: Prentice Hall, p. 255.

Stirling I 1993 Possible impacts of climatic warming on polar bears. *Canadian Wildlife Survey* **16**(3): 21–26.

Stocks BJ, Fosberg MA, Lynham TJ, Mearns L, Wotton BM, Yang Q, et al. 1998 Climate change and forest fire potential in Russian and Canadian boreal forests. *Climate Change* **38**: 1–13.

Tchebakova NM, Monserud RA, Leemans R and Nazimova DI 1995 Possible vegetation shifts in Siberia under climatic change. In: Pernetta J, Leemans R, Elder D and Humphrey S, eds *The Impact of Climate Change on Ecosystems and Species: Terrestrial Ecosystems*. Gland, Switzerland: IUCN, The World Conservation Union, pp. 67–82.

Young S 1994 Insects that carry a global warning. *New Scientist* **142**(1923): 32–35.

Zabinski K and Davis MB 1989 *Hard Times Ahead for Great Lakes Forests: A Climate Threshold Model Predicts Responses to CO_2-Induced Climate Change*.Contract No. CR-814607-01-0, Sponsored by the US EPA, Research Report. Minneapolis: University of Minnesota, pp. 5-1–5-19.

Chapter 7

Climate Change and Agriculture

"What we eat... ties us to the economic, political and ecological order of our whole planet."

Frances Moore Lappé 1982

Introduction

Productive agriculture is essential to feed a growing population and sustain modern civilization. World population will probably double over the next 100 years. Historically, cultivation of crops arose independently in several areas of the world 2,500 to 8,500 years ago. Compared to hunter-gatherers, farmers could harvest more food per area of land. Agriculture supported greater population densities than hunting and gathering, and provided the excess wealth to support skilled craftsman, governments, and armies of conquest (Diamond 1999). Agricultural productivity remains at the heart of modern economies.

Climate affects agriculture, a fact well known to every farmer. Year-to-year variations in harvest are largely due to variations in temperature and precipitation that can make the difference between bountiful "bumper" crops and economic ruin. For example, the North American Great Plains – a breadbasket of cereal crops – experienced a prolonged drought in the 1930s that turned a huge area into a "dust bowl." The economic effects were devastating – farmers, unable to meet mortgage payments, lost their farms and many migrated elsewhere in search of work. Again, droughts in the Central United States in 1980 and 1988 greatly reduced the yield of corn. Animal husbandry also depends on favorable climatic conditions. For example, the area of land needed to sustain cattle production in North America increases very rapidly in response to decreased rainfall (Figure 7.1).

Agriculture also affects climate. Forests, a major terrestrial sink for CO_2, have been greatly reduced by agricultural land clearing. Modern agriculture depends on fossil-fuel energy and contributes to greenhouse gas emissions. This is clearly evident in grain-fed livestock production, where about 20,000 cal of fossil fuel (for farm machinery, etc.) are

Climate Change: Causes, Effects, and Solutions John T. Hardy
 ISBNs: 0-470-85018-3 (HB); 0-470-85019-1 (PB)

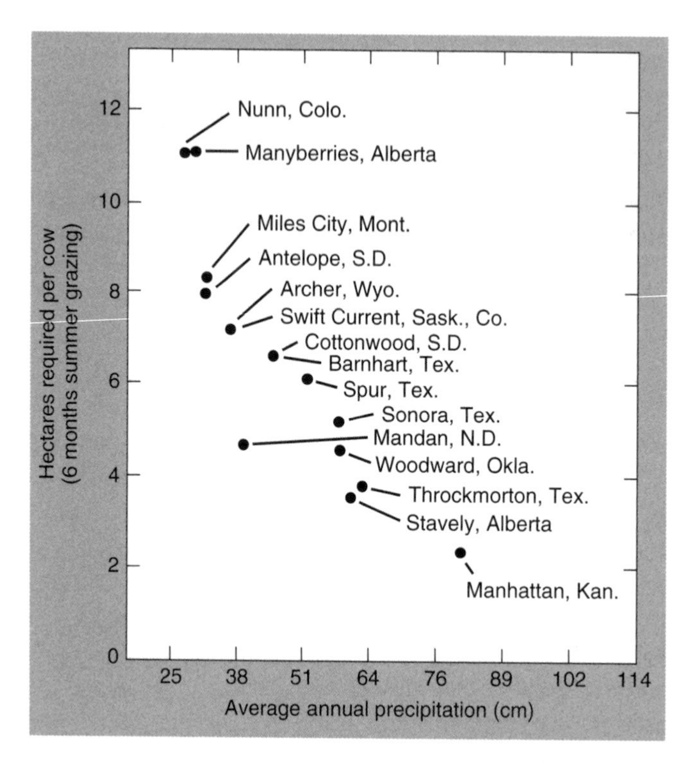

Fig. 7.1 The area of required pasturage for cattle in the US Midwest increases as annual precipitation decreases (From CIAP 1975. *Impacts of climate change on the biosphere*. Monograph 5, Part 2; September. Climate Impact Assessment Program. US Department of Transportation, Washington, DC).

necessary to produce the 500 cal contained in one pound of beefsteak.

Effects of Agriculture on Climate Change

The global flux of several greenhouse gases is influenced by agriculture. Land clearing, much of it from agriculture, is the second largest source of CO_2 emissions after fossil-fuel combustion, accounting for 10 to 30% of net global CO_2 emissions (Rosenzweig and Hillel 1998). Forests, grasslands, and soils store large quantities of carbon. Forests store 20 to 40 times more carbon per unit area than most crops and when they are cleared for cultivation, much of this carbon is released to the atmosphere. Mean estimates of carbon loss from the conversion of terrestrial ecosystems to agriculture range from 21 to 46% (Schlesinger 1986).

Some agriculture produces methane (CH_4) – the second-most important greenhouse gas. Paddy rice cultivation is responsible for about 40% of global CH_4 emissions. In flooded rice paddies, microbial decomposition of high organic aquatic sediments, under low oxygen conditions, releases CH_4 gas to the atmosphere. This source will continue to grow as rice cultivation expands in the future (Rosenzweig and Hillel 1998). Livestock production is responsible for about 15% of global CH_4 emissions. Ruminant animals (cattle, sheep, goats, camels, and buffalo) digest grasses and other cellulose forage in their stomachs and release CH_4 to the air. Cattle represent about 75% of the total livestock CH_4 emissions.

Nitrous oxide (N_2O) is another greenhouse gas closely linked to agricultural activities. Like carbon, nitrogen in vegetation and soils is lost to the atmosphere during land clearing. Also, nitrogen fertilizers are applied to crops, and generally enhance growth. However, excess nitrogen from fertilizers is leached into the soil and, through microbial denitrification, converted to volatile N_2O and released into the atmosphere. Estimates of N_2O release from agricultural fertilizers range from 0.1 to 1.5% of applied nitrogen. N_2O is produced naturally by soils, but globally, nitrogen fertilizers contribute about 0.14 to 2.4 million tons of the 8 to 22 million tons of total annual N_2O emissions (Rosenzweig and Hillel 1998).

Different agricultural practices have different consequences for greenhouse gas emissions. In general, annual crops make a net contribution to greenhouse warming. However, the high net greenhouse warming potential of conventional tillage practice can be largely eliminated by the use of no-till crop management (Figure 7.2). Intensive agriculture uses large quantities of fossil fuel for tilling and harvesting and soil carbon can be

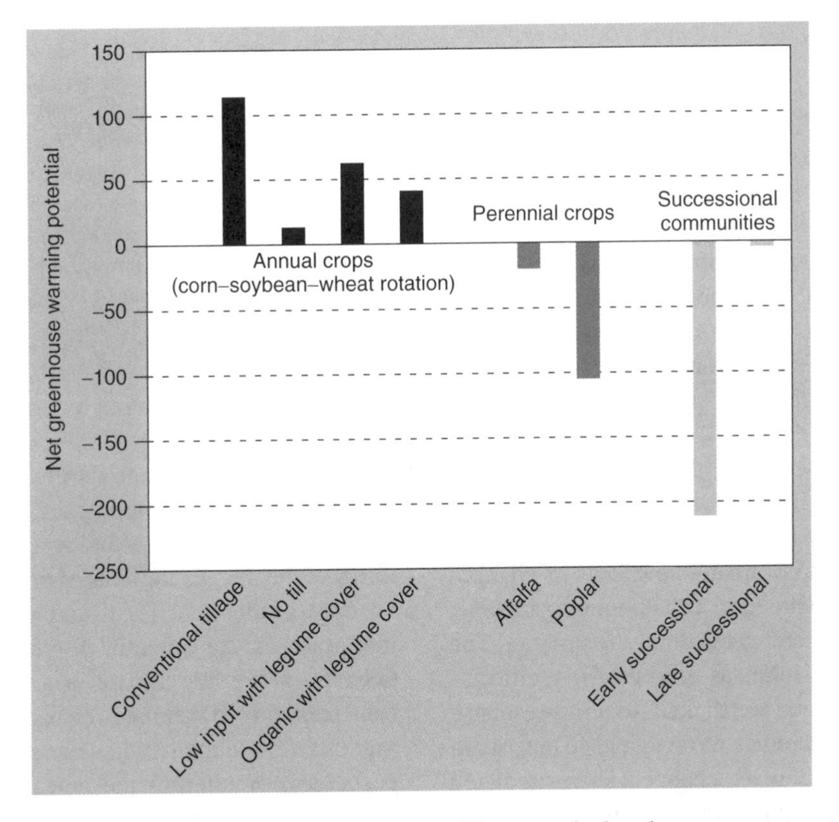

Fig. 7.2 Relative greenhouse warming potentials for different agricultural management systems based on soil carbon sequestration, agronomic inputs, and trace gas fluxes. Units are CO_2 equivalents ($g\,m^{-2}\,year^{-1}$). Negative values indicate a global warming mitigation potential (Adapted from tabular data of Robertson GP, Paul EA and Harwood RR 2000. Greenhouse gases in intensive agriculture: contributions of individual gases to the radiative forcing of the atmosphere. *Science* **289**: 1922–1925).

depleted. Low- or no-till agriculture is less energy-intensive and also maintains more carbon in the soil reservoir. Crop rotation using a legume or other nitrogen-fixing crops can reduce the need for nitrogen fertilizer. Generally, when fields are left dormant, the natural plant community that first invades (early succession community) grows rapidly, sequestering atmospheric carbon and providing a net removal of atmospheric carbon. However, as the species of plants change and the plant community matures (late succession), growth slows, and the role of natural vegetation as a carbon sink diminishes.

Effects of Climate Change on Agriculture

Changes in atmospheric CO_2, temperature, precipitation, and soil moisture, individually or together, could alter crop production. Computer models can estimate the effects of climate on agricultural production and crop prices. Dynamic crop growth models use physiological, morphological, and physical processes to predict crop growth or yield under different environmental conditions. A variety of economic models can then estimate the effects of climate change on the agricultural sector of the economy (Roesenzweig and Hillel 1998). These agricultural economic models incorporate a large number of variables to estimate the effects of changes in food production, consumption, income, employment, and gross domestic product. The general approach is as follows: First, climate-change models are linked to crop-response models, to predict crop yield changes in response to climate change. Then predicted crop yields, along with the influences of supply and demand on price, and its influence on foreign trade, are input to agroeconomic models to predict acreage change and economic consequences (Figure 7.3). However, like all

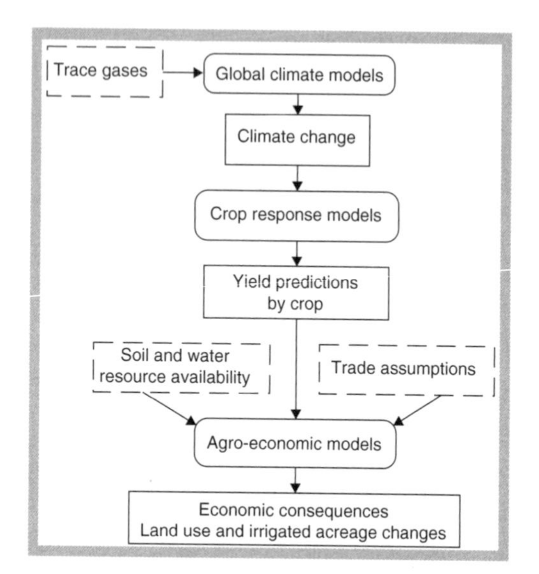

Fig. 7.3 Typical approach to modeling the effects of climate change on agriculture (From Rosenzweig C and Daniel M 1990. Agriculture. In: Smith J and Tirpak D, eds *The Potential Effects of Global Climate Change on the United States*. US EPA, Office of Research and Development, Washington, DC, New York: Hemisphere Publishing, pp. 367–417. Reproduced by permission of Routledge, Inc., part of The Taylor & Francis Group).

predictions, results contain a degree of uncertainty, and predicted economic consequences of climate change on agriculture vary widely, depending on which models are used.

There are a number of influences that could mitigate the negative effects of climate change on crop production. First, and perhaps most important, is the potential for farming practices to adapt to climate change. Farmers can respond to climate change by planting different climate-adapted species, using pesticides, or altering the dates of planting, harvesting, and irrigation. Such adaptations could minimize the impacts of climate change on crop yields. Thus, studies that assume a range of farmer adaptations predict only mild effects of climate change on crop production

(Mendelsohn et al. 1994). However, farmers, even in developed countries, are not generally influenced enough by long-term, subtle climatic changes to consciously alter their farming operations (Smit et al. 1996). Studies that assume little or no farmer adaptation suggest very negative impacts of climate change (Rosenzweig and Parry 1994).

Second, increased atmospheric CO_2 could also reduce the effects of climate change on agriculture. Higher atmospheric CO_2 levels could stimulate photosynthesis and crop production – a process called the *CO_2 fertilization* effect. However, the magnitude of such an effect is still under investigation. Greenhouse experiments on individual plant species demonstrate significant increases in yield. However, field experiments in which other factors such as water and nutrient shortages come into play often fail to show any enhanced yields.

Third, a differential day–night warming pattern would lessen the impacts of climate change on crops. In many crops, a significant increase in daytime temperature maxima during the growing season reduces photosynthesis and increases evapotranspiration, leading to a reduction in yield. However, recent trends and model predictions indicate increased cloud cover and a resulting differential warming (i.e. nighttime warming increasing faster than daytime warming) (see Chapter 4). If warming does occur primarily at night, rather than during the day, this could greatly reduce the negative impacts of climate change on crop productivity (Dhakhwa and Campbell 1998). Finally, the substitution or increased use of warmth-tolerant or drought-resistant crops could mitigate the impacts of climate change in certain areas.

However, a number of potential climate effects could have additive or synergistic effects, causing even more severe impacts than most models predict. For example, only about 20% of the world's croplands are irrigated, but this land accounts for about 40% of global crop production. Thus, any decrease in water availability would result in decreased food production in regions where water becomes critical. Also, as crops are stressed by climate change they become more vulnerable to damaging pests and diseases. The risk of crop loss in temperate regions may increase as crop pests move poleward with global warming (Porter et al. 1991). For example, aphids are a group of herbivorous insects that can, under the right environmental conditions, cause major losses to many agricultural crops. Model ecosystem studies suggest that the abundance of certain aphids can increase dramatically in response to both enhanced atmospheric CO_2 and increased temperature (Bezemer et al. 1998). A major pest of soybeans, the potato leafhopper *Empoasca fabae*, overwinters in a narrow band along the Gulf coast of the United States. Warmer winters would significantly expand the habitat of this pest northward and bring invasions earlier in the growing season (Figure 7.4).

Climate also affects animal husbandry. Indirect effects include climate-induced changes in the availability and price of feed grain and in pasture and forage crop yields. Extreme heat can affect the health of animals. For example, heat waves can kill poultry and decrease milk production in cows. Also, climate controls the distribution of livestock pests and diseases (Rötter and Van de Geijn 1999).

US Agriculture

The effect of climate change on US agriculture is particularly important and needs to be considered in the global context. The United

States, with large areas of rich soil, favorable climate, and modern technology, provides about $42 billion in food exports to the world. In fact, food production and processing is the United States's largest employer. Numerous studies on the effects of climate change on agriculture have focused on the United States.

Modern large-scale agriculture is primarily monoculture, that is, the growing of one or a few crop varieties. In the United States the economic value of just three crops – wheat, corn, and soybeans – equals the value of all other crops combined (Table 7.1). With only a few species, often with specific environmental requirements, the practice of monoculture is particularly vulnerable to stress, whether from disease, climate change, or a combination of factors.

Studies linking climate models, crop production models, and economic models predict major changes in the yield and economic value of US crops in response to climate

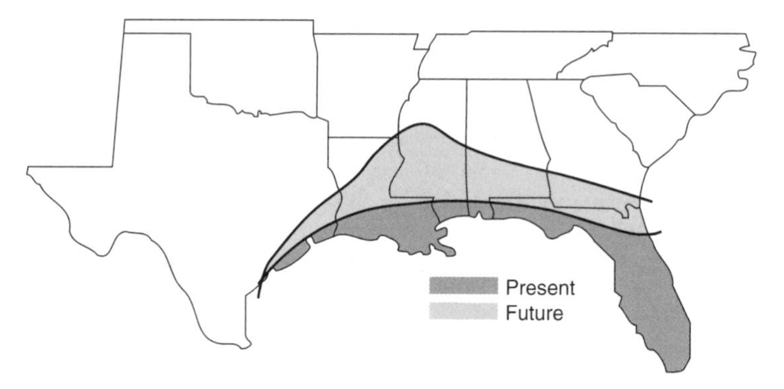

Fig. 7.4 The present and future (doubled atmospheric CO_2 GISS model) overwintering range of the potato leafhopper in the Southern United States, a major pest of soybeans (From Rosenzweig C and Daniel M 1990. Agriculture. In: Smith J and Tirpak D, eds *The Potential Effects of Global Climate Change on the United States*. US EPA, Office of Research and Development, Washington, DC, New York: Hemisphere Publishing, pp. 367–417. Reproduced by permission of Routledge, Inc., part of The Taylor & Francis Group).

Table 7.1 Major US crops (From Myers N 1979. *The Sinking Ark: A New Look at the Problem of Disappearing Species*. New York: Pergamon Press, p. 64. Reproduced with permission of Norman Myers).

	Hectares (millions 1976)	Value (millions $ 1976)	Total varieties	Major varieties	Hectarage, % of major varieties
Corn	33,664	14,742	197	6	71
Wheat	28,662	6,201	269	10	55
Soybean	20,009	8,487	62	6	56
Cotton	4,411	3,350	50	3	53
Rice	1,012	770	14	4	65
Potato	556	1,182	82	4	72
Peanut	611	749	15	9	95
Peas	51	22	50	2	96

change (Rosenzweig and Daniel 1990). Production of most crops will be reduced, with the greatest reductions in sorghum (−20%), corn (−13%), and rice (−11%). Crop prices will rise as the availability of water in some areas becomes critical and irrigation is required. If CO_2 fertilization is considered, some areas may gain in productivity. But, nonirrigated soybean and corn yields, particularly in the southeast, will drop dramatically (Figure 7.5a and 7.5b). Overall, the agricultural economy will shift northward in the Great Plains and decrease in the Southeast United States. In the US Midwestern Great Lakes region, an atmospheric CO_2 increase to 555 pmv will cause soybean yields to decrease slightly in the south, but increase in the north, with an overall beneficial 40% increase in yield for the region as a whole (Southworth et al. 2002). This shift of the optimal production area northward in the United States will have important implications for regional economies.

Studies using three different climate models and 1990 economic and agronomic conditions predict a range of economic impacts on US Agriculture (Adams et al. 1995). Assuming a beneficial CO_2 fertilization effect, overall, impacts represent only a small percentage of the economy. Regional differences are significant. Agroeconomies in the mountain and Northern Plains regions gain significantly, while Eastern and Southern States experience significant economic losses. Incorporating technologies such as higher-yielding crop varieties, fertilizers, herbicides, pesticides, and CO_2 fertilization into models offsets most-predicted negative effects of climate change on crop production. Conversely, most technological improvements could be offset by the added stress of climate change. Economically, farmers could gain from higher crop prices, while consumers could lose (Figure 7.6).

A number of environmental concerns arise from climate impacts on US agriculture. For, example, the demand for additional agricultural acreage in some more northern areas, such as Minnesota and North Dakota, will put pressure on water resources and natural habitats. As favorable climates move into areas with less desirable soil, the increased need for fertilizers and chemicals could add to watershed or groundwater pollution. As soil moisture decreases, the demand for irrigation water and the energy to pump it will increase dramatically, especially in California and the Great Plains. This will necessitate additional investment in dams, irrigation projects, and power sources. Major domestic food shortages are unlikely, as technology is applied to offset losses from climate change. However, exports will decline by about 40% as food supplies are retained for domestic consumption. Reduced US food exports will have global economic impacts, especially in developing countries.

Global Agriculture

Climate change will have serious impacts on world food supplies, especially in the less-developed countries. Global warming will probably shift growing areas by several hundred kilometers per degree increase in temperature, increasing agricultural productivity in some areas of the world, while drastically decreasing it in others.

In the developed countries of the Temperate Zone, climate change will probably have little negative impact on agricultural production. In fact, in many temperate regions, climate warming and an extended growing season could be beneficial. In Northern Europe, climate change could increase winter wheat production in Southern Sweden 10 to 20% over current levels by 2050. However, such predicted increases include many assumptions

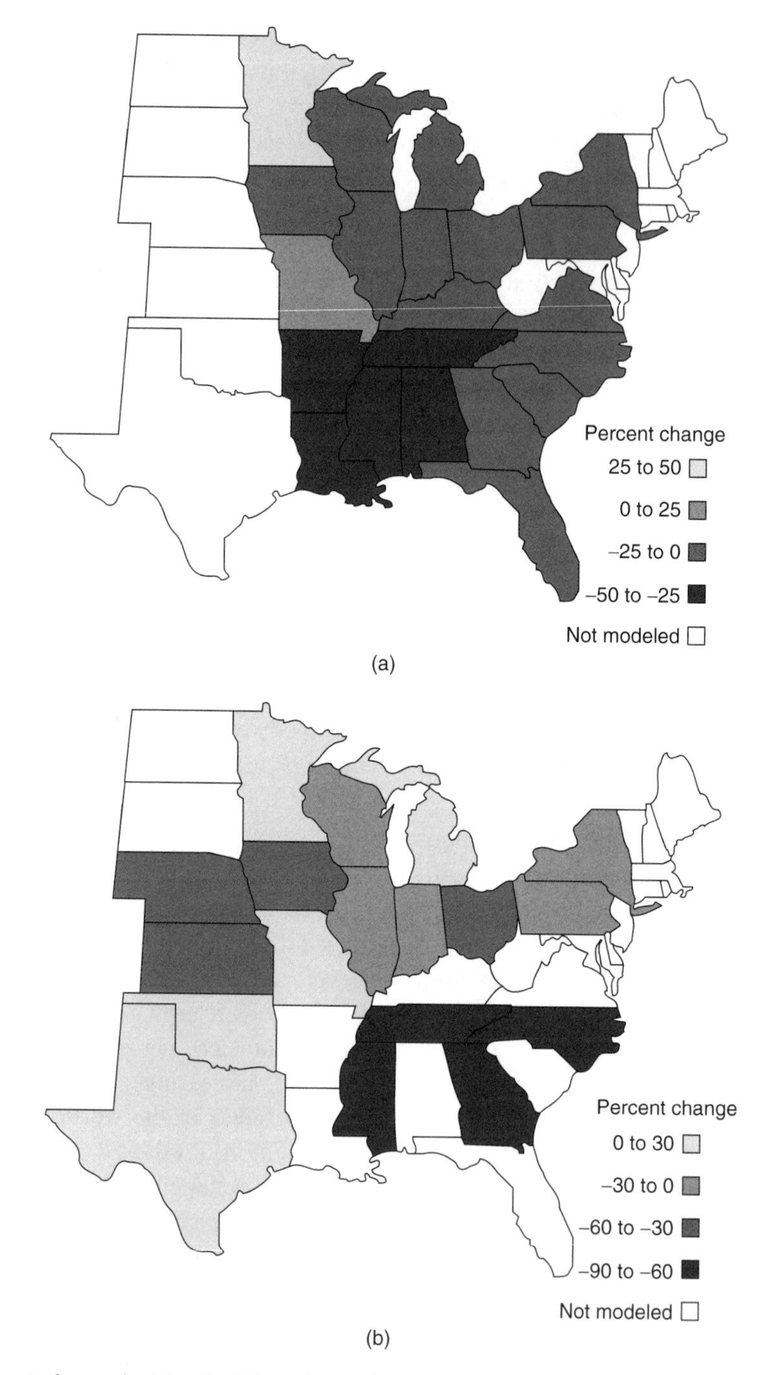

Fig. 7.5 Percent change in (a) rain-fed soybean yields and (b) dryland corn yields in response to climate change by 2060 (From Rosenzweig C and Daniel M 1990. Agriculture. In: Smith J and Tirpak D, eds *The Potential Effects of Global Climate Change on the United States*. US EPA, Office of Research and Development, Washington, DC, New York: Hemisphere Publishing, pp. 367–417. Reproduced by permission of Routledge, Inc., part of The Taylor & Francis Group).

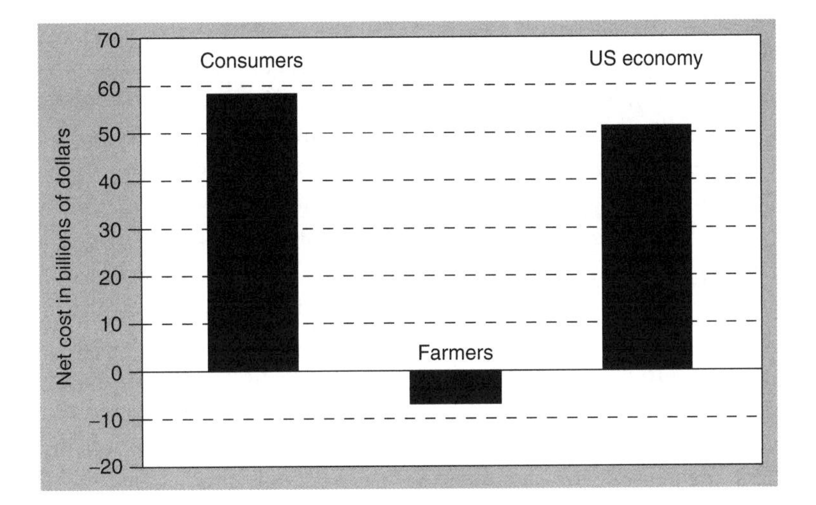

Fig. 7.6 Cost of climate change to US food-consumers, farmers, and net overall cost to economy. Negative value indicates benefit to farmers. Average predicted by two climate models. Includes changes in crop yields and supply and demand for irrigation water. Adjusted to 2001 US dollars (Adapted from data of Rosenzweig C and Daniel M 1990. Agriculture. In: Smith J and Tirpak D, eds *The Potential Effects of Global Climate Change on the United States*. US EPA, Office of Research and Development, Washington, DC, New York: Hemisphere Publishing, pp. 367–417. Reproduced by permission of Routledge, Inc., part of The Taylor & Francis Group).

about the efficiency of photosynthesis, stomatal conductance of the plants, and temperature and precipitation change (Eckersten et al. 2001). Australian wheat yields have increased over the past 50 years owing to new cultivars and changes in management practices. However, 30 to 50% of the observed increase has been attributed to climate change, with increases in minimum temperatures being the dominant influence (Neville 1997).

In Southern Quebec, Canada, researchers applied the Canadian Climate Center General Circulation Model (GCM) to examine climatic change in response to a 2X CO_2 atmosphere. They predict increases during the growing season in precipitation of 20 to 30%, in temperature of about 2.5 °C, and in growing degree-days of at least 50%. The output from the GCM was input to a crop model to estimate potential yield changes for a variety of crops. Depending on the agricultural zone and crop type, the yields increase (e.g. corn and sorghum 20%) or decrease (e.g. wheat and soybean 20 to 30%) (Singh et al. 1998).

In tropical and subtropical areas, predicted impacts on agriculture are mostly negative. For example, in the Mediterranean region of Southern Europe, grain yields will probably decrease as an increased need for irrigation places added demands on areas already suffering from acute water shortages. In poorer regions of the tropics, populations are often more directly dependent on agricultural production and more affected by its failure. Certain of these regions may be particularly vulnerable to climate change. Low-lying coastal regions, river deltas, and islands may be subject to flooding by sea-level rise (Chapter 8), in which case their agriculture would be impacted.

Water-stressed and marginal agricultural regions (e.g. in sub-Saharan Africa, Northern

Mexico, the Middle East, Northeast Brazil, and Australia) may be pushed completely out of production by climate change. In Egypt, for example, agricultural production will be greatly threatened by rising sea level in the Nile Delta and by an increased need for irrigation from the Nile River. Model simulations suggest that wheat yields may decline in the Delta by 30% and in Middle Egypt by more than 50% – all this in the face of a rapid growth in population and food demand (Eid 1994). In Trinidad, combined climate change and crop models predict significant decreases in sugarcane yields because of warming and increased soil moisture stress (Singh and El Maayar 1998).

In many global regions, human-induced deterioration of agricultural lands represents a significant negative influence on future agricultural productivity. These impacts must be added to estimates of damage from climate change. For example, Latin America has 23% of the world's arable land and 30 to 40% of the population relies on income from agriculture. However, 14% of the land suffers from moderate to extreme deterioration as a result of overgrazing, erosion, or alkalization (a buildup of alkaline minerals in soils as a result of continued irrigation and evaporation). Forty seven percent of the gazing land soils have lost their fertility. In Mexico, predicted shifts toward a warmer drier climate, in a country already stressed by low and variable rainfall, could spell economic disaster. Most studies of Latin America linking GCMs to crop models predict major decreases in yield for a variety of crops (Figure 7.7).

Many areas in the Middle East and arid Asia have rapid population growth rates and are highly dependent on grazing animals and irrigated crop production. Future predicted increases in regional water shortages, where small increases in precipitation are inadequate to counter higher evapotranspiration,

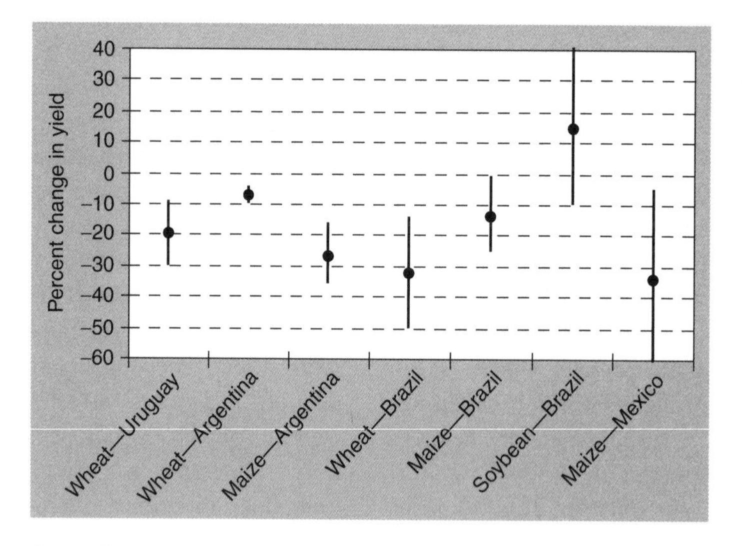

Fig. 7.7 Percent change in crop yields in Latin America in response to climate change. Range of estimates based on different GCMs under current conditions of technology and management (Adapted from Canziani O and Diaz S 1998. Latin America. In: Watson RT, Zinyowera MC and Moss RH, eds *The Regional Impacts of Climate Change: An Assessment of Vulnerability*. Special Report of IPCC Working Group II. Cambridge: Cambridge University Press, pp. 186–230).

could result in severe impacts on agriculture in the region. In Kazakstan, for example, yields of the main crop, spring wheat, are projected to decrease by 60% (Gitay and Noble 1998).

In general, studies suggest that changes in agricultural economies will be detrimental in equatorial regions, beneficial at high latitudes and mixed at temperate latitudes (Figure 7.8). Model results, incorporating flexible levels of land-use change and adaptation, indicate an overall beneficial effect on global agriculture from a 1 to 2 °C increase in global average temperature (GAT). However, if the GAT increases by 3 °C or more, we can expect severe declines in production (Darwin 1999, Brown and Rosenberg 1999).

One of the most detailed studies of climate change on global agriculture applied three different GCMs and a doubled CO_2 atmosphere to predict climate change at 122 sites in 18 countries between 1990 and 2060 (Rosenzweig and Parry 1994). The output of each GCM was linked to production models for several crops. Researchers investigated how predicted changes in climate would affect the production of wheat, maize, soybean, and rice crops that represent 70 to 75% of total world cereal production. They examined effects with and without assuming a CO_2 fertilization effect and modeled two levels of technological adaptation to climate change. Level 1 adaptation assumed little change in agricultural systems, while level 2 assumed large shifts in planting dates, increased use of fertilizers and irrigation, and development of new crop varieties. Finally, they used a world food trade and economic model to estimate changes in crop prices. Their study supports the idea that climate change will have little negative impact on (and in some cases will even benefit) agricultural production in developed

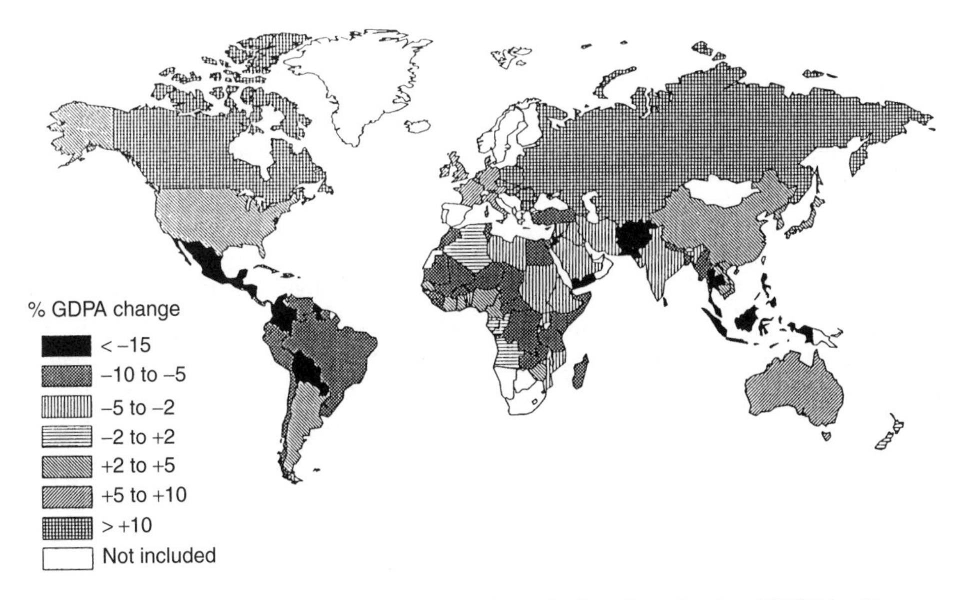

Fig. 7.8 Impact of climate change on gross domestic agricultural production (GDPA) with economic adjustment in 2060 (Reprinted from Fischer G, Frohberg K, Parry ML and Rosenzweig C 1994. Climate change and world food supply, demand and trade: who benefits, who loses? *Global Environmental Change* **4**(1): 7–23, Copyright (1994), with permission from Elsevier Science).

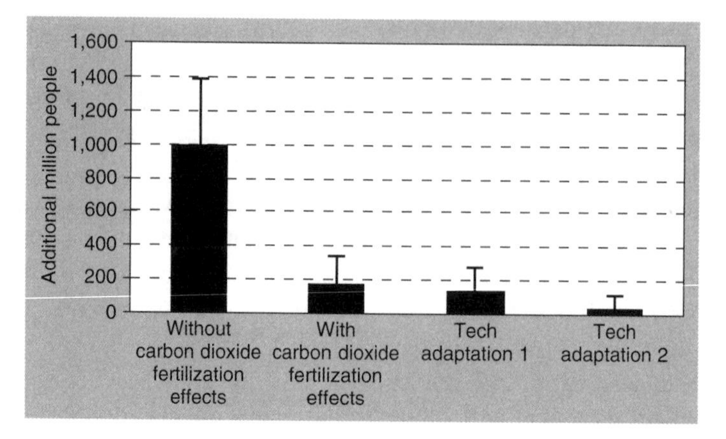

Fig. 7.9 Additional people at risk of hunger from climate change (Reprinted from Fischer G, Frohberg K, Parry ML and Rosenzweig C 1994. Climate change and world food supply, demand and trade: who benefits, who loses? *Global Environmental Change* **4**(1): 7–23, Copyright (1994), with permission from Elsevier Science).

countries. However, in the poorer nations, the lack of technological adaptation and other factors will lead to a substantial decline in agricultural production (Plate 2). Because of changes in world food supply and demand, cereal prices will increase and the number of people at risk of hunger will increase by 10s to 100s of millions (Figure 7.9).

Other case studies for different regions of the world suggest the following (Parry et al. 1988):

- a significant increase in rice production in Japan;
- increased droughts and crop stress in Australia, Brazil, India, and parts of Africa resulting from an intensified El Niño;
- dramatic losses in farm production in the central plains of the United States and Canada as it returns to "dust bowl" conditions similar to the 1930s;
- increased production in temperate countries such as Finland and the Russian Federation if technological improvements can take advantage of warmer conditions.

Summary

Large increases in agricultural production during this century will be necessary to feed the world's growing human population. Crop production is very sensitive to climate conditions. Thus, increased production in many areas will be even more difficult in the face of global climate change. In the United States, food shortages are unlikely, but investments required for irrigation and other technologies will increase food costs and lead to decreased exports. Globally, developed countries will face the same challenges as the United States, and crop prices are likely to increase. The less-developed countries will be unable to apply expensive technologies to maintain agricultural production and some will experience food shortages and increased hunger.

Many agro-climatologists cited here are pessimistic about the effects of climate change on world agriculture. However, some scientists remain confident that humans can and will adapt agricultural practices so that climate-change impacts will be minimal or even beneficial. Examples exist for several grain crops. The belt for hard red winter wheat has expanded northwestward from Central United States by hundreds of kilometers during the last part of the twentieth century through the use of different cultivars and agronomic practices. In the case of rice breeding, 40 to 50 selection cycles could be completed before atmospheric CO_2 reaches 600 ppm (Seshu et al. 1989).

Major findings concerning the effects of climate change on agriculture can be summarized as follows:

- For a doubling of atmospheric CO_2, overall global agricultural production seems sustainable. However, responses differ greatly between regions. Low-latitude, low-income areas will experience the greatest impacts.
- Sub-Saharan Africa – This arid to semi-arid region where 60% of the population depends directly on farming appears most vulnerable to climate change.
- South and Southeast Asia – More than 30% of the GDP comes from agriculture and these regions may be vulnerable.
- Pacific Island Nations – Sea-level rise and associated saltwater intrusion could negatively impact agriculture.
- Technological adaptation – Increased irrigation (where water is available), adoption of alternate crop varieties, and so on will minimize impacts for those countries that can afford it. However, climate change could seriously impact agriculture in developing countries.
- Government agricultural policies – Many present policies discourage adaptation and technological innovation and may impede adaptation to climate change.

References

Adams RM, Fleming RA, Chang C-C and McCarl B 1995 A reassessment of the economic effects of global climate change on U.S. agriculture. *Climatic Change* **30**: 147–167.

Bezemer TM, Hefin T and Knight KJ 1998 Long-term effects of elevated CO_2 and temperature on populations of the peach potato aphid *Myzus persicae* and its parasitoid *Apphidius matricariae*. *Oecologia* **116**: 128–135.

Brown RA and Rosenberg NJ 1999 Climate change impacts on the potential productivity of corn and winter wheat in their primary United States growing regions. *Climatic Change* **41**: 73–107.

CIAP 1975 *Impacts of Climate Change on the Biosphere*. Monograph 5, Part 2; September. US Department of Transportation, Washington, DC.

Canziani O and Diaz S 1998 Latin America. In: Watson RT, Zinyowera MC and Moss RH, eds *The Regional Impacts of Climate Change: An Assessment of Vulnerability*. Special Report of IPCC Working Group II. Cambridge: Cambridge University Press, pp. 186–230.

Darwin R 1999 A farmer's view of the Ricardian approach to measuring agricultural effects of climatic change. *Climatic Change* **43**: 371–411.

Dhakhwa GB and Campbell CL 1998 Potential effects of differential day-night warming in global climatic change on crop production. *Climatic Change* **40**: 647–667.

Diamond J 1999 *Guns, Germs and Steel: The Fates of Human Societies*. New York: W.W. Norton & Co.

Eckersten H, Blomback K, Katterer T and Nyman P 2001 Modelling C, N, water and heat dynamics in winter wheat under climate change in Southern Sweden. *Agriculture Ecosystems and Environment* **86**(3): 221–236.

Eid HM 1994 Impact of climate change on simulated wheat and maize yields in Egypt. In: Rosenzweig C and Iglesias A, eds *Implications of Climate Change for International Agriculture: Crop Modeling Study*. Washington, DC: US EPA, pp. 1–14.

Fischer G, Frohberg K, Parry ML and Rosenzweig C 1994 Climate change and world food supply, demand and trade: who benefits, who loses? *Global Environmental Change* **4**(1): 7–23.

Gitay H and Noble IR 1998 Middle East and Arid Asia. In: Watson RT, Zinyowera MC and Moss RH eds, *The Regional Impacts of Climate*

Change: An Assessment of Vulnerability. Special Report of IPCC Working Group II. Cambridge: Cambridge University Press, pp. 231–252.

Lappé FM 1982 *Diet for a Small Planet*. New York: Ballantine Books, p. 8.

Mendelsohn R, Nordhaus RW and Shaw D 1994 The impact of global warming on agriculture: a Ricardian analysis. *American Economic Review* **84**(4): 753–771.

Myers N 1979 *The Sinking Ark: A New Look at the Problem of Disappearing Species*. New York: Pergamon Press, p. 64.

Neville N 1997 Increased Australian wheat yield due to recent climate trends. *Nature* **387**: 484, 485.

Parry ML, Carter TR and Konijn N, eds 1988 *The Impact of Climatic Variations on Agriculture. Vol. 1: Assessments in Cool, Temperate and Cold Regions; Vol. 2: Assessments in Semi-Arid Regions*. Boston: Kluwer Academic Publishers.

Porter JH, Parry ML and Carter TR 1991 The potential effects of climate change on agricultural insect pests. *Agricultural/Forest Meteorology* **57**: 221–240.

Robertson GP, Paul EA and Harwood RR 2000 Greenhouse gases in intensive agriculture: contributions of individual gases to the radiative forcing of the atmosphere. *Science* **289**: 1922–1925.

Rosenzweig C, Daniel M and US EPA, Office of Research and Development 1990 Agriculture. In: Smith J and Tirpak D, eds *The Potential Effects of Global Climate Change on the United States*. Washington, DC: Hemisphere Publishing, pp. 367–417.

Rosenzweig CR and Hillel D 1998 *Climate Change and the Global Harvest: Potential Impacts of the Greenhouse Effect on Agriculture*. Oxford: Oxford University Press.

Rosenzweig C and Parry M 1994 Potential impact of climate change on world food supply. *Nature* **367**: 133–138.

Rötter R and Van de Geijn SC 1999 Climate change effects on plant growth, crop yield and livestock. *Climatic Change* **43**: 651–681.

Schlesinger WH 1986 Changes in soil carbon storage and associated properties with disturbance and recovery. In: Trabalka JR and Reichle DE, eds *The Changing Carbon Cycle: A Global Analysis*. New York: Springer-Verlag, pp. 194–220.

Seshu DV, Woodhead T, Garrity DP and Oldeman LR 1989 Effect of weather and climate on production and vulnerability of rice. *Climate and Food Security. International Symposium on Climate Variability and Food Security in Developing Countries*, 5–9 February 1987, New Delhi, India, Manila: International Rice Research Institute, pp. 93–113.

Singh B and El Maayar M 1998 Potential impacts of greenhouse gas climate change scenarios on sugar cane yields in Trinidad. *Tropical Agriculture* **75**(3): 348–354.

Singh B, El Maayar M, André P, Bryant CR and Thouez J-P 1998 Impacts of a GHG-induced climate change on crop yields: effects of acceleration in maturation, moisture stress and optimal temperature. *Climatic Change* **38**: 51–86.

Smit B, McNabb D and Smithers J 1996 Agricultural adaptation to climatic variation. *Climatic Change* **33**: 7–29.

Southworth J, Pfeifer RA, Habeck M, Randolph JC, Doering OC, Johnston JJ, et al. 2002 Changes in soybean yields in the midwestern United States as a result of future changes in climate, climate variability, and CO_2 fertilization. *Climatic Change* **53**: 447–475.

Chapter 8

Climate Change and the Marine Environment

"We have been sustained by the ocean for two millennia... this harmony may be interrupted by the action of nations very distant from our shores. I hope that the peoples of the Pacific [Islands] can help convince the industrialized nations to discontinue their profligate contamination of the atmosphere."

Amata Kabua, President of the Marshall Islands, 1989

Introduction

Humans depend on the sea. Climate change could affect sea-level ocean–atmosphere interactions, ocean heat transport, biogeochemical cycles, and marine ecosystems including fisheries. Two-thirds of the Earth's surface is covered by ocean. As it warms, the ocean's volume expands. This expansion, together with additional water from melting glaciers, will raise the sea level. Higher sea level will affect shallow water marine communities and greatly impact human coastal populations.

Physical interactions between the atmosphere and hydrosphere (ocean), which play a key role in determining climate, will probably be altered themselves by climate change (Bigg 1996). Increasing water temperature increases the movement of water vapor from the ocean to the atmosphere (evaporation) and decreases the solubility of atmospheric greenhouse gases (Chapter 1). Changes in ocean temperature and/or salinity will alter density-driven ocean currents and heat transport. Marine biogeochemical processes in the ocean such as photosynthesis, organosulfide production, and calcium carbonate formation are influenced by climate. Changes in these processes, through complicated feedbacks, will significantly add to climate-change impacts.

Finally, global climate change will alter the available habitat and population distribution of marine plankton, invertebrates, fish, and marine mammals. Warmer temperatures may increase the incidence of toxic algal blooms and marine diseases. Here we examine the effects of greenhouse warming on the world's oceans and describe how marine biota and human communities will be affected by changing ocean conditions.

Climate Change: Causes, Effects, and Solutions John T. Hardy
 ISBNs: 0-470-85018-3 (HB); 0-470-85019-1 (PB)

Sea-Level Rise

Greenhouse warming will lead to sea-level rise (Figure 8.1). As seawater warms, its volume expands. Also, as freshwater stored in alpine and polar continental regions melts and flows to the sea, it contributes further to sea-level rise. Melting of sea ice or icebergs per se does not add to sea level (just as ice melting in a glass does not cause it to overflow). Determining average sea level is not a simple task, but a variety of methods yield data on sea level from thousands of years ago to the present (Box 8.1).

Sea level is already rising and will continue to rise. Tide-gauge records, in some cases covering the last 100 years, show a general increase in sea level of 2.4 ± 0.9 mm per year (Peltier and Tushingham 1989). Although predictions of future sea-level rise vary (Titus et al. 1991), there is little doubt that sea level will rise substantially over the next 100 to 200 years. On the basis of ranges encompassed by seven different climate models and 35 different SRES, the Intergovernmental Panel on Climate Change predicts a global sea-level rise between 1990 and 2100 of 0.09 to 0.88 m with a central value of 0.48 m (Figure 8.2). Probability distributions of sea-level rise estimates from different researchers suggest that greenhouse-induced sea-level rise has a 50% chance of exceeding 34 cm and only a 1% chance of exceeding 1 m by the year 2100. There is a 65% chance that sea level will rise 1 mm year^{-1} more rapidly in the next 30 years (after 1996) than in the previous century (Titus and Narayanan 1996). About 70% of the predicted sea-level rise will result from the thermal expansion of ocean water

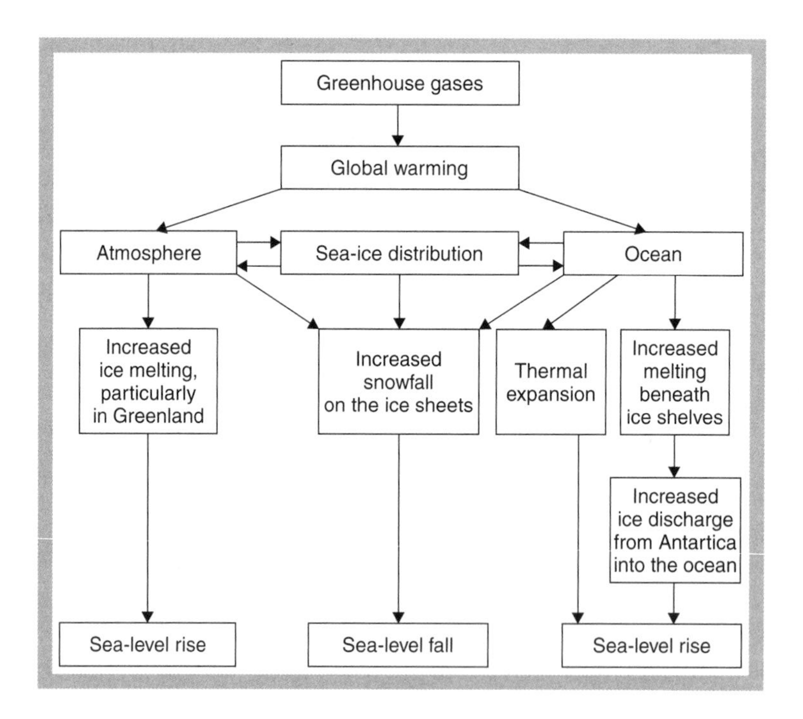

Fig. 8.1 Major processes relating greenhouse warming to sea level (From Titus JG 1986. Greenhouse effect, sea-level rise, and coastal zone management. *Coastal Zone Management Journal* **14**(3): 147–171).

Box 8.1 Determining sea level

Sea level is not constant or uniform – it varies daily and monthly with tidal cycles and differs spatially over the globe in response to ocean currents and atmospheric pressure differences. Thus, determining the average or "true" sea level is not an easy task, but it can be estimated using a number of different techniques. These include releveling surveys (i.e. geological studies of past levels of coastal terraces, beaches, marshes, and archeological sites), tide-gauge measurements, and satellite altimetry measurements (TOPEX/Poseidon).

Tide gauges, attached to the sea bottom in shallow coastal areas (e.g. the piling of a pier), continuously measure rising and falling sea level, often producing records over decades. However, the land itself (to which such gauges are fixed) can move vertically. The Earth's crust can be uplifted by the tectonic movement of crustal plates. Land areas covered by heavy layers of thick ice during the last glaciation continue to rise since the weight was removed. Such upward land movements may appear in tide-gauge records as a decrease in sea level. In other areas, sinking land margins may appear as a sea-level rise. Therefore, true sea-level changes must include a correction for land movement, that is, an isostatic adjustment. This is done by using a network of geodetic reference points and historical measurements compared to stable areas, for example, in the interior of continents and by global positioning system (GPS) measurements in relation to satellite sensors. Changes in sea level can also be detected by carbon-14 age dating of historic biological markers (e.g. the location of past populations of specific marine and coastal vegetation).

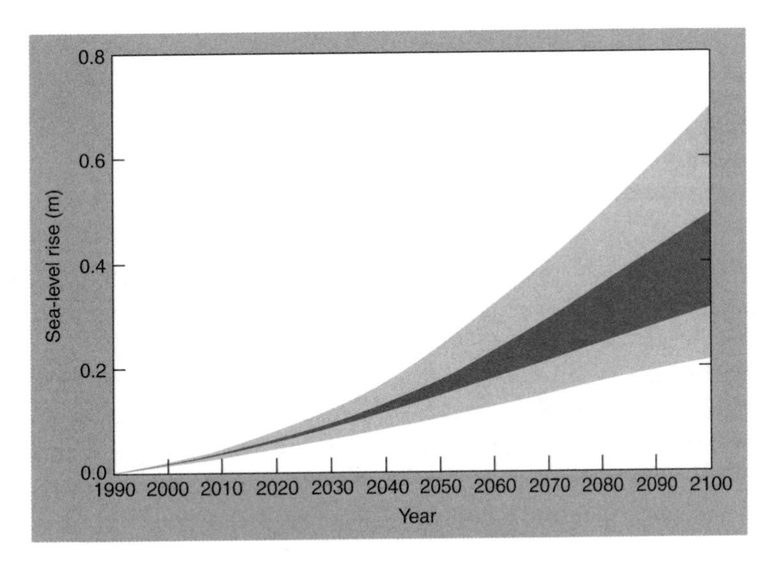

Fig. 8.2 Global average sea-level rise from 1990 to 2100. The region in dark shading is the average of seven climate models for all IPCC SRES. Light shading shows the range of all models for all 35 scenarios. The range does not include the uncertainty surrounding the possible collapse of the West Antarctic ice sheet (Adapted from Church JA and Gregory JM 2001. Changes in sea level. In: Houghton JT, Ding Y, Griggs DJ, Noguer M, van der Linden PJ, Dai X, et al., eds *Climate Change 2001: The Scientific Basis. Intergovernmental Panel on Climate Change, Working Group I.* Cambridge: Cambridge University Press, pp. 639–693).

as it warms and 30% from melting glaciers and ice caps that add freshwater to the sea. Sea-level rise will lag the global temperature increase; thus sea level could continue to rise for several centuries beyond 2100 to several meters above its current level. However, predictions of sea-level rise contain several uncertainties. If the West Antarctic ice sheet melts and collapses (a less likely, but possible, scenario), sea level could increase by 6 m. Also, any changes in atmospheric circulation in the North Atlantic could alter precipitation and snow and ice accumulation in Greenland (Bromwich 1995).

Sea-level rise will have a number of important impacts on ecosystems and humans. About half the world's population lives within 200 km of the ocean, and many millions live in coastal areas that are less than 5 m above sea level. Sea-level rise impacts include increased beach erosion, saltwater intrusion into groundwater, and flooding of coastal habitats. Generally, beach loss from erosion will far exceed that expected from direct inundation. The area eroded will depend upon the average slope of the beach out to a water depth where waves cease to impact the bottom (generally about 10 m deep) (Brunn 1962). This means that in many areas a 1-cm rise in sea level would result in a shoreline retreat of 1 m, that is, sea-level rise to beach loss ratio of 1 : 100. In areas of stronger wave action, the shore retreat would be even greater, and in areas of low beach slope, for example, parts of Florida, it could mean a sea-level rise to beach retreat ratio of 1 : 1000. In such an area, a 1-m rise in sea level would mean a 1-km retreat of the shore. For a property owner with 2 km of beachfront, this would mean a loss in land area of 50%.

In many areas, beaches represent a valuable economic resource based on tourism. The protection, stabilization, and replenishment of these beaches will require massive investments of capital to retain their economic value. Financial institutions are already weighing the risk of investing in and insuring low-lying coastal properties (Chapter 9).

Many coastal and island communities draw their freshwater from groundwater wells. Because it is less dense than saltwater, fresh groundwater generally collects beneath coastal areas or islands as a "freshwater lens" floating on top of intruding subsurface seawater. As sea level rises, such freshwater lenses will be squeezed into smaller volumes, that is, saltwater will intrude into the fresh groundwater.

Certain regions are particularly vulnerable to the effects of sea-level rise. For island nations of the Pacific, the Indian Ocean, and the Caribbean, erosion, flooding, and saltwater intrusion into water supplies all threaten the social and economic viability of island states (Roy and Connell 1991). For example, most of the land area of the atoll island states of Kiribati, Maldives, Marshall Islands, Tokelau, and Tuvalu, with a combined population of greater than 300,000, is less than 3 m above sea level. These islands could lose much or all of their land area to the sea. One study recommends that plans be drawn up for the relocation of major Pacific island settlements (Nunn 1988). Island people may become the first environmental refugees of the greenhouse era.

Other areas sensitive to sea-level rise are low-lying river deltas, for example, the Mississippi, Nile, Ganges, Orinoco, and numerous others. A 1-m sea-level rise would dramatically change the physical character of the San Francisco Bay and delta. Assuming levee constructions were only able to protect urban areas, the area of the bay–delta system would approximately double by 2100, creating a large inland sea area. Tidal circulation patterns would change, and salinity would increase, drastically affecting California's irrigated agriculture in the San Joaquin Valley and Southern California's water supply (Williams 1985).

In the United States, a 1-m sea-level rise would have serious consequences (Titus et al. 1991):

- 30,000 km^2 of coastal area will be inundated;
- 26 to 82% of the US coastal wetlands will be eliminated;
- if unprotected, 13,000 to 26,000 km^2 of dryland shores will be inundated;
- the loss of wetlands and undeveloped lowlands, together with the costs of protecting areas with beach (sand) replenishment, bulkheads, and levees (ignoring future development) would be $270 to 475 billion.

Faced with rising sea level, and its potential effects on major urban areas (London, Amsterdam, New York, and Washington, DC to name a few), developed countries will spend trillions of dollars to construct seawalls and dikes to hold back the advancing sea. For example, the very existence of Venice, Italy, has depended for centuries on a series of engineering solutions to exclude floods from the sea. An estimated local 30-cm rise in sea level from greenhouse warming during this century means that, unless something is done, the central Piazza San Marco would be under water almost everyday during the flooding season. One solution being pursued at a cost of $2.5 to $3 billion is the construction of a series of movable floodgates separating the city and the lagoon from the sea (Harleman et al. 2000). In the Netherlands, seawall protection against a 0.5-m sea-level rise could cost $3.5 trillion (WWF 2002).

However, poorer developing countries will not be able to afford such gigantic engineering projects. The Nile Delta, South China, and Bangladesh are examples of regions with very dense human populations that could be displaced by a rising sea. Bangladesh, in particular, has already experienced mortalities in the millions from flooding and cyclone-driven storm surges. A 1.5-m sea-level rise there would flood about 16% of the land area and displace 17 million people (UNEP 2002) (Figure 8.3). Even a 0.5-m sea-level rise (expected by 2100) in the Nile Delta of Egypt would displace 3.8 million people and inundate 1,800 km^2 of cropland (UNEP 2002).

Ocean Currents and Circulation

Changes in global precipitation and temperature patterns could alter large-scale oceanic circulation patterns (Weaver 1993). Oceanic water masses flow from areas of high density (e.g. colder and/or more saline) to areas of lower density (e.g. warmer and/or less saline). Thus, density differences drive the large-scale ocean currents. Model predictions suggest that greenhouse warming will be greater at lower latitudes, that is, further north and south. As a result, latitudinal temperature gradients, along with the intensity of large-scale atmospheric and ocean surface circulation driven by these gradients, should lessen. However, the recent increasing temperature contrasts between coastal regions and adjacent continental landmasses, together with the alongshore winds they generate (Chapter 3), should continue to strengthen (Bakun 1992). In addition, global warming may increase the frequency and intensity of El Niño Southern Oscillation (ENSO) cycles in the central Pacific Ocean (Box 8.2, Plate 7).

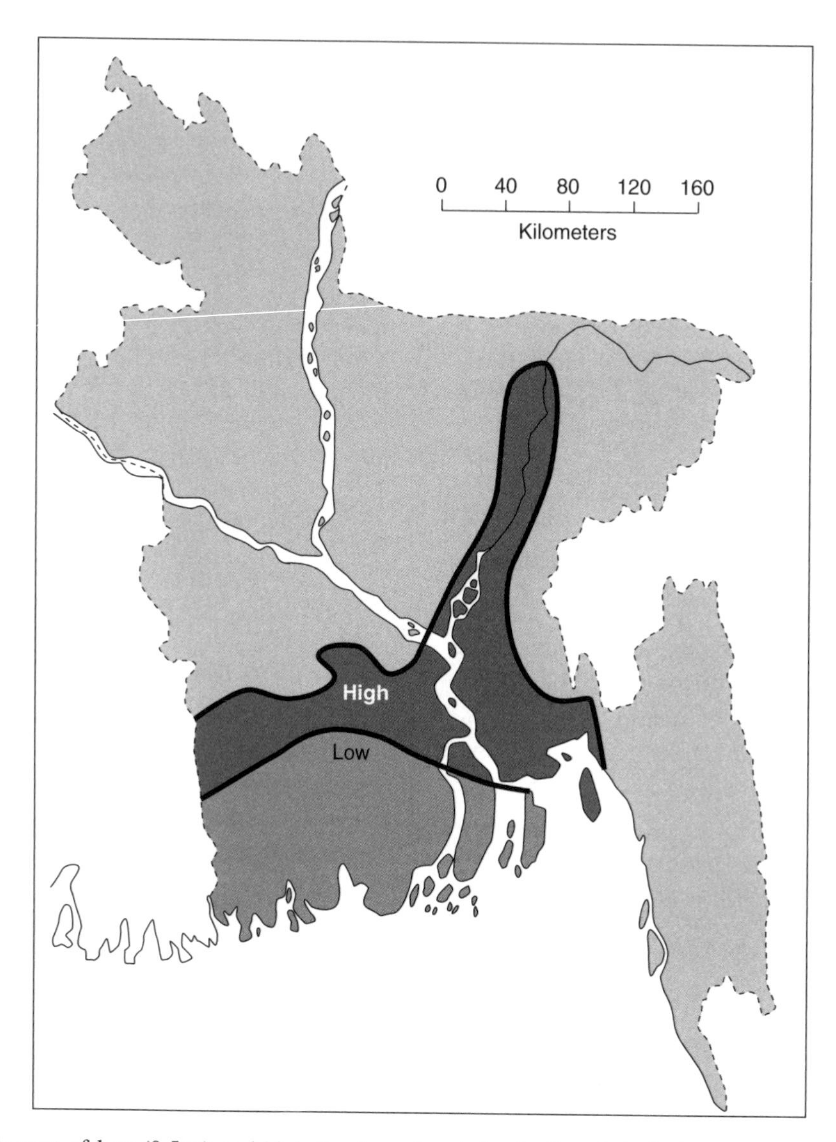

Fig. 8.3 Impact of low (0.5 m) and high (3 m) sea-level rise in Bangladesh (From Broadus J, Milliman J, Edwards S, Aubrey D and Gable F 1986. Rising sea level and damming of rivers: possible effects in Egypt and Bangladesh. In: Titus J *Effects of Changes in Stratospheric Ozone and Global Climate*; Vol. 4: Washington, DC: Sea-Level Rise. US EPA, pp. 165–189).

Ironically, some studies suggest that greenhouse warming could trigger a rapid cooling, at least in the North Atlantic region, along with severe drought in the low latitudes of Mexico and North Africa. The large-scale thermohaline ocean circulation (the conveyer belt) is a crucial part of the atmosphere/ocean climate system (Chapter 1). Historically, periods of decreased salinity in the North Atlantic portion of this conveyer belt led to a slowdown in the sinking of water, that is, in North Atlantic Deep Water Formation (NADWF). This, in turn, led to decreased heat transport to the North Atlantic and a rapid cooling

Box 8.2 Decadal oceanographic oscillations

Changes resulting from natural cycles can provide information on how the Earth and humans may respond to anthropogenic greenhouse warming. A number of natural periodic cycles link the atmosphere and the ocean and greatly influence regional and even global climate. Cycles on periods of decades or so (decadal oscillations) alter the transport of heat and greenhouse gases from the surface to the deep ocean or between the ocean surface and the atmosphere, affect regional evaporation and precipitation patterns, and alter the productivity of plankton, invertebrate, and fish populations. Oscillation-induced changes in drought and flood affect many human activities, particularly fisheries, forestry, and agriculture. Global climate change could alter the frequency and the intensity of these periodic events.

The best-known cycle is the El Niño Southern Oscillation (ENSO) – a naturally occurring cyclic change in ocean currents and heat transport in the Equatorial Pacific Ocean (NOAA 2002). ENSO is an alternation between a high atmospheric pressure center in the Eastern Pacific and a low-pressure center over Indonesia and Northern Australia. In normal periods, this pressure difference drives trade winds from east to west along the Equator (Plate 7a). This forces warm water to pile up (sea level actually increases by about 40 cm) in the west and depresses the thermocline (the depth where warm water above transits to more colder water below) to about 200 m deep. The warm water and evaporation results in heavy rains in the Western Pacific and dry air over South America. Also, along the west coast of South America, water is driven offshore and replaced by the upwelling of cool nutrient-rich deep water.

In contrast, in El Niño periods, the east–west pressure difference becomes low and the trade winds weaken in the Western Pacific. The warm water piled up in the Western Pacific now flows back toward the east and subsurface (Kelvin) waves travel across the Pacific to depress the thermocline off South America (Plate 7b). The result is a general warming and increase in precipitation in the Central to Eastern Pacific and dry conditions in the west. La Niña is the cold counterpart of El Niño – the other extreme of the ENSO cycle where sea-surface temperatures in the tropical Pacific drop below normal (Plate 7c).

During the past 5,000 years, ENSO events occurred on average once or twice per decade, but since the mid-1970s, they have occurred more often and persisted longer (Rodbell et al. 1999). Some scientists believe the increasing frequency and intensity of the ENSO is a result of anthropogenic global greenhouse warming (Trenberth and Hoar 1996). Some changes associated with El Niño include reduced productivity of the giant kelp (seaweed) in California, increased diseases in marine organisms, coral bleaching, and release of large quantities of CO_2 from the Central Pacific into the atmosphere. The economic cost from floods, hurricanes, drought, and fire as a result of the 1982 to 1983 ENSO is estimated at $8.1 billion (UCAR/NOAA 1994).

In the North Atlantic region, a decadal scale oscillation occurs known as the North Atlantic Oscillation (NAO). In a high NAO state, a high-pressure center near the Azores (west of Portugal) and a low-pressure center in the North Atlantic near Greenland and Iceland create winds that blow from North America to Europe. In a low NAO cycle, the Icelandic low pressure moves south off Newfoundland and a high pressure forms over Northern Greenland.

This causes dry polar air to blow across Northern Europe and then westward toward North America. Northern Europe experiences much cooler summers and more severe winters in low NAO compared to that in high NAO periods. Similarly, a North Pacific Oscillation has significant effects on the climate and fisheries productivity of that region (Francis et al. 1998). All these cycles probably interact in ways not yet understood.

trend known as the *Younger Dryas Event* (Chapter 2). The cooler surface water also led to a decrease in sea-surface evaporation and a resultant decrease in precipitation over Europe and North Africa. In fact, precipitation in Mexico and the Sahel region of Africa still exhibits rapid (decadal) change in response to salinity changes in the North Atlantic (Street-Perrott and Perrott 1990). Also, natural fluctuations with a frequency of 30 to 40 years in the thermohaline circulation of the North Atlantic are correlated with temperature cycles in Northwestern Europe (Stocker 1994). If greenhouse warming leads to increased precipitation at high latitudes or increased ice-sheet disintegration (both recent trends, see Chapter 3), the salinity of the North Atlantic will decrease. This could lead to a collapse of the oceanic conveyer belt, a rapid cooling in the North, and severe droughts in areas such as Mexico and the Sahel. An increase of atmospheric CO_2 to 750 ppmv within 100 years (the actual recent growth rate) could lead to a permanent shutdown of the thermohaline circulation (Stocker and Schmittner 1997). Such a scenario has been called the *Achilles heel* of our climate system (Broecker 1997).

Marine Biogeochemistry

The ocean is important in the global biogeochemical cycles of carbon, sulfur, and other elements. These cycles interact closely with climate in several ways. Organosulfur compounds such as Dimethyl Sulfide (DMS) are produced by marine phytoplankton, particularly a group of photosynthetic organisms called *coccolithophores*. These sulfur compounds escape to the atmosphere and form sulfate aerosols that serve as condensation surfaces for water vapor, that is, they promote cloud formation (Charlson et al. 1987). Because of their indirect role in cloud formation (and hence radiation balance and precipitation), any change in ocean temperature or chemistry that affects coccolithophores (e.g. by changing their growth and abundance) could affect climate.

The ocean is a sink for carbon dioxide. The pH of seawater is buffered against change by dissolved carbonate. Atmospheric CO_2 dissolves in seawater to form bicarbonate and hydrogen ions, thus acidifying (lowering the pH) seawater. The solubility of CO_2 decreases with increasing temperature. Therefore, as the ocean warms, its ability to absorb CO_2 from the atmosphere will decrease. This will act as a positive feedback (see Chapter 4), that is, warming will decrease the oceanic CO_2 sink, leading to more CO_2 remaining in the atmosphere and additional warming.

The living and dead organic biomass of the ocean contains 700 Gt of carbon, an amount almost equal to that in the atmosphere. Photosynthetic marine plankton (phytoplankton) constitute the base of the open ocean marine food web. Like land plants, they use solar energy to take up carbon from seawater and form organic compounds and oxygen (Figure 8.4). The rate of photosynthetic carbon fixation (primary production) differs greatly by geographic region over the ocean. A significant change in phytoplankton productivity as a result of climate change could affect this part of the global carbon cycle.

Many marine organisms, from plankton to coral, remove calcium from seawater and deposit it as solid calcium carbonate (limestone) in their cell wall or exterior skeleton. Thus, these organisms form an integral part of the global carbon cycle. Calcium carbonate formation represents both a sink for dissolved carbon and a source of CO_2 to the atmosphere (Figure 8.4). The reaction is temperature- and pH-dependent. The net effect of increasing ocean temperature and decreasing pH on $CaCO_3$ formation is crucial in terms of the ocean's role in regulating atmospheric CO_2 (Elderfield 2002).

A business-as-usual scenario of global warming could decrease the pH of the oceans by 0.35 units by 2100. This change could be important since ocean pH normally differs by only a few tenths. By the middle of this century, increased CO_2 concentrations could decrease biological carbonate formation in tropical oceans by 14 to 30%, threatening coral reefs and many other marine organisms (Kleypas et al. 1999).

Dissolved oxygen is essential for respiration and survival of aquatic animals. The solubility of oxygen in water decreases with increasing

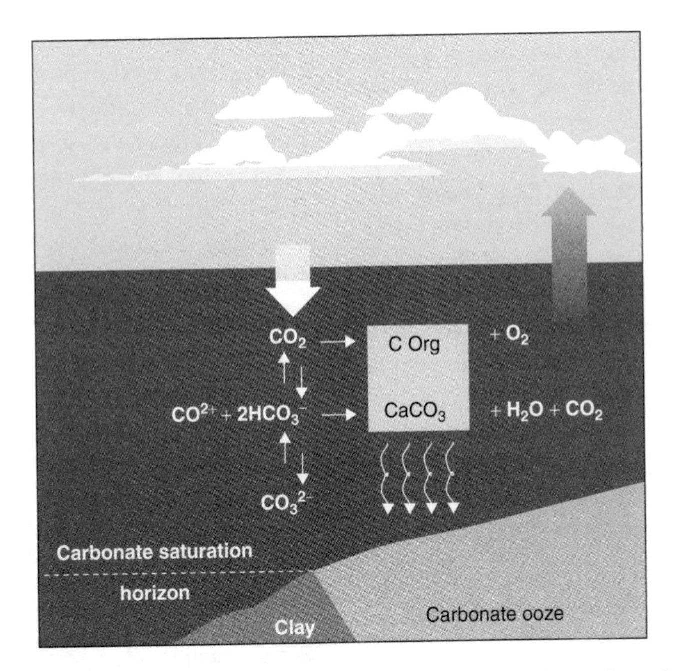

Fig. 8.4 The biogeochemistry of the ocean carbon pump is complex. Formation of organic carbon by photosynthesis is a sink for atmospheric CO_2, whereas biological $CaCO_3$ deposition by certain plankton and reef-building corals is a source of CO_2 to the atmosphere on short timescales. Thus, depending on the ratio of these processes in a particular region, biological carbon sequestration can serve as either a source or sink to the atmosphere. However, when organisms with calcium carbonate skeletons die, much of the carbonate goes into long-term ocean bottom or reef deposits (Reprinted with permission from Elderfield H 2002. Carbonate mysteries. *Science* **296**: 1618–1621. Copyright (2002) American Association for the Advancement of Science).

temperature and salinity. Also, oxygen tends to become depleted by bacterial respiration in waters loaded with organic material. Some coastal waters may experience oxygen depletion as a result of greenhouse warming. For example, discharge of organic material and freshwater from the Mississippi River creates a large (16,500 km^2) zone of hypoxia (low oxygen) below the surface mixed layer of the Northern Gulf of Mexico. Model simulations for a doubled-CO_2 climate predict a 30 to 60% decrease in oxygen below this mixed layer, leading to enlargement of the oxygen-depleted (dead) zone of the Gulf of Mexico (Justic' et al. 1996).

Finally, research is only beginning to explore the complicated interactions between global warming, oceanic circulation, biota, and carbon cycle feedbacks. According to model simulations, the North Atlantic Deep Water Formation will weaken, and at high levels of atmospheric CO_2, it will collapse. Weakened heat transport to deep water will result in warmer surface water and thus lower CO_2 solubility. The net effect of this feedback will be an additional increase in atmospheric CO_2 of 4% by 2100 and 20% by 2500, although a range of possible scenarios could occur (Joos et al. 1999).

Marine Ecosystems

Plankton

Each plankton species has its own optimal environmental conditions for photosynthesis, growth, and reproduction. A 2 °C rise in average ocean surface temperature would probably bring about major shifts in the abundance and distribution of individual plankton species (Fogg 1991) and affect the abundance of herbivores and fish further up the food web. In offshore waters of Southern California, measurements from 1951 to 1993 indicate an 80% decrease in the abundance of animal plankton concurrent with a rise in surface water temperatures of 1.2 to 1.6 °C (Roemmich and McGowan 1995).

Fish

Global warming could devastate some of the world's most productive fisheries. Four primary factors control fish populations: the number of fish spawning, the availability of food for young fish larvae, the physical transport of larvae in the water, and predation on larvae and juveniles. All these processes are affected by temperature. The availability of plankton food is critical to young developing larvae, and fisheries yield decreases exponentially with decreases in plankton production (Figure 8.5). This means that a small decrease in plankton production, for example, in response to climatic change, could lead to a large decrease in fish production.

Since the success of fish populations is closely linked to water temperature, it is not surprising that most fish populations are distributed regionally or globally, within a characteristic temperature range (thermal habitat) that can be mapped as temperature isotherms (lines of equal temperature). Changes in ocean temperature are likely to lead to changes in fish populations. Generally, increasing temperatures are favorable to fish stocks that exist at the higher-latitude end of their temperature range and detrimental to stocks that live at the lower-latitude end of their temperature range.

Although overfishing can devastate fish stocks, major changes in fish populations also often result from changing environmental conditions including water temperature. The catch from several widely separated populations of sardines in the Pacific increases during periods of cold water and decreases dramatically during periods of warm water (Figure 8.6). Norwegian scientists have also

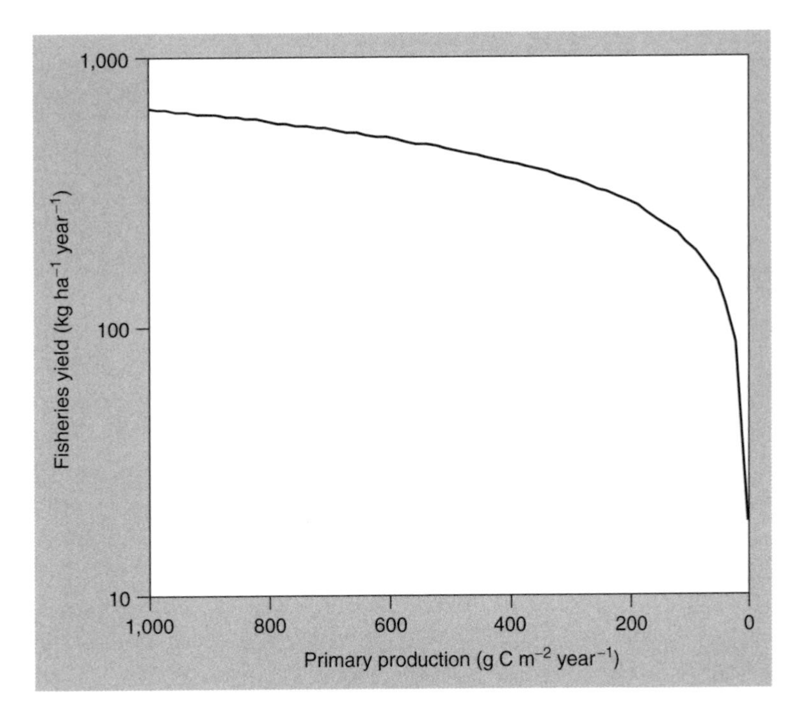

Fig. 8.5 Fisheries yield decreases exponentially with decreasing production of phytoplankton. Data is based on field measurements from a wide variety of geographic locations and different types of ecosystems (Reprinted from Nixon SW 1983. Nutrient dynamics, primary production and fisheries yields of lagoons. In: *Proceedings of the International Symposium on Coastal Lagoons*; Bordeaux, Sept 1981. Oceanologia Acta: 357–371, Copyright (1983), with permission from Elsevier Science).

documented the close relationship between climate and the success of fish populations. They found that during the 1960s and the 1970s, cold, less saline waters from the polar region expanded into the North Atlantic. In response, populations of herring, cod, haddock, and blue whiting declined sharply (Blindheim and Skjoldal 1993).

Cod in the North Sea exist at the southern end of their range, and their production decreases during warm periods (Figure 8.7). Fishing pressure, combined with unusually warm temperatures during the past decade, now threaten this fishery with collapse. To sustain the population, harvests will need to be reduced much more than they would in the absence of warming (O'Brien et al. 2000).

Canadian researchers used a coupled atmosphere/mixed-layer ocean climate model to predict the effects of a doubled CO_2 atmosphere on sea-surface temperatures in the Northeast Pacific Ocean. They predict that warmer temperatures will lead to a shrinking habitat and a 5 to 9% decrease in the production of food for sockeye salmon, resulting in smaller and less abundant fish (Figure 8.8).

With a doubling of atmospheric CO_2, temperature patterns along the west coast of North America will shift markedly northward. Some important commercial fish stocks, formerly occupying a "thermal habitat" extending from Northern California to the Gulf of Alaska will become restricted to a small area in the Bering Sea (Figure 8.9). The end result could

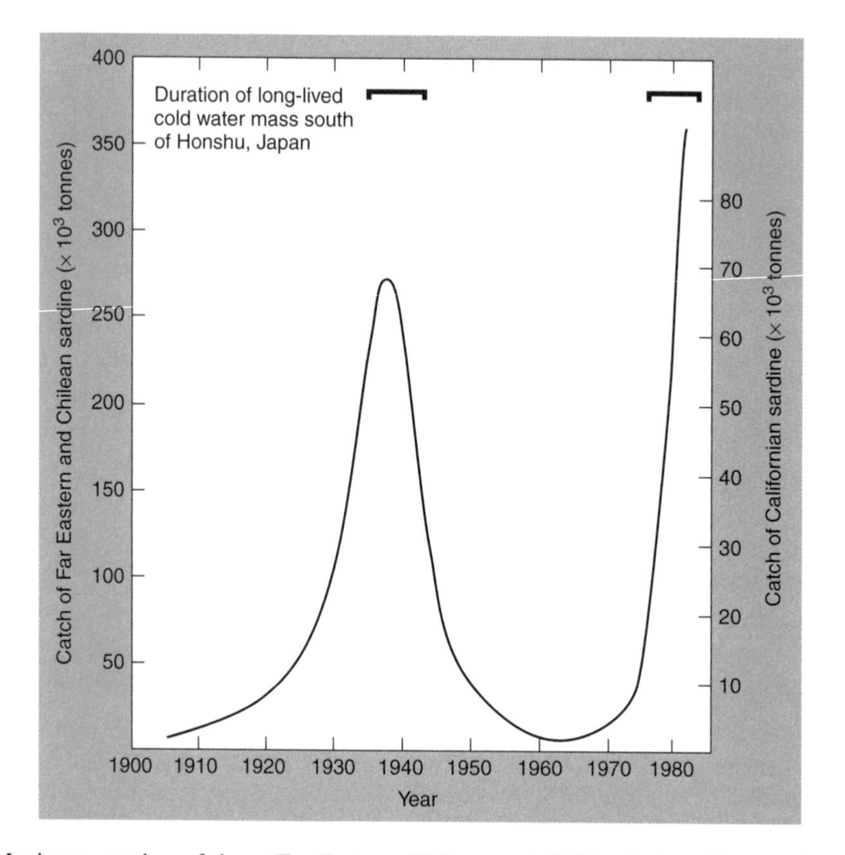

Fig. 8.6 Maximum catches of three (Far Eastern, Chilean, and Californian) sardine species coincide with the long-lived cold-water mass events in the Western Pacific (Adapted from Kawasaki T 1985. Fisheries. In: Kates RW, Ausubel JH and Berberian M, eds *Climate Impact Assessment. SCOPE.* John Wiley & Sons, pp. 131–153).

be a catastrophic loss of available habitat for many species and a likely negative economic impact on the fishing industry of Western North America (Francis and Sibley 1991).

Coastal biota

Climate change could alter the species composition and distribution of coastal marine ecosystems. The growth and reproduction of most marine organisms is closely linked to a specific optimum temperature and salinity for that species. Climate change can affect ocean temperatures and, through changes in land runoff, alter coastal and estuarine salinity. Long-term and recent changes serve to illustrate how marine communities respond to increased temperature. For example, the intertidal community of Monterey Bay, California, was described in detail between 1931 and 1933, and then examined again in 1993 to 1994. There was a significant shift in 32 of 45 invertebrate species with an increase in more southern species and a decrease in more northern species, while average shoreline temperatures increased by 0.75 °C (Barry et al. 1995).

Coral reefs seem particularly vulnerable to climate change. Widespread "bleaching" as

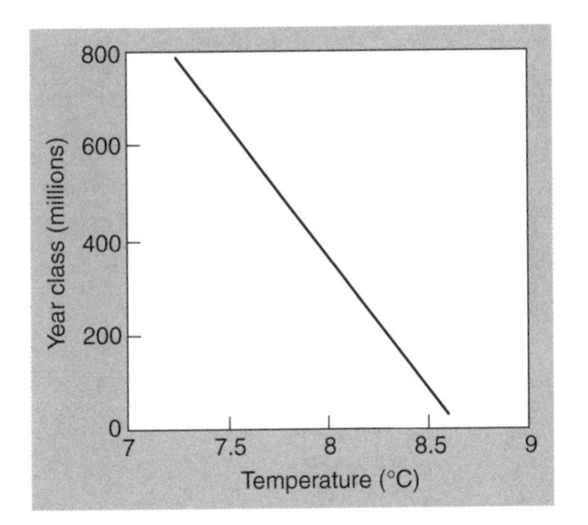

Fig. 8.7 For North Sea cod, year-class strength (millions of one year old fish) decreases sharply with increases in water temperature. Data shown is for the high level (240,000 tonnes) of spawning-stock biomass (Reprinted with permission from O'Brien CM, Fox CJ, Planque B and Casey J 2000. Climate variability and North Sea cod. *Nature* **404**: 142. Copyright (2000) Macmillan Magazines Limited).

well as mortality in recent decades is linked to the occurrence of warm water and oceanic hotspots (Box 8.3, Figure 8.10).

Flooding and saltwater intrusion from projected sea-level rise threatens the viability of many biologically rich coastal wetlands. In the United States, between 1948 and 1993, the vegetation community of New England tidal salt marshes changed dramatically in response to increasing rates of sea-level rise, possibly combined with other factors (Warren and Niering 1993).

Mangrove forests inhabit extensive coastal areas in the subtropics to tropics, supplying firewood fuel, lumber, shelter from storm erosion, and a habitat and nursery ground for numerous species of fish and shellfish, many of commercial importance. In Florida and the Caribbean region, climate change will probably increase sea level, decrease

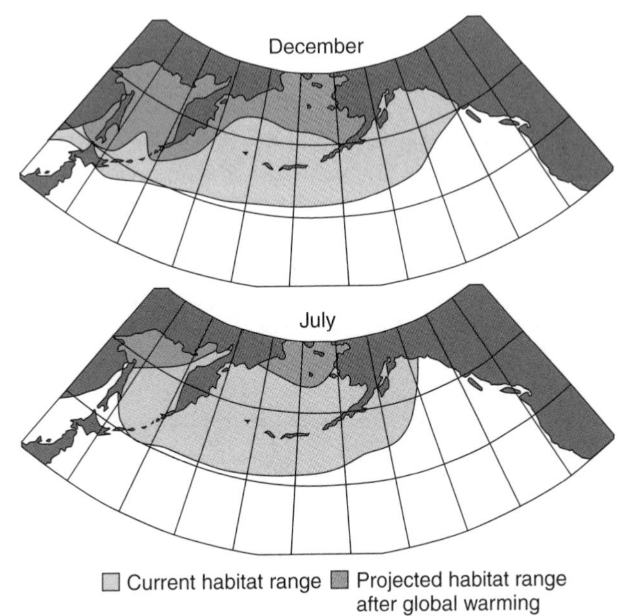

Fig. 8.8 Projected loss of marine habitat for sockeye salmon due to global warming by the middle of this century (From Welch DW, Ishida Y and Nagasawa K 1998. Thermal limits and ocean migrations of sockeye salmon (*Onocorhynchus nerka*): long-term consequences of global warming. *Canadian Journal of Fisheries and Aquatic Sciences* **55**: 937–948. Reproduced by permission of NRC Research Press).

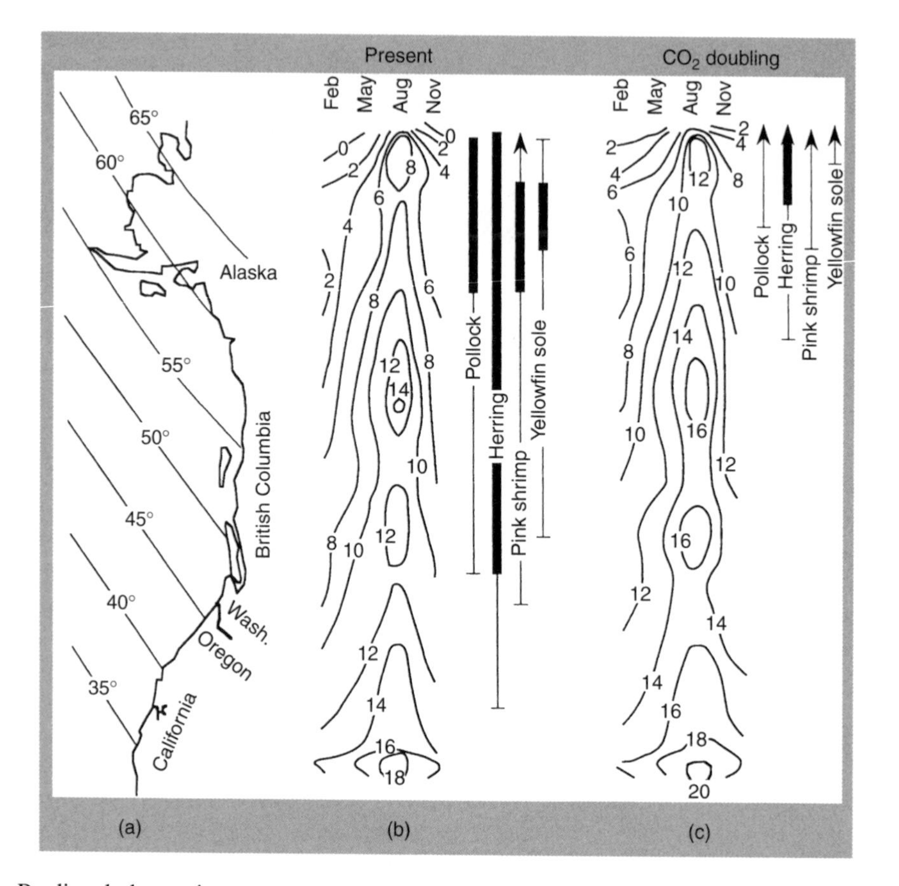

Fig. 8.9 Predicted change in temperature patterns and fish distribution off Western North America in response to a doubling of atmospheric CO_2. (a) West coast of North America; (b) present range of four species; (c) predicted range after a double-CO_2 global warming. Isoline values from 0 to 20 enclose the temperature range vertically over latitudes and horizontally over seasons (From Strickland RM, Grosse DJ, Stubin AI, Ostrander GK and Sibley TH 1985. Definition and characterization of data needs to describe the potential effects of increased atmospheric CO_2 on marine fisheries of the Northeast Pacific Ocean. Virginia U.S. Department of Energy, Office of Energy Research. DOE/NBB-075. TR028. NTIS Springfield, p. 139).

Box 8.3 Coral reefs and climate change

Coral reefs, roughly between latitudes 30 °N and S, serve multiple functions. They provide a habitat for a great diversity of plants and animals, protect shorelines from storms and erosion, and serve as an economic resource for tourism and fisheries. Reef-forming corals deposit solid calcium carbonate (limestone). These limestone structures can be massive and form the base of entire islands, archipelagos, and large landmasses. Coral reefs are among the world's most biologically diverse communities. Reef-building coral animals contain symbiotic algae (dinoflagellates) in their tissue. These algae enhance coral growth rates and aid in nutrient recycling. They are vital to the survival of coral reefs. If stressed by higher than normal temperature, the coral lose their algae or "bleach." The coral may recover or, if the stress is prolonged, may die. Since the 1980s, the frequency and extent of "coral bleaching" has grown alarmingly (Brown and Ogden 1993, Wellington et al. 2001). In some areas, large expanses of reef have died. Satellite sea-surface temperature maps demonstrate that the large-scale bleaching events are almost always associated with anomalous "oceanic hotspots" – areas of the ocean that exceed the long-term mean monthly maximum temperature by 1 °C or more (Goreau and Hayes 1994, Strong 1989). The 1998 ENSO was particularly strong and led to increased temperatures and severe coral bleaching in many areas of the Indo-Pacific. Over 90% of the corals bleached and, in some areas, entire island reefs died. The area and number of species of coral infected with debilitating or lethal diseases is growing and may be related to temperature stress. For example, in the Florida Keys between 1996 and 1998 the proportion of sampled sites with diseased coral and the proportion of species affected increased from 16 to 82% and 27 to 85%, respectively (Harvell et al. 1999). Reefs in some areas have died from a variety of human impacts (Figure 8.10). At least one-quarter of known reefs are seriously degraded and over half are seriously threatened. Concern is growing that climate change, through temperature-induced bleaching, increased disease incidence, elevated sea level, and lower ocean pH, threatens the very global survival of coral reefs.

precipitation and runoff, and lead to increased salinity, which will reduce mangrove production (Snedaker 1995).

Estuaries are enclosed bodies of water with a direct link to the sea and an input of freshwater. They are among the most biologically productive areas of the world's ocean and often serve as sheltered nursery grounds for the early development of fish and shellfish. However, they may be especially vulnerable to climate change. For example, changes in rates of freshwater inflow or evaporation can alter salinity. Warmer or less saline water, because of its lower density, remains in the upper water column. Such stratification reduces mixing and can promote oxygen depletion in deeper waters. Also, if the temperature of exterior coastal water warms above a particular species' preferred or tolerable range, it may be blocked from its normal migration route outside the estuary. Finally, changes in seasonal temperature patterns can lead to a mismatch between the plankton blooms and the arrival of juvenile fish that depend on these blooms as their food source.

Case studies in the United Kingdom and United States suggest major impacts of climate change on estuaries. The Thames Estuary, in the United Kingdom, exemplifies the close links between large-scale climatic events, estuaries, and fisheries. There, the abundance of most commercially important fishery species is closely linked to the large-scale oceanographic-climatic variation, the North Atlantic Oscillation (NAO) (Box 8.2). During warm years, populations of southern species, such as bass, increase, whereas in cold years, northern species, such as herring, thrive (Attrill and Power 2002). In the United States major changes may be in store for the fisheries of Chesapeake Bay, one of the largest and most productive estuaries (Kennedy 1989).

(a)

(b)

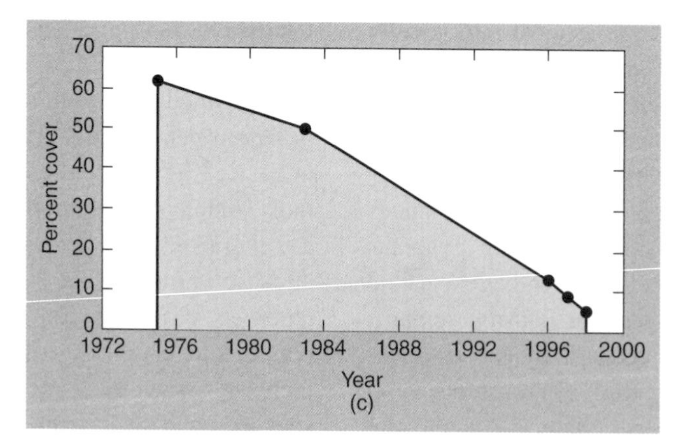

(c)

Fig. 8.10 Carysfort Reef is the largest and once was the most luxuriant reef in the Florida Keys: (a) 1975 showing extensive live coral, mostly *Acropora palmata*; (b) 1995 showing degraded and mostly dead coral; (c) between 1975 and 1998 Carysfort lost over 92% of its living coral cover from pollution, disease, and physical damage. Ocean warming adds additional stress to already threatened reefs (Photos and data from Philip Dustan).

Marine mammals

Many marine mammals are threatened by habitat destruction and fragmentation. Global warming will put additional pressure on their ability to survive (Harwood 2001). Many marine mammals serve as keystone species in the food web. Depletion of their numbers can result in a cascade of changing abundance in other species. The extent of sea ice has declined and will probably decline another 40% or more by 2100 (Hadley Centre 2002). This will shrink the available habitat of pinnipeds (seals and sea lions) and polar bears that rely on sea ice for resting and pupping and of cetaceans such as the bowhead whale that rely on high levels of plankton productivity associated with the ice edge. Earlier breakup of sea ice in Western Hudson Bay may already be responsible for the declining condition of polar bears in that region (Stirling et al. 1999). However, the most threatened aquatic mammals may be certain species of seals that are endemic to inland seas and large lakes such as the Caspian Sea in Southwestern Asia and Lake Baikal in Southern Siberia. If adequate ice does not develop, these seals will not have enough habitat for breeding. In fact, this situation may already be under way in the Caspian, where recent small ice areas have led to crowding during the breeding season and mass mortality from disease (Kennedy et al. 2000).

Marine diseases

Environmental stress, whether from pollution or from climate change, can weaken organisms, making them more susceptible to disease. The incidence of disease in many marine species is increasing around the world (Harvell et al. 1999). These disease outbreaks often result in mass mortalities of marine plants, invertebrates, and vertebrates. For example, the pathogenic infection and die-off of the dominant sea urchin and keystone herbivore *Diadema antillarum* in the Caribbean in the 1980s led to a massive phase shift. In the absence of herbivorous urchins, algae thrived and covered and killed the coral. The result was a shift from coral- to algal-dominated reefs. Coral reefs are also dying from temperature stress and more frequent and widespread marine diseases (Box 8.2).

An almost 25-year warming trend on the east coast of the United States may be responsible for the spread of several diseases and mass mortalities in oyster populations. Also, mass mortalities of seals in recent years in Northern Europe resulted from infection with a pathogenic virus known as phocine distemper virus. As temperatures rise, seals spend more time on the beach as opposed to the water and congregate in dense aggregations. This sets the stage for the rapid spread of the opportunistic pathogen (Lavigne and Schmitz 1990).

Diseases, spread through the marine environment, are not restricted to marine species. Many human pathogens such as cholera are naturally active in coastal waters, or following introduction via sewage or storm water outfalls, remain dormant for a period of time. Warm water may trigger them to emerge in an infectious state.

Blooms of algae, particularly toxic and undesirable species, have increased in many regions of the world (Smayda and Shimizu 1993) and seem linked to warming ocean temperatures (Epstein et al. 1994). These blooms are implicated in mass mortalities of marine mammals and fish. For example, a toxic phytoplankton (*Gymnodinium catenatum*) killed 70% of the Mediterranean monk seal population on the coast of the Western Sahara in 1997 (Forcada et al. 1999). In addition, changing temperature and ocean currents along the northwest coast of Spain are expected to lead to an increase in blooms of the toxic single-celled algal dinoflagellates responsible for paralytic shellfish poisoning (Fraga and Bakun 1993).

Summary

Recent changes in the marine environment in response to greenhouse warming will probably continue and even accelerate. Predicted changes, by the end of this century, include the following:

A global average sea-level rise of about 0.2 to 0.7 m resulting in

- increased beach erosion and coastal flooding;
- loss of coastal ecosystems such as mangroves and wetlands;
- displacement of human populations away from low-lying areas;
- saltwater intrusion into coastal aquifer water supplies.

Possible changes in large-scale ocean circulation patterns:

- a possible collapse of the oceanic conveyer belt, followed by rapid climate cooling and decreased precipitation in the Mediterranean and the African Sahel region;
- intensified upwelling of cool water along the western coasts of the Americas, Southern Europe, and Africa.

Alteration of important oceanic biogeochemical cycles:

- a warming-induced decrease in the ability of the ocean to absorb CO_2;
- a decrease in bio-carbonate formation necessary for reef-forming corals and other marine species;
- lower concentrations of life-supporting dissolved oxygen in some marine waters.

Direct impacts on marine ecosystems:

- changes in the species population abundance of plankton that form the base of the marine food web;
- changes in the geographic distribution and species composition of coastal ecosystems;
- additional temperature stress on communities already threatened by human impacts, for example, coral reefs and mangroves;
- decreased productivity and altered distributions of fish and marine mammals.

References

Attrill MJ and Power M 2002 Climatic influence on a marine fish assemblage. *Nature* **417**: 275–278.

Bakun A 1992 Global greenhouse effects, multidecadal wind trends, and potential impacts on coastal pelagic fish populations. *International Council for Exploration of the Sea Marine Science Symposium: Hydrobiological Variability in the ICES Area 1980–1989*, Mariehamn 1991: 316–325.

Barry JP, Baxter CH, Sagarin RD and Gilman SE 1995 Climate-related, long-term faunal changes in a California rocky intertidal community. *Science* **267**: 672–675.

Bigg GR 1996 *The Oceans and Climate*. Cambridge. Cambridge University Press.

Blindheim J and Skjoldal HR 1993 Effects of climatic change on the biomass yield of the Barents Sea, Norwegian Sea, and West Greenland large marine ecosystems. In: *Proceedings of the International Conference on Large Marine Ecosystems: Stress, Mitigation and Sustainability*. Monaco, IUCN: The World Conservation Union, pp. 185–198.

Broadus J, Milliman J, Edwards S, Aubrey D and Gable F 1986 Rising sea level and damming of rivers: possible effects in Egypt and Bangladesh.

In: Titus J, ed. *Effects of Changes in Stratospheric Ozone and Global Climate; Vol. 4: Sea Level Rise*. Washington, DC: US EPA, pp. 165–189.

Broecker WS 1997 Thermohaline circulation, the Achilles heel of our climate system: will man-made CO_2 upset the current balance? *Science* **278**: 1582–1588.

Bromwich D 1995 Ice sheets and sea level. *Nature* **373**: 18,19.

Brown B and Ogden JC 1993 Coral bleaching. *Scientific American* **January**: 63–70.

Brunn P 1962 Sea level rise as a cause of shore erosion. *Journal of Waterways and Harbors Division (ASCE)* **1**: 116–130.

Charlson R, Lovelock J, Andrae M and Warren S 1987 Oceanic phytoplankton, atmospheric sulphur, cloud albedo and climate. *Nature* **326**: 655–661.

Church JA and Gregory JM 2001 Changes in sea level. In: Houghton JT, Ding Y, Griggs DJ, Noguer M, van der Linden PJ, Dai X, et al., eds *Climate Change 2001: The Scientific Basis. Intergovernmental Panel on Climate Change, Working Group I*. Cambridge: Cambridge University Press, pp. 639–693.

Elderfield H 2002 Carbonate mysteries. *Science* **296**: 1618–1621.

Epstein PR, Ford TE and Colwell R 1994 Marine ecosystems. In: *Health and Climate Change*. The Lancet, London, pp. 14–17.

Fogg GE 1991 Changing productivity of the oceans in response to a changing climate. *Annals of Botany* **67**(Suppl. 1): 57–60.

Forcada J, Hammond P and Aguilar A 1999 The status of the Mediterranean monk seal in the Western Sahara and the implications of a mass mortality. *Marine Ecology Progress Series* **188**: 249–261.

Fraga S and Bakun A 1993 Global climate change and harmful algal blooms: the example of *Gymnodinium catenatum* on the Galacian coast. In: Smayda TJ and Shimizu Y, eds *Toxic Phytoplankton Blooms in the Sea: Proceedings of the Fifth International Conference on Toxic Marine Phytoplankton; Newport, Rhode Island; 28 October-1 November 1991*. Amsterdam. Elsevier, pp. 59–65.

Francis RC and Sibley TH 1991 Climate change and fisheries: what are the real issues? *Northwest Environmental Journal* **7**: 295–307.

Francis RC, Hare SR, Hollowed AB and Wooster WS 1998 Effects of interdecadal climate variability on the oceanic ecosystems of the NE Pacific. *Fisheries Oceanography* **7**: 1–21.

Goreau TJ and Hayes RL 1994 Coral Bleaching and Ocean "Hot Spots". *Ambio* **23**(3): 176–180.

Hadley Centre 2002 Meteorological Office, United Kingdom *http://www.meto.govt.uk/research/hadley centre/models/modeldata.html*

Harleman DRF, Bras RL, Rinaldo A and Malanotte P 2000 Blocking the tide. *Civil Engineering* **October**: 52–57.

Harvell CD, Kim K, Burkholder JM, Colwell RR, Epstein PR, Grimes DJ, et al. 1999 Emerging marine diseases-climate links and anthropogenic factors. *Science* **285**: 1505–1510.

Harwood J 2001 Marine mammals and their environment in the twenty-first century. *Journal of Mammalogy* **82**(3): 630–640.

Joos F, Plattner G-K, Stocker TF, Marchal O and Schmittner A 1999 Global warming and marine carbon cycle feedbacks on future atmospheric CO_2. *Science* **284**: 464–467.

Justic' D, Rabalais NN and Turner RE 1996 Effects of climate change on hypoxia in coastal waters: a doubled CO_2 scenario for the northern Gulf of Mexico. *Limnology and Oceanography* **41**: 992–1003.

Kawasaki T 1985 Fisheries. In: Kates RW, Ausubel JH and Berberian M, eds *Climate Impact Assessment. SCOPE*. Chichester, UK, John Wiley & Sons, pp. 131–153.

Kennedy V 1989 Potential effects of climate change on Chesapeake Bay animals and fisheries. In: Topping JC, ed. *Coping With Climate Change: Proceedings of the Second North American Conference on Preparing for Climate Change*. December 6–8, Washington, DC: The Climate Institute, pp. 509–513.

Kennedy S, Kuiken T, Jepson PD, Deaville R, Forsyth M, Barrett T, et al. 2000 Mass die-off of Caspian seals caused by canine distemper virus. *Emerging Infectious Diseases* **6**(6): 637–639.

Kleypas JA, Buddemeier RW, Archer D, Gattuso J-P, Langdon C and Opdyke BN 1999 Geochemical consequences of increased atmospheric carbon dioxide on coral reefs. *Science* **284**: 118–120.

Lavigne DM and Schmitz OJ 1990 Global warming and increasing population densities: a prescription for seal plagues. *Marine Pollution Bulletin* **21**: 280–284.

Nixon SW 1982 Nutrient dynamics, primary production and fisheries yields of lagoons. In Proceedings

International Symposium on Coastal Lagoons. Bordeaux, Sept 1981. *Oceanologia Acta* 357–371.

El Niño 2002 *National Oceanic and Atmospheric Administration*. Available from: *http://www.pmel.noaa.gov/tao/elnino/nino-home.html*

Nunn PD 1988 *Future sea-level rise in the Pacific: Effects on selected parts of Cook Islands, Fiji, Kiribati, Tonga and Western Samoa. Technical Report*. Suva, Fiji, School of Social and Economic Development, University of the South Pacific, p. 46.

O'Brien CM, Fox CJ, Planque B and Casey J 2000 Climate variability and North Sea cod. *Nature* **404**: 142.

Peltier WR and Tushingham AM 1989 Global sea level rise and the greenhouse effect: might they be connected? *Science* **244**: 806–810.

Rodbell DT, Seltzer GO, Anderson DM, Abbott MB, Enfield DB and Newman JH 1999 A similar to 15,000-year record of El Nino-driven alluviation in southwestern Ecuador. *Science* **283**: 516–520.

Roemmich D and McGowan J 1995 Climatic warming and the decline of zooplankton in the California Current. *Science* **267**: 1324–1326.

Roy P and Connell J 1991 Climatic change and the future of atoll states. *Journal of Coastal Research* **7**(4): 1057–1075.

Smayda TJ and Shimizu Y, eds 1993 *Toxic Phytoplankton Blooms in the Sea: Proceedings of the Fifth International Conference on Toxic Marine Phytoplankton; Newport, Rhode Island; 28 October-1 November 1991*. Amsterdam, Elsevier.

Snedaker SC 1995 Mangroves and climate change in the Florida and Caribbean Region: scenarios and hypotheses. *Hydrobiologia* **295**: 43–49.

Stocker TF 1994 The variable ocean. *Nature* **367**: 221–222.

Stocker TF and Schmittner A 1997 Influence of CO_2 emission rates on the stability of the thermohaline circulation. *Nature* **388**: 862–865.

Stirling I, Lunn N and Iacozza J 1999 Long-term trends in the population ecology of polar bears in western Hudson Bay in relation to climate change. *Arctic* **52**: 294–306.

Street-Perrott FA and Perrott RA 1990 Abrupt climate fluctuations in the tropics. The influence of Atlantic ocean circulation. *Nature* **343**: 607–612.

Strickland RM, Grosse DJ, Stubin AI, Ostrander GK and Sibley TH 1985 *Definition and characterization of data needs to describe the potential effects of increased atmospheric CO_2 on marine fisheries of the Northeast Pacific Ocean*. Virginia, U.S. Department of Energy, Office of Energy Research. DOE/NBB-075. TR028. NTIS Springfield, p. 78.

Strong AE 1989 Greater global warming revealed by satellite-derived sea-surface temperature trends. *Nature* **338**: 642–645.

Titus JG 1986 Greenhouse effect, sea level rise, and coastal zone management. *Coastal Zone Management Journal* **14**(3): 147–171.

Titus JG and Narayanan V 1996 The risk of sea level rise. *Climatic Change* **33**: 151–212.

Titus JG, Park RA, Leatherman SP, Weggel JR, Greene MS, Mausel PW, et al. 1991 Greenhouse effect and sea level rise: the cost of holding back the sea. *Coastal Management* **19**: 171–204.

Trenberth KE and Hoar TJ 1996 The 1990–1995 El Nino-Southern Oscillation event: Longest on record. *Geophysical Research Letters* **23**(1): 57–60.

UCAR/NOAA 1994 University Center for Atmospheric Research and National Oceanic and Atmospheric Administration. Boulder Colorado and Washington, DC. *http://www.ucar.edu/ucar/index.html*

UNEP 2002 Available from: *http://www.grida.no/climate/vital/impacts.htm*

Warren RS and Niering WA 1993 Vegetation change on a Northeast tidal marsh: interaction of sea-level rise and marsh accretion. *Ecology* **74**(1): 96–103.

Weaver AJ 1993 The oceans and global warming. *Nature* **364**: 192–193.

Welch DW, Ishida Y and Nagasawa K 1998 Thermal limits and ocean migrations of sockeye salmon (*Onocorhynchus nerka*): long-term consequences of global warming. *Canadian Journal of Fisheries and Aquatic Sciences* **55**: 937–948.

Wellington GM, Glynn PW, Strong AE, Navarrete SA, Wieters E and Hubbard D 2001 Crisis on coral reefs linked to climate change. *EOS, Transactions of the American Geophysical Union* **82**(1): 1–5.

Williams PB 1985 *An overview of the impact of accelerated sea level rise on San Francisco Bay. Report on Project 256. December 20*. San Francisco, Phillip Williams & Associates.

WWF 2002 Available from: *http://www.panda.org/climate/*

SECTION III

Human Dimensions of Climate Change

Chapter 9

Impacts on Human Settlement and Infrastructure

"The insurance business is the first in line to be affected by climate change ... it could bankrupt the industry."

Franklin Nutter, President of the Reinsurance Association of America

Introduction

Climate change will have wide-ranging impacts on society and the infrastructure that supports civilization. Global warming could impact not only agriculture (Chapter 7) and human health (Chapter 10) but also patterns of human settlement, energy use, transportation, industry, environmental quality, and other aspects of infrastructure that affect our quality of life (IPCC 1990).

Numerous examples from history illustrate how the success of civilization and human welfare is intimately linked to climate (Gore 1993). Natural warming or cooling periods of only 1 or 2 °C have impacted human activities, resulted in population migrations, or altered settlement patterns. For example, the warm period around 950 AD allowed the settlement of Greenland, and briefly North America, by Nordic people; but at about the same time, severe droughts in Central America contributed to the collapse of the Mayan Civilization. During the Little Ice Age (1550 to 1850 AD), global average temperatures 1 to 2 °C lower than now contributed to fishing and crop failures and repeated famine in Europe. Also, during the same period (in 1815), a large volcanic eruption in Indonesia discharged huge quantities of dust and soot into the atmosphere. The resultant cooling in 1816 became known as, "the year without a summer," and crop failures in Europe led to widespread food riots, political unrest, and migration.

Fossil-fuel use will affect future climate. Fossil fuels, currently the mainstay of economically developed countries, supply energy, either directly as fuel or indirectly as generated electricity, for manufacturing, agriculture, transportation, and space heating. Future greenhouse gas (GHG) emissions and resultant climate change will depend largely

Climate Change: Causes, Effects, and Solutions John T. Hardy
 ISBNs: 0-470-85018-3 (HB); 0-470-85019-1 (PB)

on future rates of fossil-fuel consumption. Many complex and interacting factors determine the consumption rate of fossil fuels. Demand is a result of population growth rate, availability of fossil fuel, energy efficiency, conservation measures, use of nonfossil energy sources, general industrial productivity, and energy policy. All these factors will affect fossil-fuel utilization rates and future climate.

Future climate, in turn, will affect fossil-fuel use. As climate changes, patterns of energy use will change. Humans living in cold climates require large quantities of energy for space heating of residential and commercial buildings. These requirements will decrease in response to warmer winters. In warm climates, energy is needed for air conditioning, and in arid regions, irrigated agriculture requires energy for pumping water. Energy demand for these activities could increase.

The impacts of climate change on human settlement patterns and infrastructure will differ regionally and could range from insignificant to catastrophic. The costs of mitigating these impacts will vary greatly (Chapters 11 and 12), but are likely to be felt most by developing nations. Important links exist between global climate, extreme climate events, energy use, environmental quality, human settlement patterns, and the transportation and industrial infrastructure.

Energy

The effects of energy use on climate

The United Nations Framework Convention on Climate Change (UNFCCC 2002) calls for "stabilization of greenhouse gas concentrations in the atmosphere at a level that would prevent dangerous anthropogenic interference with the climate system. . . ." Can this goal be met? The actual level at which atmospheric CO_2 stabilization is achieved will depend on the product of several factors, known together as the Kaya identity (Hoffert et al. 1998):

$$Mc = N(GDP/N)(E/GDP)(C/E);$$

where

Mc = CO_2 emitted from fossil-fuel combustion
N = population
GDP = gross domestic product
E/GDP = energy intensity in W year $\$^{-1}$
C/E = carbon intensity = the weighted average of the carbon-to-energy emission factors of all energy sources in kg C W^{-1} $year^{-1}$.

Thus, global 1990 fossil-fuel CO_2 emissions can be estimated as

$$5.3 \times 10^9 \text{ persons} \times 4{,}100\$ \text{ per person year}^{-1}$$
$$\times\ 0.49 \text{ W year } \$^{-1} \times 0.56 \text{ kg C W}^{-1}\text{year}^{-1}$$
$$= 6 \times 10^{12} \text{ kg C (6.0 gigatons of carbon)}$$

The level of atmospheric CO_2 stabilization that can be achieved in this century will depend on all these factors (Hoffert et al. 1998). The IPCC developed a number of possible GHG emission scenarios based on socioeconomic projections. The business as usual scenario (IS92a) assumes current rates of population and economic growth with no new climate-change policies. On the basis of such a scenario, population and GDP per capita will increase, while energy intensity and carbon intensity should decrease (i.e. improve) (Figure 9.1). Nevertheless, despite improvement in these last two factors, the rate of atmospheric CO_2 emissions will double by 2050 and more than triple by 2100 (Figure 9.2a). To meet the economic assumptions of "business as usual," total power will need to increase greatly during this century. In the developing world, because of the projected rapid growth rate in energy use, achievable increases in energy efficiency will have

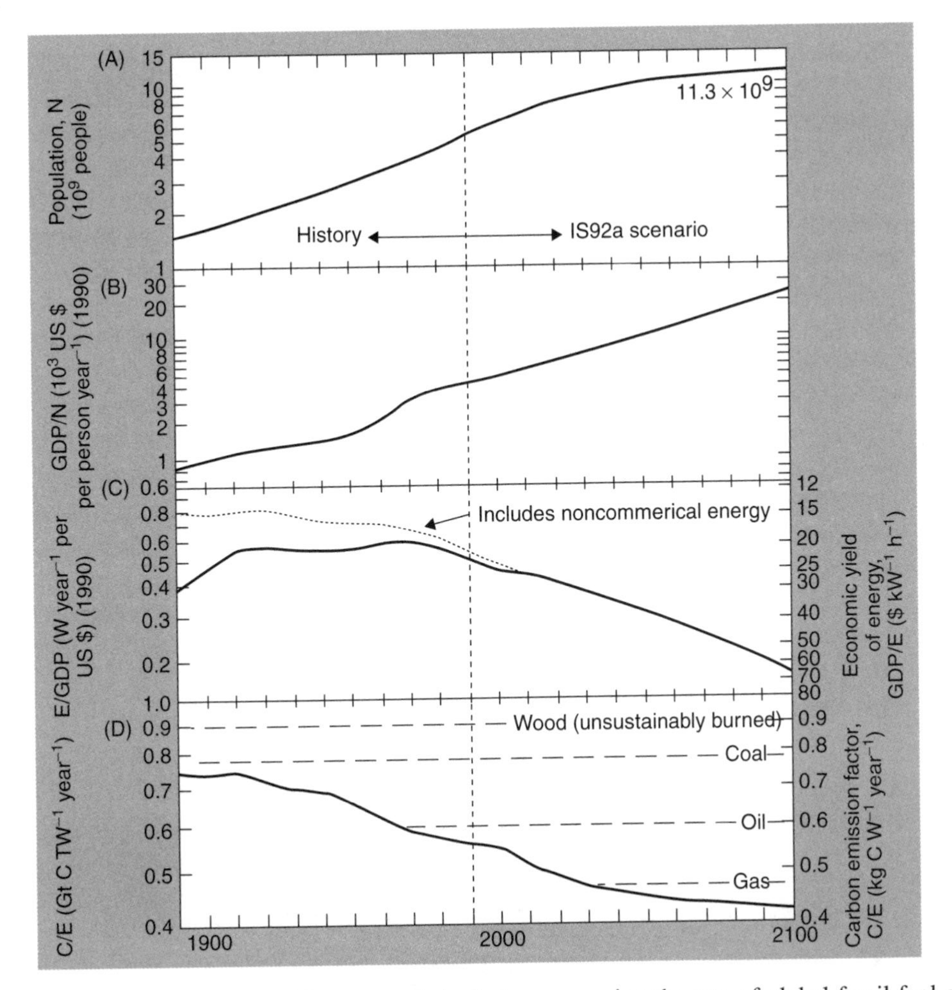

Fig. 9.1 Historical and predicted future trends in factors governing the rate of global fossil-fuel carbon emissions. (A) Population and (B) per capita income increase, while (C) energy consumed/gross domestic product decreases, that is, energy efficiency increases. (D) Carbon intensity (carbon emissions/energy consumed) continues to decrease. Note that in (D), although carbon emissions per energy consumed decreases, because of growth in population and total global energy consumption, greenhouse gas emissions will increase (see Figure 1.10) (Reprinted with permission from Hoffert MI, Caldeira K, Jain AK, Haites EF, Danny Harvey LD, Potter SD, et al. 1998. Energy implications of future stabilization of atmospheric CO_2 content. *Nature* **395**: 881–884. Copyright (1998) Macmillan Magazines Limited).

little impact in reducing total GHG emissions (Pearson and Fouquet 1996).

Improvements in energy efficiency alone will not be sufficient to stabilize CO_2 at reasonable target values. Meeting CO_2 stabilization goals will require a simultaneous decrease in carbon fuels as a proportion of total energy (Figure 9.2b). New carbon-free sources of energy will be required to decrease carbon intensity (Hoffert et al. 1998). In fact, stabilizing atmospheric CO_2 at twice preindustrial levels while maintaining "business

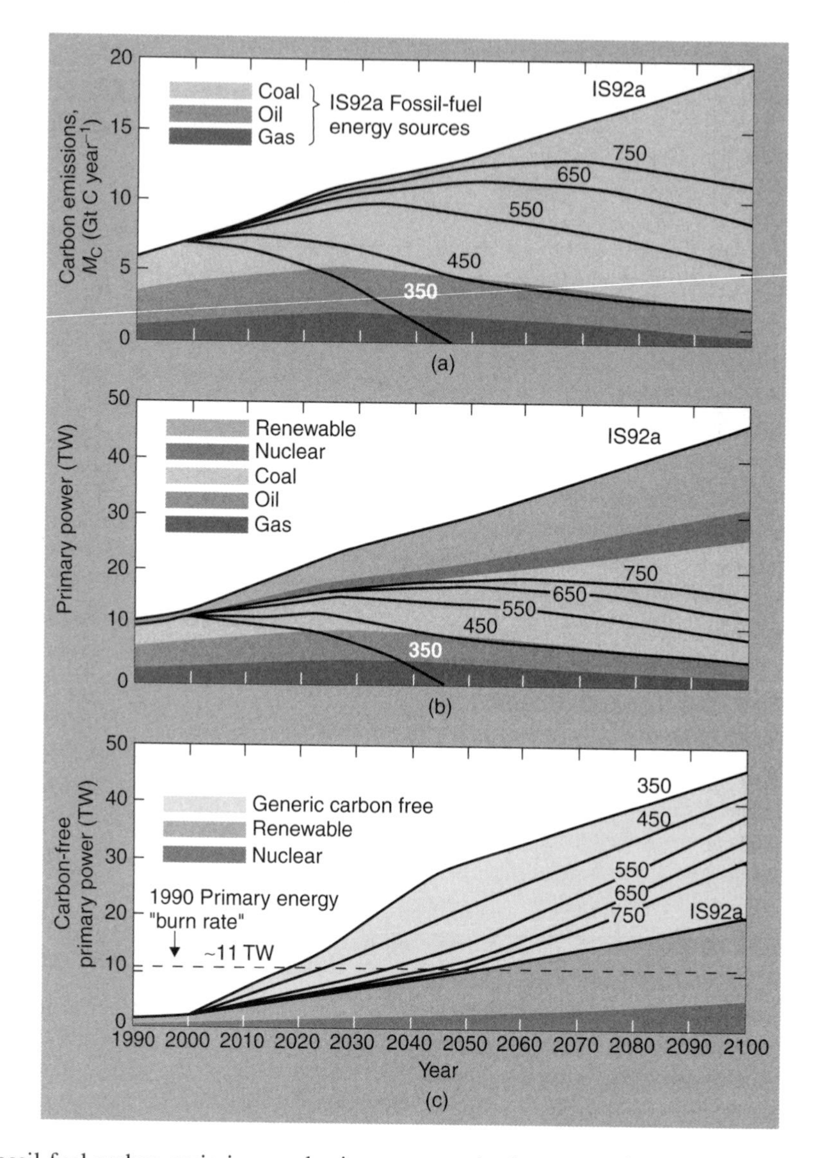

Fig. 9.2 Fossil-fuel carbon emissions and primary power in the twenty-first century for IPCC IS92a and WRE (1996) stabilization scenarios. Predicted allowable emission levels over time that will ultimately stabilize atmospheric CO_2 at 750, 650, 550, 450, and 350 ppmv. (a) In the business as usual scenario (IS92a), carbon emissions continue to grow and the proportion from coal increases. Stabilizing CO_2 at 350 ppmv will require a complete phase-out of coal by 2020, oil by 2040, and gas by 2050 – highly unlikely. Even stabilization at 450 ppmv will require a complete phase-out of coal by 2050. (b) In the absence of carbon-free energy, the power needed to meet economic goals of IS92a will need to come from a mixture of energy sources. Stabilization of CO_2 will require decreases in the fossil-fuel contribution to total primary power. (c) Stabilizing CO_2 at lower (below 750 ppmv) levels will require rapid development of carbon-free power (Reprinted with permission from Hoffert MI, Caldeira K, Jain AK, Haites EF, Danny Harvey LD, Potter SD, et al. 1998. Energy implications of future stabilization of atmospheric CO_2 content. *Nature* **395**: 881–884. Copyright (1998) Macmillan Magazines Limited).

as usual" will require a massive transition to carbon-free power systems (Figure 9.2c). Carbon-free technologies exist (Chapter 11), but some are still experimental, while others have limited potential. Energy production and consumption must make the transition rapidly to a largely carbon-free global economy; otherwise civilization will indeed experience "a dangerous anthropogenic interference with the climate system."

The effects of climate change on energy supply and demand

Population and economic growth will lead to future increases in energy demand in most countries, but the impacts of climate change on supply and demand will vary greatly by region. For example, in the United Kingdom and Russia a 2 to 2.2 °C warming by 2050 will decrease winter space-heating needs, thus decreasing fossil-fuel demand by 5 to 10% and electricity demand by 1 to 3% (Moreno and Skea 1996). By 2050 in the Southern United States, summertime electrical demand will increase greatly because of air-conditioning demands. In the Northeastern United States, summertime decreases in stream flow will reduce hydropower generation during that season (Linder 1990).

Electrical generation must meet average demands, but it must also be sufficient to meet peak demands. Energy demand is greatest at certain times, generally showing daily and seasonal peak periods. For example, in temperate regions electrical demand for space heating and lighting peaks in winter, while that for air conditioning peaks in summer. In many areas during summer, electrical pumps are needed to draw water from wells or pump water through irrigation systems. Also, different demands peak at particular times of the day – industrial during working hours and residential during the early evening.

Model studies, assuming a 3 to 5 °C increase in average US temperature by 2055, suggest that electrical demand and fuel costs will increase significantly because of climate change (Linder 1990). Annual electrical energy demands will increase slightly more than 1% per degree centigrade, or a total of 4 to 6% by 2055. As a result of climate change, peak national demand will increase 16 to 23% above base case values, that is, above the increased demand due largely to population growth without climate change (Figure 9.3).

The costs of increasing electrical capacity to meet the increased demand due to climate change will be large. By 2055, the annual costs for capital, fuel, and climate-induced modifications in utility operations will be 7 to 15% greater than costs without climate change. Regional needs will include increased peak winter capacity in the Northeastern United States and an increase in summer peak capacity of greater than 20% in the southeast, the Southern Great Plains, and the southwest (Linder 1990). However, with milder winters in the north, some research suggests that a 1.8 °C warming would actually result in a net reduction in overall US energy demand of 11% and a cost reduction (1991 dollars) of $5.5 billion by 2010 (Rosenthal et al. 1995).

An increase in electrical demand (much of it generated by fossil-fuel combustion) would make policies that limit GHG emissions more difficult to achieve. Because of the long lead time required to plan and build new power plants, electric utility managers need to plan with climate change in mind. With increased demand, the need to import power could affect the balance of payments of a country's foreign trade.

Because of its effect on runoff and stream flows, climate change will also affect

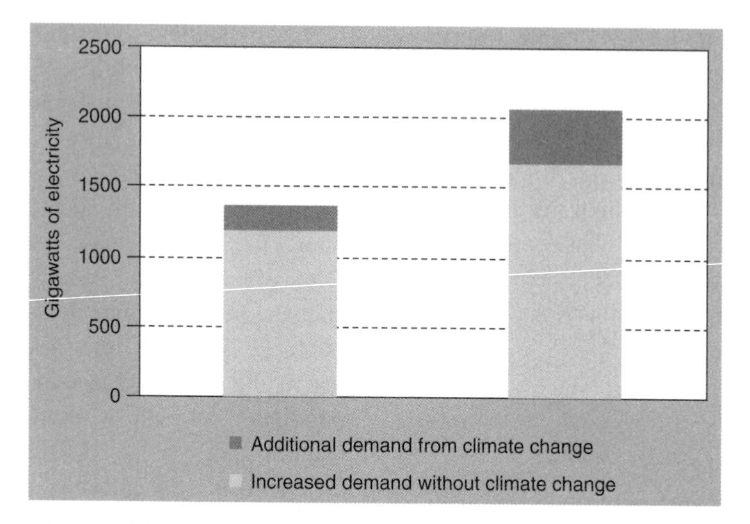

Fig. 9.3 Climate change will add to US electrical demand by the year 2055. Assumed GNP is lower in projection on left and higher in right (From Linder KP 1990. National impacts of climate change on electric utilities. In: Smith JB and Tirpak DA, eds *The Potential Effects of Global Climate Change on the United States.* New York: Hemisphere Publishing Corporation, pp. 579–596. Reproduced by permission of Routledge, Inc., part of The Taylor & Francis Group).

hydroelectric power generation. Hydropower supplies about 2.3% of the world's total energy and 18% of the world's electricity. In Latin America it meets over 50% of electrical needs. The African drought of 1991 to 1992 led to a significant decrease in hydropower, including losses of 30% from the Kariba Dam that supplies power to Zambia and Zimbabwe.

Climate change will also affect biomass (trees or other vegetation) energy, which currently provides 11% of global energy (IEA 1998). In sub-Saharan Africa in 1990, biomass fuel (mostly wood) provided 53% of the total energy (in Sudan and Ethiopia it was 80% and 90%, respectively). If climate change, as expected, decreases rainfall in North Central Africa, forests will suffer from drought and fuel wood will become scarce. The poor will be most vulnerable to reductions in the fuel wood supply. Biomass is currently used or proposed as a future source of energy for vehicles. Substitution of gasoline with alcohol derived from agricultural products such as maize represents a viable renewable energy source and has been quite successful in Brazil. Climate change could impact grain production (Chapter 7) and this, in turn, would affect alcohol fuel production and cost.

Environmental Quality

Global warming will add to environmental quality and resource depletion problems. If, as predicted, the lower atmosphere becomes more stratified and temperature inversions become more frequent, the atmosphere will be less mixed. Summertime smog, the photochemical buildup of nitrogen oxides (NO_x) and tropospheric ozone (O_3), will increase, adding to human health problems. Also, negative effects of ozone and/or acidic deposition on North American forests will increase at higher temperatures (McLaughlin and Percy 1999). Some trees, already sensitive to urban ozone pollution, could face even greater stress when ozone formation is enhanced by increased temperature and moisture from greenhouse warming (McLaughlin and Downing 1995). In regions where climate change decreases precipitation, the incidence and intensity of brush and forest fires will increase.

If, as models suggest, climate change increases the need for electricity, and additional power capacity is required, then negative environmental impacts could include the following:

- decreased air quality (e.g. additional emissions of sulfur dioxide, nitrogen oxides, and other pollutants);
- increased land use for new power plant sites, fuel extraction and storage, and solid waste disposal;
- increased water demand (e.g. for power plant cooling and fuel processing);
- resource depletion of nonrenewable fuels such as natural gas.

On the other hand, reduction in fossil-fuel combustion, if achieved, would have the added benefit of avoiding the costs of these environmental impacts.

Extreme Climatic Events

Climate model projections for the twenty-first century predict more extreme high temperatures, fewer extreme low temperatures, a reduced diurnal temperature range, increased intensity of precipitation events, and midcontinental reductions in soil moisture. Changes in the frequency of occurrence and intensity of temperature and precipitation extremes can be more important than changing long-term averages. During the twentieth century, the incidence of climate extremes changed significantly (Easterling et al. 2000). Some areas experienced fewer days of extreme cold or precipitation, and some more days of extreme heat or precipitation, but these trends were not universal. One widespread trend was an increase in the average minimum temperature and a decrease in the number of frost days. Also, one-day and multiday heavy precipitation events increased in many regions. In Europe, the warming trend between 1946 and 1999 was generally accompanied by a slight increase in wet extremes (ECAP 2002). In the United Kingdom, heavy precipitation events increased in the winter and decreased in the summer. Elsewhere, for example, the Sahel region of Africa and China, the frequency and area affected by drought increased. Worldwide economic losses from storms increased greatly during the twentieth century. However, these growing losses result more from population growth and demographic shifts to more storm-prone locations rather than from greater storm intensities.

Natural systems are vulnerable to increases in climate extremes and the occurrence of climatic disturbances. The development or life cycle of numerous organisms will be affected by climate change (Box 6.2). For example, the maximum temperature experienced during embryonic development determines the adult sex of many turtle species. In Britain and Scandinavia, populations of birds, amphibians, and deer are affected by the periodicity and severity of the North Atlantic Oscillation (Box 8.2).

Some climate models suggest that a warmer atmosphere and ocean will add momentum to the sea–air exchange of energy and, thus, increase the frequency of tropical cyclones, thunderstorms, tornadoes, hailstorms, droughts, and wildfires. For example, a doubling of CO_2 may increase the intensity of tropical hurricanes and cyclones by as much as 40% (Emanuel 1987). In areas like the Caribbean, the economic impacts of such storms could increase significantly. These events could be made even more severe in coastal areas by rising sea level.

Some models suggest an increase in the intensity of El Niño events, but others do not. Global models generally show little agreement when predicting changes in the frequency or location of midlatitude storms or tropical cyclones. However, more high-resolution regional models do suggest increases in tropical cyclone intensities. A review of numerous observed and predicted climate extremes (Easterling et al. 2000) indicates the probability of changes during this century, which include the following:

- *very likely* – higher maximum temperatures, more hot summer days, increase in the heat index (combined temperature and humidity), higher minimum temperatures, more heavy one-day and multiday precipitation events, more heat waves, droughts, and reduced soil moisture at midlatitudes;
- *likely* – higher minimum temperatures and fewer frost days;
- *possible* – more intense tropical storms and El Niño events.

Heavy precipitation events can cause flooding, erosion, and in mountainous regions, mudslides. Model predictions suggest that, for many temperate countries, heavy summer precipitation events will increase by about 20% during this century – nearly four times the average overall precipitation increase (Groisman et al. 1999). Increased precipitation and runoff in some regions will lead to an increased frequency and/or intensity of flooding with consequent economic costs. For example, model studies of New South Wales, Australia, predict that a CO_2 doubling will result in an increase in extremely heavy rainfall events by a factor of 2 to 4. If the 1-in-400 year flood becomes the 1-in-100 year flood, combined residential and commercial damage in New South Wales would increase by \$54 million to \$313 million and the number of residences flooded would double to quadruple (Smith and Hennessy 2002). In mountain regions people often live in settlements on steep and potentially unstable hillsides. Landslides from extreme rainfall events can have devastating consequences.

Extreme climate events could significantly increase property insurance costs (Baker 2002). Insurance companies are among the world's largest investors and have extensive real-state holdings. In 1992, total financial losses from weather-related disasters cost the insurance industry a record \$23 billion. With climate change, costs to insure against extreme events and sea-level rise will surely escalate. Some insurance companies, in response to climate-change forecasts, are decreasing their investments in coastal real estate, wildfire areas like Southern California, and flood-prone valleys. Eighty insurers from around the world signed the 2001 United Nations Environment Program, "Statement of Environmental Commitment by the Insurance Industry" concerning global change and the environment. Skeptics argue that mitigative policy responses should wait until we are more certain about the reality of climate change. However, disaster insurers believe that these uncertainties make a compelling argument for taking climate change seriously. Without insurance, taxpayers and/or individuals will be called upon to pay the price of climate-induced disasters.

Human Settlements

Climate change will alter regional agricultural and industrial potential and could trigger large-scale migrations and redistributions of people. Such population displacements can result in serious socioeconomic disruptions, negative health impacts, and increased human suffering (similar to the mass movement of war-related refugees). The lifestyle of most human populations is adapted to a very narrow range of climatic conditions. Human settlements generally concentrate in areas of high industrial or agricultural potential, that is, areas with hospitable climates, near coastlines, in river and lake basins, or close to major transportation routes.

Living patterns and technologies of particular populations have evolved to cope with occasional storms or disasters or slow natural climate change, but not with rapid climate change. Tens of thousands of years ago, North Africa contained numerous large lakeside human settlements. As the climate changed after the last glaciation, these populations gradually adapted by assuming a nomadic way of life. However, a more rapid "desertification" of some semitropical regions due to climate change (Chapter 6) will not provide time for such a gradual adaptation. Desertification from tree removal, overgrazing, and other detrimental land-use practices is already a global crisis. In sub-Saharan Africa, millions suffer from frequent drought-related crop failures. Climate change will probably accelerate this crisis. In other regions, as increased precipitation results in flooding, populations will need to abandon long-inhabited floodplains or construct expensive dam or levee systems.

Climate change, according to most scenarios, will place added demands on urban infrastructures (Box 9.1). Urban populations in much of the world are already experiencing explosive growth. Climate change could accelerate urbanization, as people migrate away from low-lying coastal to interior areas or from drought-stricken farms to cities (IPCC 1990). Unabated, sea-level rise will have devastating consequences for densely populated river delta areas in Egypt, India, Bangladesh, and elsewhere. Inhabitants will need to migrate to mainland interior areas to escape flooding. For example, a 1-m sea-level

Box 9.1 Modeling infrastructure effects

The US EPA sponsors research on regional climate change in the United States, including effects on urban infrastructures. One project examines the future evolution of infrastructure in the Boston Metropolitan Area. The project, Climate's Long-Term Impacts on Metro Boston (CLIMB), attempts to document the present infrastructure and determine how climate change will affect such things as flooding and drainage, water supply, water quality, built environment, energy, transportation, communication, and public health. A research team is formulating a computer model to provide a tool for scenario analysis and policy development. The model examines effects over the period 2000 to 2100 and calculates three types of costs associated with climate-induced changes in infrastructure systems and services: loss of service cost, repair or replacement cost, and adaptation cost (CLIMB Project 2002). Preliminary results suggest a number of significant effects of climate change on infrastructure.

Another study, the Metropolitan East Coast Assessment (MEC), covers areas of New York, Connecticut, and New Jersey with a regional population of 19.6 million. The MEC Project focuses on seven sectors: wetlands, transportation, infrastructure, water supply, public health, energy, and land-use decision-making (MEC 2002).

rise would seriously affect nearly a hundred million people along the coasts of China alone (Han 1989). In some sea-level rise scenarios, low-lying island nations virtually cease to exist (Chapter 8). Migrating populations would create infrastructure problems for regions suddenly faced with large numbers of "climate-change immigrants." Additional infrastructure requirements would include more housing, medical facilities, and other essential urban services.

Infrastructure

Transportation

Climate change will impact the transportation sector in a number of ways. Industrial or agricultural relocations in response to climate change will require additional investments in transportation – from new highways or rail links to new shipping ports. In many developing countries the percentage of paved roads is very low (the average for 15 different African countries is only 23%). In areas of predicted precipitation increases, landslides and road erosion will raise maintenance costs. Also, long-distance power and pipelines will be threatened in areas of increased precipitation owing to slope instability and landslides, and in arctic regions by warming and melting of the supporting permafrost (Nelson et al. 2001).

Ship and barge transport are affected by climate. Regional decreases in precipitation and runoff could reduce transport and navigation on rivers. For example, during the US drought of 1988, low water in the Mississippi River impeded barge traffic for several months. The reverse can also be true; increased flooding can lead to more siltation and impede river navigation. Marine ports in high-latitude areas, such as the oil fields of Prudhoe Bay Alaska or Siberia, may benefit from a longer ice-free season following global warming. In the US Great Lakes, models suggest a longer ice-free shipping period. However, at the same time, lower precipitation and runoff will lead to lower lake levels and increased costs for dredging of Great Lakes ports (Smith 1990).

The growing number of automobiles in the world, an expected increase from 400 million to over 1 billion by the year 2030, will greatly add to GHG emissions, unless transportation alternatives are adopted soon. The average car emits 50 to 80 tons of CO_2 over its full life, and the transportation infrastructure is responsible for a large percentage of CO_2 emissions (35% in the United States and about 27% globally). The quantity of these emissions per vehicle is a function of fuel efficiency. Automobile gasoline efficiency will need to increase to 25 km L^{-1} (60 miles $gallon^{-1}$) by 2030 to maintain present fuel consumption rates (Sierra Club 2002). Individual choices and government policies regarding transportation options have significant impacts on GHG emissions and resultant climate change. For example, a person traveling in a single occupant vehicle emits over five times the quantity of CO_2 per distance traveled as a passenger using a train or a bus (Figure 9.4).

Automobile commuting relates to urban planning and human settlement patterns. In the United States and Australia, decades of town planning have led to suburban sprawl and creation of the "Car City" where high intensities of auto commuting result in high levels of GHG emissions. In contrast, cities in Europe and Asia have generally remained more compact, and with greater public transport, more energy-efficient. Generally, commuters tend to use automobiles less frequently when there are more people and jobs per area (Figure 9.5). However, some city planners in the United States and Australia are taking a new approach – designing urban centers that

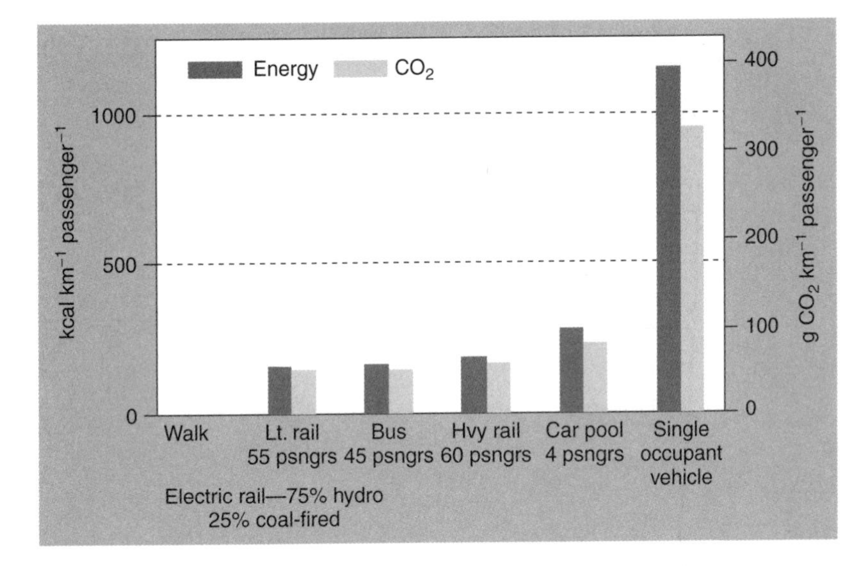

Fig. 9.4 Public transport options, settlement patterns, and transportation choices all influence CO_2 emissions from the transportation sector (From Watson RH, Byers R and Lesser J 1991. Energy efficiency – a "no regrets" response to global climate change for Washington State. *The Northwest Environmental Journal* **7**: 309–328).

incorporate the living and working environments into a mixed use area that facilitates walking, cycling, and public transport such as light rail (Stocker and Newman 1996).

Greenhouse warming represents a serious threat to the infrastructure over widespread areas of Northern Canada, Alaska, and Siberia. In high-latitude areas, such as Northern Canada and Siberia, winter transport is over roads on ice. Climate warming could substantially decrease the length of the ice road season (Lonergan et al. 1993). Increased thawing of arctic and subarctic permafrost will threaten the stability of structures and force changes in construction practices. For example, permafrost covers about 18% of the Tibet Plateau and Inner Mongolia regions of China. A 2 °C warming over a 10- to 20-year period would thaw 40 to 50% of the permafrost (IPCC 1990). Areas at high risk from melting permafrost and subsidence extend discontinuously around the Arctic Ocean. They include population centers (Barrow, Inuvik), river terminals on the Arctic Coast of Russia (Salekhard, Igarka, Dudinka, Tiksi), natural gas production complexes in Northwest Siberia, the trans-Siberian railway, the Bilibano nuclear power station in the Russian Far East, and pipeline corridors across Northwestern North America (Figure 9.6).

Industry

Many industries, from basic manufacturing to consumer goods and services, will be affected by climate change. Energy-intensive industries such as steel, aluminum, and cement production will be negatively impacted by climate changes that reduce power production or increase the cost of power (above). Likewise, agro-industry in developed countries requires large amounts of power and is sensitive to the supply and cost of energy. Many less-developed countries, heavily dependent on

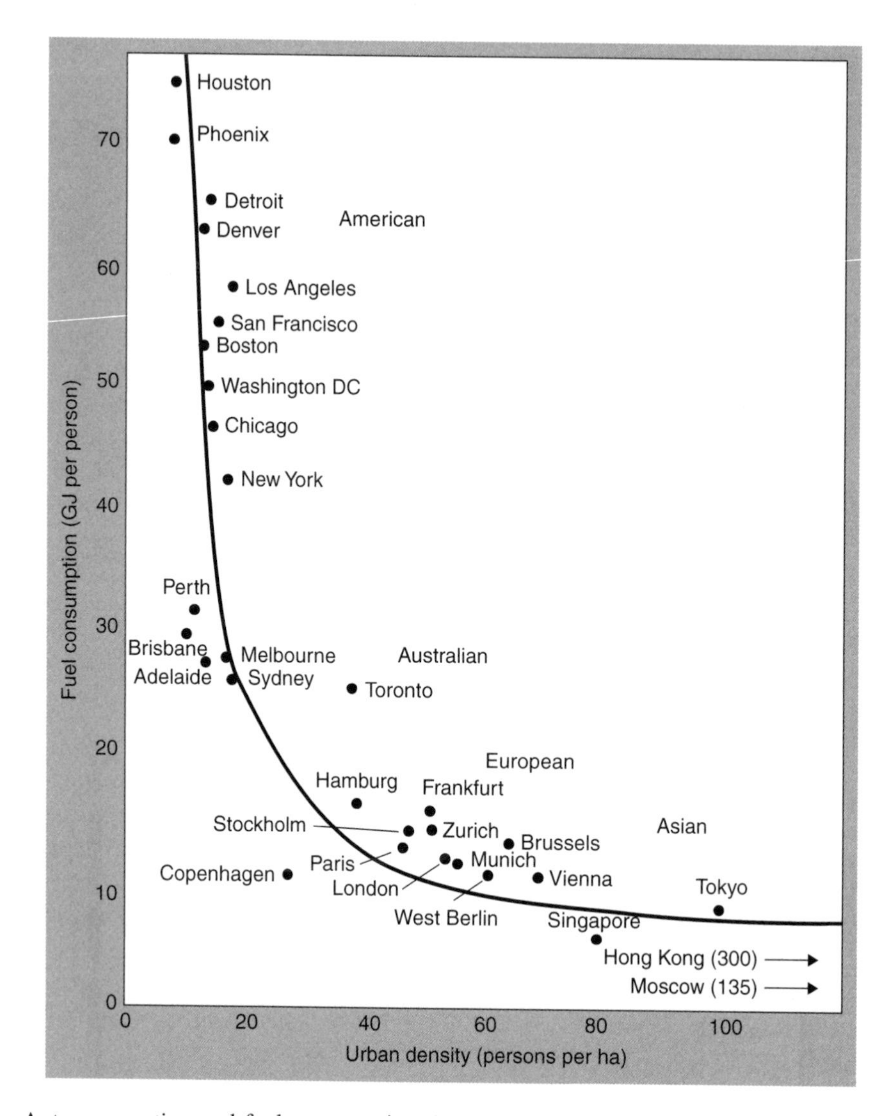

Fig. 9.5 Auto commuting and fuel consumption decreases with increasing urban population density (Reproduced from Stocker L and Newman P 1996. Urban design and transport options: strategies to decrease greenhouse gas emissions. In: Bouma WJ, Pearman GI and Manning MR, eds *Greenhouse: Coping With Climate Change*. Collingwood, VIC 3006, Australia: CSIRO Publishing, by permission of CSIRO Publishing).

subsistence food and fiber production, are particularly vulnerable to climate change. In contrast, climate change may have little direct effect on food supplies in oil-rich countries of the Middle East, where domestic agricultural production is small (Pilifosova 1998).

Industries, in some areas, will need to relocate to avoid flooding from sea-level rise or drought-induced water shortages. In high latitudes, such as Canada and Siberia, climate warming could lead to an increase in agriculture, forestry, mining, industry, and human

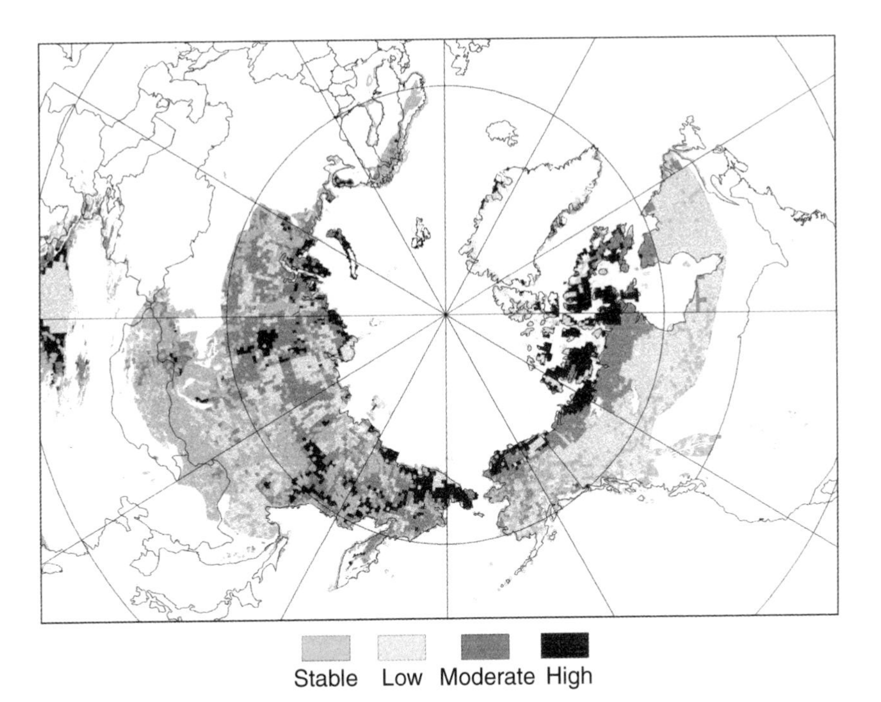

Fig. 9.6 Permafrost hazard potential in the Northern Hemisphere (polar projection) (Reprinted with permission from Nelson FE, Anisimov OA and Shiklomanov NI 2001. Subsidence risk from thawing permafrost. *Nature* **410**: 889–890. Copyright (2001) Macmillan Magazines Limited).

settlement. Additional infrastructure, in the form of ports, roads, railroads, and airports, would be required.

Tourism, important to the economies of many countries, will be impacted by climate change (Viner and Agnew 1999). Impacts will include more frequent periods of extreme heat in many eastern Mediterranean resorts and loss of beaches or reefs from sea-level rise. In the United States, tourism in New England's White Mountain region is related to the colorful fall foliage season, ski season, and recreational fishing – all potentially affected by climate change. In many regions of the world, decreasing snowfall could negatively impact winter recreation. Studies of ski resorts suggest a shortening of the ski season and a significant drop in income (Figure 9.7). For example, a 4 to 5 °C increase throughout the ski season in Southern Quebec would lead to a 50 to 70% decrease in the number of ski days (Lamother and Périard 1988). In contrast, the summer recreation season in some temperate or boreal areas could become longer.

Ecotourism is a growth industry. However, ecosystems such as coral reefs, tropical rainforests, and wildlife are already threatened by human activities. The additional stress of anthropogenic climate change, in some regions, could lead to extinction of such valuable ecosystems. Forest preserves, wildlife protected areas, and wetlands are often part of a fragmented landscape surrounded by human settlements or agriculture. As climate changes, wildlife inhabiting such isolated preserves, without corridors to other natural areas, will be unable to migrate and will either adapt to the new climate or die. For

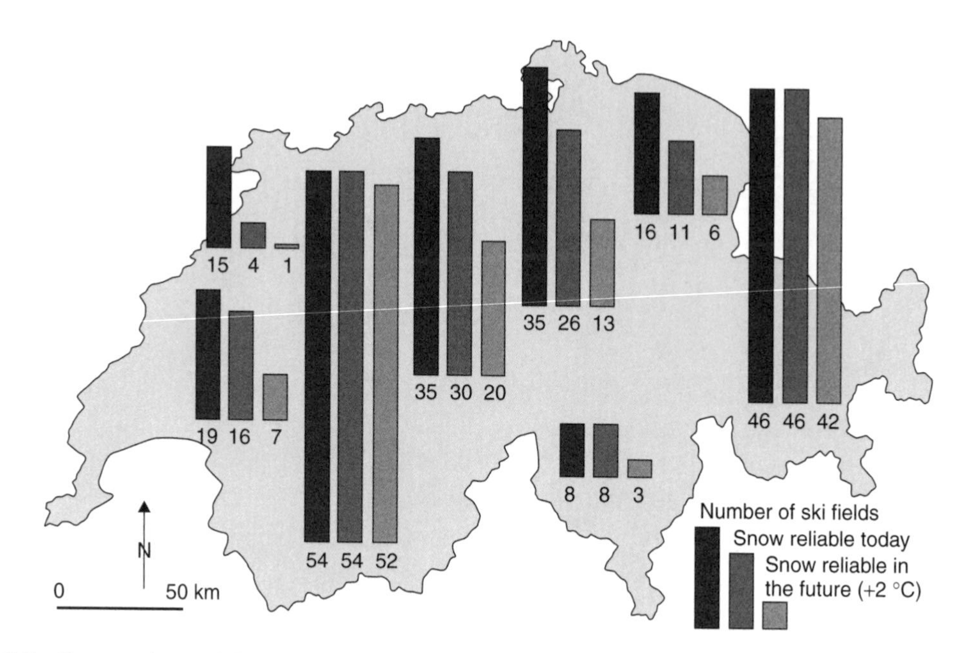

Fig. 9.7 Present-day and future snow conditions in ski resorts of Switzerland. The limit for safe snow conditions is set at 100 days with a snow depth >30 cm (Reproduced from Abegg B and Elsasser H 1996. Klima, wetter und tourismus in den Schweizer Alpen. *Geographische Rundschau* **12**: 737–742).

example, climate models of the wildlife-rich African savanna predict reduced precipitation and runoff (Hulme 1996). Water shortages and resultant wildlife losses there could translate into significant tourist dollar losses.

Coral reefs and beaches provide the base of important tourist economies in many countries. For example, in the Commonwealth of the Bahamas, tourism employs 50% of the labor force directly and 25% in related services. It accounts for more than 50% of government revenues and earns $1.3 billion in foreign exchange. Sea-level rise and beach erosion or coral reef mortality due to warmer ocean temperatures or increased CO_2 (Chapter 8) could have serious economic impacts.

Climate change presents new opportunities for some industries. To stabilize climate, new carbon-free and low-impact energy technologies need to be developed. Manufacturers of solar energy systems, fuel cells, wind generators, and other forms of low GHG technologies should benefit. In fact, some authors suggest a global research and development effort in alternative fuels on a massive scale, and with the urgency of the Manhattan atom bomb development project or the Apollo space program of the last century (Hoffert et al. 1998).

The relative impact of climate change on human infrastructure will vary by region. Compared to affluent developed countries, developing countries have less ability to adapt technologically to climate change. Also, developing countries, because of their high dependence on climate-sensitive industries such as forestry, agriculture, and fisheries, are more economically vulnerable to climate change. Nomadic peoples of the Arctic and Middle East who rely on the environment for their subsistence may be especially vulnerable. In the United States, while Cleveland, Ohio, may expect to save $4.5 million a year in reduced snow and ice removal costs, Miami, Florida, may need to invest $600 million during this century to deal with problems of sea-level rise (Miller 1989). Venice, Italy, is again reviewing a proposed $2 billion floodgate system, as well as other measures, to factor in sea-level rise (Chapter 8). Also, urban areas, because of pavement cover, buildings, and air pollution, suffer from the "urban heat island effect." Cities such as Shanghai, China, can experience temperatures of 5 °C or more greater than the surrounding countryside. Climate change may enhance this effect and add to potential health problems related to heat stress (Chapter 10).

Summary

Climate will probably lead to shifts in regional patterns of energy use. These include some reduction in space-heating requirements in temperate regions and increased energy requirements for air conditioning and irrigation in less temperate regions. Additional energy consumption could further compromise environmental quality. To avoid a dangerous anthropogenic interference with the climate system, a massive shift to carbon-free power systems must be accomplished.

Human populations are heavily concentrated in urban centers and more than 80% live within 300 km of coastlines. Models predict increases in extreme climatic events, such as windstorms, heavy precipitation and flooding, heat waves, and droughts. This could result in even greater impacts on human settlement patterns and infrastructure with serious economic consequences. Settlement patterns will probably shift in some regions in response to decreasing rainfall (desertification) and in other regions in response to increasing rainfall (flooding). Millions living near low-lying coastal areas will be displaced if the projected 100-year rise in global sea level is not averted.

Transportation depends on highways, bridges, and rivers. Shifting settlement patterns, more frequent extreme weather events, or melting permafrost could challenge transport systems. Industries, especially those heavily dependent on resources or the environment, such as agriculture, forestry, fisheries, and tourism (often the case in developing countries), are especially vulnerable to climate change. Many industries, including electric utilities, insurance, resource harvesting (forestry and fisheries), and tourism are analyzing and planning their possible options for adapting to or reducing their vulnerability to climate change.

References

Abegg B and Elsasser H 1996 Klima, wetter und tourismus in den Schweizer Alpen. *Geographische Rundschau* **12**: 737–742.

Baker JD 2002 Climate change and the re-insurance industry. Presentation to the Reinsurance Association of America, May 24, 2000. Available from: *http://www.noaa.gov/baker/reinsure/speech1.html.*

CLIMB Project 2002 Available from: *http://www.tufts.edu/tie/climb/.*

Easterling DR, Meehl GA, Parmesan C, Changnon SA, Karl TR and Mearns LO 2000 Climate extremes: observations, modeling, and impacts. *Science* **289**: 2068–2074.

Emanuel KA 1987 The dependence of hurricane intensity on climate. *Nature* **326**: 483–485.

ECAP 2002 Available from: *http://www.knmi.nl/voorl/.*

Gore A 1993 Chapter 3, Climate and civilization, *Earth in the Balance: Ecology and the Human Spirit*. New York: Plume Publishing, pp. 56–80.

Groisman PY, Karl TR, Easterling DR, Knight RW, Jamason PF, Hennessay KJ, et al. 1999 Changes in the probability of heavy precipitation: important indicators of climatic change. *Climatic Change* **42**: 243–283.

Han M 1989 Global warming induced sea level rise in China: response and strategies. *Presentation to World Conference on Preparing for Climate Change* December 19, Cairo, Egypt.

Hoffert MI, Caldeira K, Jain AK, Haites EF, Danny Harvey LD, Potter SD, et al. 1998 Energy implications of future stabilization of atmospheric CO_2 content. *Nature* **395**: 881–884.

Hulme M 1996 Chapter 5, Climatic change within the period of meteorological records. In: Adams WM, Goudie AS and Orme AR, eds *The Physical Geography of Africa*. Oxford: Oxford University Press, pp. 88–102.

IEA 1998 *Biomass Energy: Data, Analysis, and Trends*. Paris: International Energy Agency, p. 339, *http://www.iea.org/index.html*.

IPCC 1990 *Potential impacts of climate change: human settlement; energy, transport and industrial sectors; human health; air quality, and changes in ultraviolet-B radiation*. Intergovernmental Panel on Climate Change, Working Group II (A. Izrael, Chair). Geneva WMO/UNEP, p. 34.

Lamother AM and Périard G 1988 *Implications of climate change for downhill skiing in Quebec. Climate Change Digest: CCD 88-03. Downsview*. Ontario: Environment Canada, p. 12.

Linder KP 1990 National impacts of climate change on electric utilities. In: Smith JB and Tirpak DA, eds *The Potential Effects of Global Climate Change on the United States*. New York: Hemisphere Publishing Corporation, pp. 579–596.

Lonergan S, DiFrancesco R and Woo M 1993 Climate change and transportation in Northern Canada: an integrated impact assessment. *Climatic Change* **24**(4): 331–351.

McLaughlin SB and Downing DJ 1995 Interactive effects of ambient ozone and climate measured on growth of mature forest trees. *Nature* **374**: 252–254.

McLaughlin S and Percy K 1999 Forest health in North America: some perspectives on actual and potential roles of climate and air pollution. *Water and Soil Pollution* **116**(1–2): 151–197.

MEC 2002 Metropolitan East Coast Assessment. New York: Center for International Earth Science Information Network (CIESIN) Columbia University, Available from: *http://metroeast_climate.ciesin.columbia.edu/*.

Miller TR 1989 Impacts of global climate change on metropolitan infrastructure. In: Topping Jr JC, ed. *Coping With Climate Change. Proceedings of the Second North American Conference on Preparing for Climate Change*. Washington, DC: Climate Institute, 6–8 December, pp. 366–376.

Moreno RA and Skea J 1996 Industry, Energy, and Transportation: Impacts and Adaptation, Chapter 11. In: Houghton JT, Meira LG, Callander BA, Harris N, Kattenberg A, Maskell K, eds *Climate Change 1995: The Science of Climate Change*. Intergovernmental Panel on Climate Change. Cambridge: Cambridge University Press, pp. 365–398.

Nelson FE, Anisimov OA and Shiklomanov NI 2001 Subsidence risk from thawing permafrost. *Nature* **410**: 889,890.

Pearson PJG and Fouquet R 1996 Energy efficiency, economic efficiency and future CO_2 emissions from the developing world. *The Energy Journal* **17**(4): 135–159.

Pilifosova O 1998 Middle East and Arid Asia. Chapter 7. In: Watson RT, Zinyowera MC and Moss RH, eds *The Regional Impacts of Climate Change: An Assessment of Vulnerability. Intergovernmental Panel on Climate Change*. Cambridge: Cambridge University Press, pp. 231–252.

Rosenthal DH, Gruenspecht HK and Moran E 1995 Effects of global warming on energy use for space heating and cooling in the United States. *Energy Journal* **16**(2): 77–96.

Sierra Club 2002 Available from: *http://www.sierraclub.org/global-warming*.

Smith DI and Hennessy K 2002 Climate change, flooding and urban infrastructure. Australian National University and CSIRO Atmospheric Research. Available from: *http://dar.csiro.au/res/cm/c7.htm*.

Smith J 1990 Great lakes. In: Smith JB and Tirpak DA, eds *The Potential Effects of Global Climate Change on the United States*. US EPA, Washington, DC: Hemisphere Publishing Corporation, pp. 121–182.

Stocker L and Newman P 1996 Urban design and transport options: strategies to decrease greenhouse gas emissions. In: Bouma WJ, Pearman GI and Manning MR, eds *Greenhouse: Coping With Climate Change*. Collingwood, VIC 3006, Australia: CSIRO Publishing, pp. 520–537.

UNFCCC 2002 Available from: *http://unfccc.int/*.

Viner D and Agnew M 1999 Climate change and its impact on tourism. Report prepared for the World Wildlife Fund. UK: Climate Change Research Unit, University of East Anglia, p. 50.

Watson RH, Byers R and Lesser J 1991 Energy efficiency – a "no regrets" response to global climate change for Washington State. *The Northwest Environmental Journal* **7**: 309–328.

Chapter 10

Effects of Climate Change on Human Health

"... climate change is likely to have wide-ranging and mostly adverse impacts on human health, with significant loss of life."

Intergovernmental Panel on Climate Change 1995

Introduction

In the summer of 1995, a heat wave struck the Eastern and Midwestern United States, leaving more than 500 dead in Chicago alone. Clearly, extreme heat can kill humans, but also virtually every aspect of predicted climate change has implications for human health. Changes in temperature and precipitation, sea level, fisheries, agriculture, natural ecosystems, and air quality will all directly or indirectly affect human morbidity (illness) or mortality (Figure 10.1). Climate change could negatively impact human health in economically developed countries in North America (US EPA 2002a) and Europe (Kovats et al. 1999). But in general, these countries should have sufficient resources to reduce climate-change impacts on human health (Balbus and Wilson 2000). On the other hand, the less-developed poorer nations with very rapid population growth, poverty, poor health care, economic dependency, and isolation will be quite vulnerable to the human health effects of climate change (Githeko et al. 2000, Woodward et al. 1998).

Climate change can directly affect health because high temperatures place an added stress on human physiology. Changes in temperature and precipitation including extreme weather events and storms can cause deaths directly, or by altering the environment, result in an increased incidence of infectious diseases. Air pollution can be exacerbated by higher temperature and humidity. Finally, virtually all effects of global climate change, ranging from sea-level rise to impacts on agriculture and human infrastructures are linked at least indirectly to human health.

Climate Change: Causes, Effects, and Solutions John T. Hardy
 ISBNs: 0-470-85018-3 (HB); 0-470-85019-1 (PB)

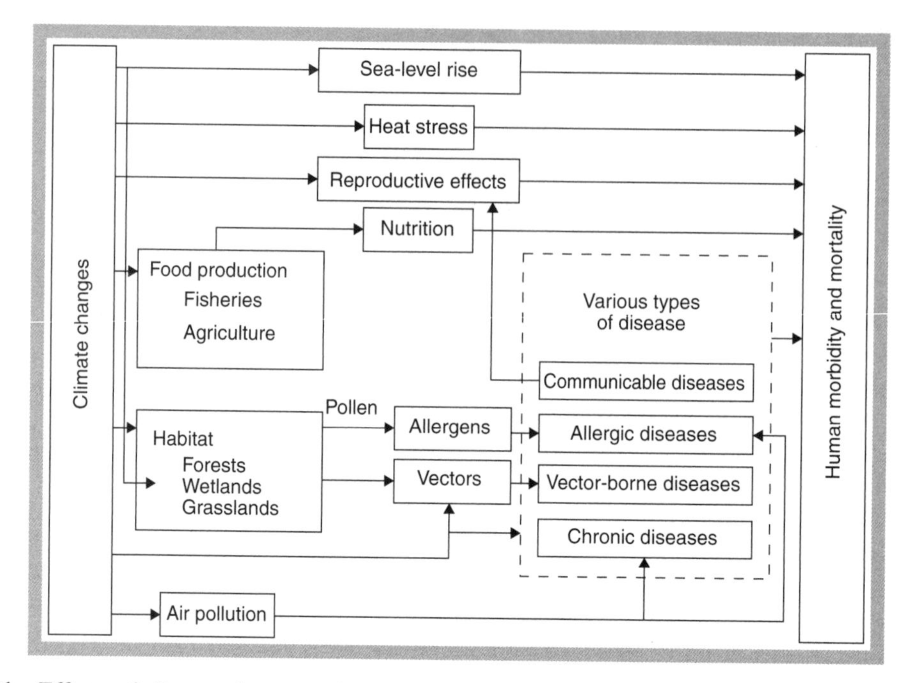

Fig. 10.1 Effects of climate change on human health (Adapted from Longstreth JA 1990. Human health. In: Smith JB and Tirpak DA, eds *The Potential Effects of Climate Change on the United States.* US EPA, Washington, DC, New York: Hemisphere Publishing Corporation, pp. 525–556. Reproduced by permission of Routledge, Inc., part of The Taylor & Francis Group).

Direct Effects of Heat Stress

The incidence and severity of many health problems increase with increasing temperature. As temperatures increase, the body expends added energy to keep cool. The most immediate consequence, if the body's temperature rises above 41 °C, is heat stroke. This disturbance to the temperature-regulating mechanism of the body results in fever, hot and dry skin, rapid pulse, and sometimes progresses to delirium and coma. Also, temperature stress can exacerbate many existing health conditions including cardiovascular and cerebrovascular disease, diabetes, chronic obstructive pulmonary disease, pneumonia, asthma, and influenza. Mortality from such diseases, especially among children and the elderly, increases dramatically during periods of unusually hot weather.

Quantitative algorithms based on historical data that relate morbidity and mortality to weather conditions suggest that global warming will increase heat-related morbidity and mortality (Listorti 1997). The incidence of mortality from heart disease and stroke increases dramatically when average daily temperatures exceed 27 °C to 30 °C (Figures 10.2a and 10.2b). Also, fetal and infant mortality is generally higher in summer and lower in winter, probably as a result of increases in infectious diseases in summer. Because of the expanses of concrete and blacktop and the resulting "heat island effect," urban areas are particularly subject to extreme temperature events called *heat waves*. For example, in 1966 the mortality rate in New York more than doubled for a brief period during a *heat wave* (Figure 10.3). A heat wave in

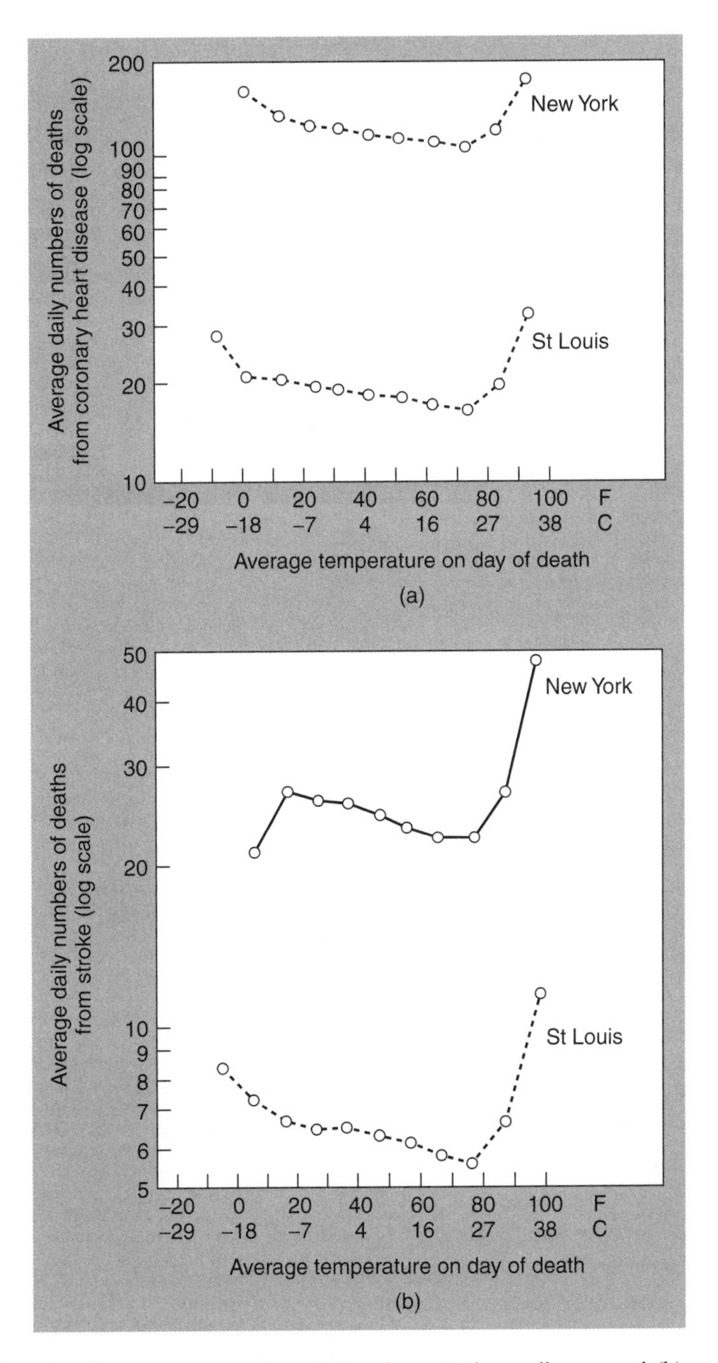

Fig. 10.2 Relationship of temperature and mortality from (a) heart disease and (b) stroke (From Rogot E and Padgett SJ 1976. Associations of coronary and stroke mortality with temperature and snowfall in selected areas of the United States 1962–1966. *American Journal of Epidemiology* **103**: 565–575, by permission of Oxford University Press).

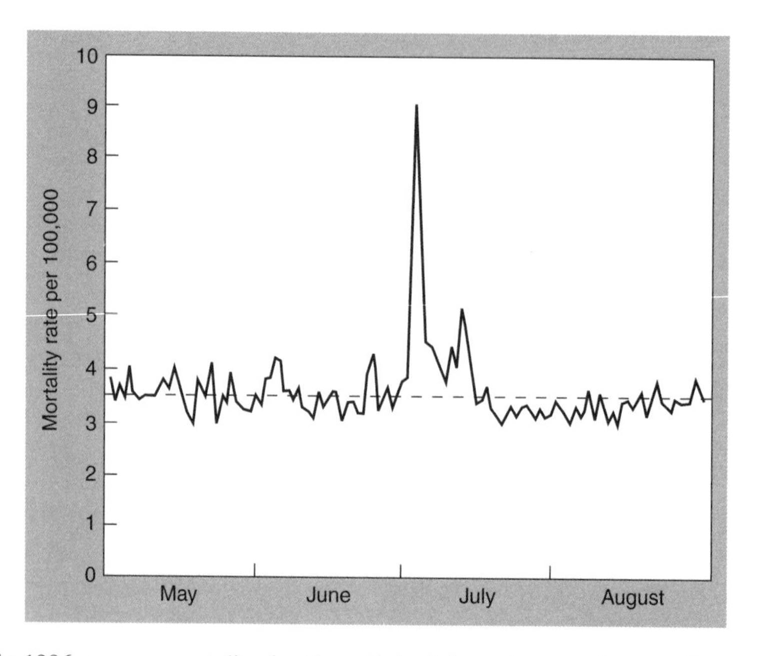

Fig. 10.3 Daily 1996 summer mortality for New York. A large increase in mortality occurred during the heat wave in early July (From Kalkstein LS 1994. Direct impacts on cities. Health and Climate Change. Special Issue. *The Lancet* 26–28).

London in 1995 was associated with a 16% increase in mortality, and one in Athens in 1987 with 2,000 additional deaths.

Acclimatization to direct heat stress is possible, at least in developed countries where investments in air conditioning, changed workplace habits, and altered home construction can all contribute to reducing heat stress. In the United States, warming due to a doubling of atmospheric CO_2 will increase heat-stress-related mortality 400% for acclimatized and 700% for unacclimatized scenarios, respectively (Figure 10.4). In developing countries where air conditioning is lacking and housing is poorly insulated or ventilated, acclimatization to higher temperatures is unlikely.

Negative impacts of heat stress will occur in both tropic and temperate regions. In temperate regions, however, warmer winters could offset some of the negative summertime heat effects. At-risk groups such as the elderly will have less risk of cold exposure, if winter months become milder in the wake of climate change. Whether milder winters in temperate regions will offset increases in summer heat-related mortality remains uncertain and will probably depend on location and specific adaptive responses. In the United States, only about 1,000 people die each year from cold weather, while twice that number die from excess heat (US EPA 2002a). By some estimates, a 2 to 2.5 °C increase in temperature in the European Union would increase heat-related summer deaths by thousands, but might also reduce winter deaths by just as much (Beniston and Tol 1998).

Infectious Diseases

Global warming represents a new stimulus to the spread of infectious diseases (Last 1997). According to Jonathon Patz of the Johns Hopkins School of Hygiene and Public Health,

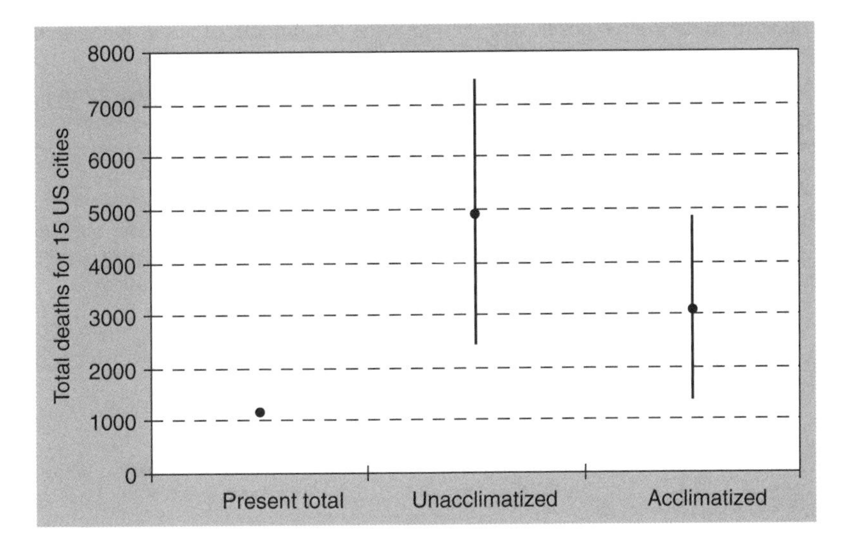

Fig. 10.4 Estimates of present and predicted (double atmospheric CO_2 warming) heat-related mortality. Total of 15 US cities. Acclimatized scenario includes some use of air conditioning, fans, and altered life habits to accommodate heat (Based on data in Kalkstein LS 1994. Direct impacts on cities. Health and Climate Change. Special Issue. *The Lancet* 26–28).

"The spread of infectious diseases will be the most important public health problem related to climate change." Infectious diseases include those spread through intermediate hosts such as rats, flies, mosquitoes, ticks, or fleas (vector-borne) and those spread directly between humans (nonvector-borne). Chronic noninfectious diseases account for the vast majority of deaths in economically developed countries, but in developing nations, climate-sensitive infectious diseases are among the leading causes of death. Many of the most widespread vector-borne diseases will probably spread even further in response to global climate change (Table 10.1). In general, diseases whose hosts are temperature-sensitive and currently restricted to the tropics (e.g. malaria, schistosomiasis, yellow fever, and dengue fever) will spread poleward into more temperate regions.

Malaria, one of the most serious global health problems, claims millions of lives each year. Its distribution is sensitive to climate conditions. It was once common in many parts of Europe, but there, as in many other developed regions, insecticides and improved public health practices have rendered it uncommon. However, cool temperatures currently limit malaria to only a fraction of its potential worldwide distribution. The anopheline mosquito, a major malaria vector, thrives between 20 and 30 °C and the malaria parasite *Plasmodium spp.* develops more rapidly within the mosquito as the temperature rises. Also, increased rainfall from climate change would provide additional surface water breeding points for the malaria mosquito.

Diseases such as malaria are closely linked to climate conditions. Rwanda, Africa, is close to the altitude and latitude range of the *Anopheles* mosquito. In 1987, the El Niño in the Pacific (Box 8.1) was responsible for record high temperatures and rainfall in Rwanda. The mosquito habitat expanded,

Table 10.1 Major tropical vector-borne diseases and the likelihood of their increase in response to climate change (From Kovats RS, Menne B, McMichael AJ, Corvalan C and Bertollini R 2000. *Climate Change and Human Health: Impact and Adaptation.* WHO/SDE/OEH/00.4, Geneva and Rome: World Health Organization, p. 22).

Disease	Likelihood of increase	Vector	Present distribution	People at risk (millions)
Malaria	+++	Mosquito	Tropics/subtropics	2,020
Schistosomiasis	++	Water snail	Tropics/subtropics	600
Leishmaniasis	++	Phlebotomine sandfly	Asia/Southern Europe/Africa /Americas	350
American trypanosomiasis (change disease)	+	Triatomine bug	Central and South America	100
African trypanosomiasis (sleeping sickness)	+	Tsetse fly	Tropical Africa	55
Lymphatic filariasis	+	Mosquito	Tropics/subtropics	1,100
Dengue fever	++	Mosquito	All tropical countries	2,500–3,000
Onchocerciasis (river blindness)	+	Black fly	Africa/Latin America	120
Yellow fever	+	Mosquito	Tropical South America and Africa	–
Dracunculiasis (guinea worm)	?	Crustacean (copepod)	South Asia/Arabian peninsula/Central West Africa	100

Note: +++ = highly likely; ++ = very likely; + = likely; ? = unknown.

and the incidence of malaria rose significantly compared to previous years (Lovinsohn 1994). In Pakistan, both temperatures and malarial occurrence have increased significantly since the 1980s (Bouma et al. 1994).

Different models of projected climate change all suggest an increase in the potential occurrence zone for malaria (Martens et al. 1995). For example, on the basis of the Hadley Centre HadCM3 model, the global area of medium to high risk for malaria transmission will expand significantly by 2050 (Plate 8). Seasonal malaria, the type most likely to lead to epidemics among unexposed nonimmune populations, is predicted to increase.

Malaria could become a serious problem for temperate developed countries within decades (Martin and Lefebvre 1995). Global warming will probably increase the acceptable temperate area habitat for anopheline mosquitoes hundredfold (Montague 1995). This, combined with increased air travel, could affect the spread of diseases such as malaria into currently temperate areas of the United States, Southern Europe, Australia, and elsewhere (Montague 1995). Models linking climate change with the environmental/physiological requirements of anopheline mosquitoes predict possible increases in malarial infections in Southern Europe. In fact, recent isolated cases of malaria in Italy and the Eastern United States are consistent with this hypothesis (Martens 1999).

In many areas, malaria is limited to warmer lowland elevations and excluded by the cooler temperatures of highland elevations. In theory, a small increase in temperature could open highland areas of many countries to the malaria parasite. However, no correlation seems to exist between recent climate change and malaria resurgence at high-altitude sites in East Africa (Hay et al. 2002).

Climate-change models (HadCM3) predict an additional 300 million cases of *P. falciparum* and 150 million cases of *P. vivax* types of malaria worldwide by 2080 (Martens et al. 1999). The greatest increase in potential transmission for malaria will be in Central Asia, North America, and Europe, where the vector mosquitoes are present, but where current climates are too cold for transmission. In South America, the number of persons at risk of infection from year-round malaria transmission will double from 25 million to 50 million between 2020 and 2080 (Martens et al. 1999).

The range of dengue fever will probably expand in response to climate change. Dengue fever, spread by the mosquito *Aedes aegypti*, is the most important vector-borne viral disease in the world with up to 100 million cases per year (Kovats et al. 1999). Higher temperatures increase the efficiency of *Aedes* mosquitoes in transmitting the dengue virus (Watts et al. 1987). Model predictions suggest that in response to climate change, the mosquito will expand its latitudinal distribution during the warmer months (Hopp and Foley 2001). Freezing temperatures kill the mosquito eggs, but as the minimum temperature – for overwintering its eggs and larvae – moves poleward with climate change, much of Central and Southern Europe (Ward and Burgess 1993) as well as the United States could be at risk. Climate-change models predict increases in rainfall and temperature for parts of Australia that could increase the threat of several mosquito-borne viruses, including dengue fever, by extending the season of the mosquito vectors (Russell 1998).

Increasing global travel and trade has led to the spread of disease vectors to new areas. The major vector of dengue fever, *Aedes albopictus*, was introduced into the United States from East Asia in shipments of used car tires and has now spread through much of North America. In Italy, the same mosquito has spread to 22 northern provinces since its appearance in about 1992. West Nile viral encephalitis, another sometimes deadly disease carried by *Culex* mosquitoes, and hosted by birds, broke out in France in the 1960s. Perhaps in response to a very warm summer along the east coast of North America, it appeared in New York in 1999. It is now spreading across the United States. Its link to climate change remains unclear, but certainly warmer temperatures could enhance its spread.

Schistosomiasis (Bilharzia), another vector-borne disease, infects perhaps as many as 200 million people worldwide. It is a water-borne parasitic worm whose eggs enter water supplies by way of human feces or urine. The parasitic worm and its intermediate host, a snail, thrive in warm water. In countries like Egypt, the snail population declines during the cooler months, but climate warming will extend the season and incidence of infection.

Hantavirus is spread though the saliva and feces of deer mice. In the Southwestern United States in 1993, heavy rains following a six-year drought provided an abundant food supply of pine nuts and ideal conditions for a rodent vector population explosion. This was followed by an outbreak of the deadly

Hantavirus (Stone 1995). Scientists worry that climate change could hasten the spread of similar deadly diseases (Brown 1996).

Tick-borne diseases, such as Rocky Mountain spotted fever and Lyme disease, are likely to increase in northern areas. Tick-borne encephalitis as well as Lyme disease increased in Sweden during the 1980s and 1990s, as Northern European winter temperatures became milder and the spring vegetation season advanced an average 12 days in latitudes between 45 to 70 °N (Lindgren and Gustafson 2001). However, a greater abundance of deer (hosts to the tick) could also be at least partially responsible for the spread of tick-borne diseases.

In Africa, climate change may greatly increase the habitat area for the tsetse fly, carrier of African trypanosomiasis (sleeping sickness) (Figure 10.5).

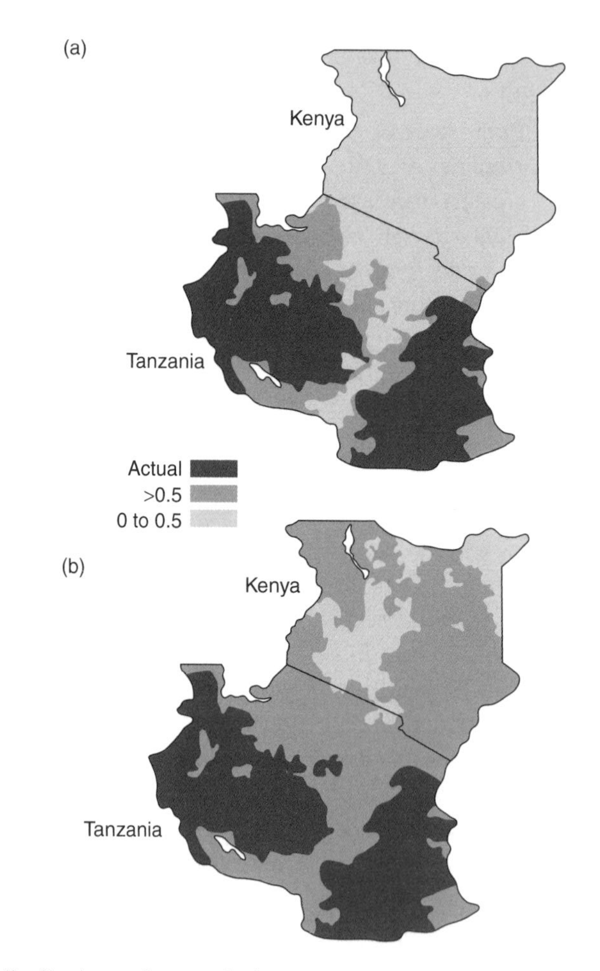

Fig. 10.5 Predicted distributions of tsetse fly in Kenya and Tanzania: (a) distribution in the absence of climate change; (b) distribution following a 3 °C rise in average temperature. Black = current distribution, shaded = high probability of occurrence, white = low probability of occurrence (Adapted from Rogers DJ and Packer MJ 1994. Vector-borne diseases, models, and global change. *The Lancet* **342**(8882): 1282. Reprinted with permission from Elsevier Science).

Some nonvector-borne diseases could also increase in response to climate change. Population explosions of certain toxin-containing marine algae, "toxic algal blooms," are increasing globally. In fact, the world seems to be experiencing a "global epidemic" of harmful coastal algal blooms (Epstein et al. 1994, Van Dolah 2000). Humans are poisoned from eating algal-contaminated shellfish or from breathing marine aerosols containing such algae. Cholera (*Vibrio cholerae*) is a waterborne disease, which in its dormant state is quite salt-tolerant. Some evidence suggests that cholera outbreaks in several coastal areas may be linked to warm-water events and associated plankton blooms that concentrate the pathogen (Colwell and Huq 1994). In 1992, a new strain of cholera emerged in coastal areas of India and began spreading in Asia. Finally, food-related infections, such as food poisoning from salmonellosis, thrive at warmer temperatures and could become more common (UKDH 2001).

Air Quality

Air pollution, an exacerbating factor for pulmonary health conditions such as asthma and cardiorespiratory disorders, will probably increase as a result of climate change. Increased energy demand (Chapter 9) will add to fossil-fuel combustion. This will increase emissions of particulates (a cause of respiratory problems), toxic aromatic hydrocarbons (carcinogens), and sulfur dioxide (acid rain). Air quality is a function not only of pollutant emissions but also of atmospheric circulation and mixing. Thus, if climate change causes increased atmospheric stratification, pollutants will tend to accumulate near the Earth's surface.

Climate change will increase ozone pollution. Increased temperature and water vapor, along with increased ultraviolet-b (UV-b) radiation from stratospheric ozone depletion, will accelerate the chemical transformations that form tropospheric ozone (smog) (Box 10.1). Ozone and the photochemical products formed, such as peroxides, are lung irritants. Global climate models predict about a 20-ppbv (parts per billion by volume) or 40% increase in the atmospheric concentration of ozone in midsummer in Central England between 1990 and 2100 (UKDH

Box 10.1 Ozone and photochemical smog formation (Adapted from US EPA 2002b. *Air Quality Planning and Standards*. Unites States Environmental Protection Agency. Available from: *http://www.epa.gov/oar/oaqps*)

Tropospheric Ozone – Tropospheric ozone, "bad ozone," is an air pollutant. It should not be confused with "good ozone," present in the stratosphere, which protects life from the damaging effects of ultraviolet radiation (Box 1.4). Motor vehicle exhaust, industrial emissions, gasoline vapors, and chemical solvents release nitrogen oxides (NO_x) and volatile organic compounds (VOC). Strong sunlight and hot weather promote the transformation of these chemicals and the formation of tropospheric ozone. Many urban areas tend to have high levels of "bad" ozone, but other areas are also subject to high ozone levels as winds carry NO_x emissions hundreds of miles away from their original sources. Ozone concentrations can vary from year to year. Changing weather patterns (especially the number of hot, sunny days), periods of air stagnation, and other factors that contribute to ozone formation make long-term predictions difficult.

Repeated exposure to ozone pollution may cause permanent damage to the lungs. Even when ozone is present in low levels, inhaling it triggers a variety of health problems, including chest pains, coughing, nausea, throat irritation, and congestion. It can also worsen bronchitis, heart disease, emphysema, and asthma, and reduce lung capacity. Healthy people also experience difficulty in breathing when exposed to ozone pollution. Because ozone pollution usually forms in hot weather, anyone who spends time outdoors in the summer may be affected, particularly children, the elderly, outdoor workers, and people exercising. Ground-level ozone damages plant life and is responsible for 500 million dollars in reduced crop production in the United States each year. It interferes with the ability of plants to produce and store food, making them more susceptible to disease, insects, other pollutants, and harsh weather. "Bad" ozone damages the foliage of trees and other plants, ruining the landscape of cities, national parks and forests, and recreation areas.

2001). In New York City, a 5 °C increase in temperature together with increased UVB would more than double the June level of hydrogen peroxide in the air. And, in the San Francisco Bay area of California, the area and number of people exposed to ozone in excess of air quality standards would greatly expand. In the Bay area, a 4 °C increase in average temperature in August would increase the ozone concentration from 15 pphm (parts per hundred million) to 18 pphm (Figure 10.6).

Hot, dry summers in some areas will contribute to wider distributions of spores and pollen, adding to asthma and allergic disorders in humans. Higher humidity and precipitation in other areas would promote populations of molds and house dust mites, whose feces are a powerful allergen (Martens 1999).

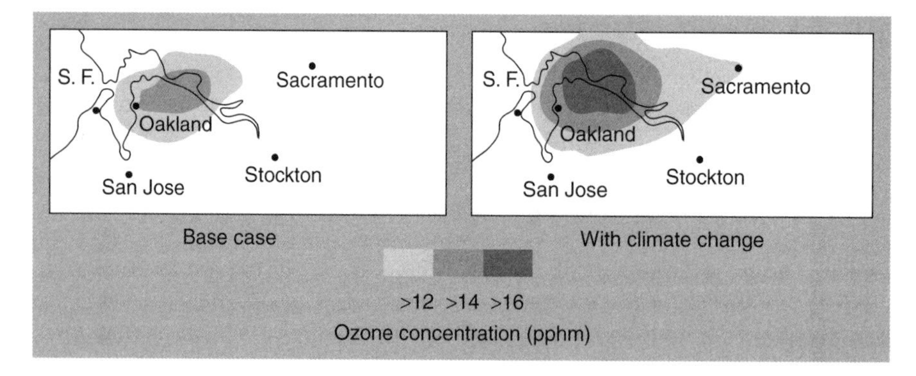

Fig. 10.6 In the San Francisco Bay area, a regional air pollutant transport model compares present-day August conditions (base case) with a climate change including a 4 °C increase in temperature and attendant increase in water vapor. Although the study includes simplifying assumptions, the trend is clear. Ozone concentrations (pphm = parts per hundred million) increased and the area where the ozone air quality was exceeded almost doubled from 3,700 to 6,600 km^2 (From Bufalini JJ, Finkelstein PL and Durman EC 1990. Impact of climate change on air quality. In: Smith, JB and Tirpak DA, eds *The Potential Effects of Global Climate Change on the United States*. New York: Hemisphere Publishing Corporation, pp. 485–524. Reproduced by permission of Routledge, Inc., part of The Taylor & Francis Group).

Reduction in greenhouse gas emissions could, in addition to reducing global warming, have many ancillary benefits for human health. Each year about 700,000 deaths worldwide result from air pollution, and the World Health Organization has ranked air pollution as one of the top 10 causes of disability. Reduction of greenhouse gas emissions will also result in reductions in associated copollutants such as particulate matter and ozone that negatively affect human health. Thus, adoption of effective greenhouse gas mitigation technologies over two decades in only four cities in the Americas (Mexico City, New York, Santiago, and Sao Paulo) could prevent about 64,000 premature deaths, 65,000 chronic bronchitis cases, and 37 million person-days of restricted activity or work loss (Cifuentes et al. 2001).

Interactions and Secondary Effects

Many of the effects of climate change discussed elsewhere in this volume have implications for human health or well-being. Several aspects of predicted climate change could indirectly place additional burdens on public health systems. Climate change will have numerous direct effects on water and food supplies, ecosystems, sea level, extreme weather events, and human infrastructure (see other chapters). In fact, the direct effects of increased heat stress and air pollution on human health will probably be outweighed by impacts resulting from complex changes in ecosystems and altered patterns of disease (McMichael and Haines 1997). More severe storms, droughts, or floods would increase human deaths and provide beneficial conditions for the spread of certain diseases. For example, during the summers of 1997 and again in 2002, record-breaking floods in Central Europe killed hundreds, forced the evacuation of hundreds of thousands, and caused billions of dollars in damage.

Malnutrition and unsafe drinking water, the two greatest risk factors enhancing disease in the developing world, could increase in response to climate change. Crop production in some regions will suffer from climate change (Chapter 7) and subsequent malnutrition could increase human susceptibility to infectious diseases. In 1990 about 1.1 billion people lacked access to safe drinking water (Kovats et al. 2000). Poor water quality accounts for less than 1% of the deaths in the developed economies, but nearly 11% in sub-Saharan Africa and 9% in India (Murray and Lopez 1996). A decrease in water availability as a result of climate change (Chapter 5) will further compromise sanitary standards and allow waterborne pathogens to spread.

Sea-level rise and flooding of coastal areas would lead to intrusion of saltwater into groundwater supplies and could interfere with coastal wastewater treatment plants. The saltwater malaria mosquito, *Anopheles sundaicus*, would be able to move further inland. Climate change could also trigger a mass movement of environmental refugees (e.g. from coastal areas flooded by sea-level rise) and place an added burden on the public health system (Chapter 8). Potential increases in extreme weather events, such as storms, floods, or drought, could lead to increased loss of human life.

Summary

Human health will suffer from many aspects of climate change (Box 10.2). Direct impacts include increasing incidences of thermal stress, leading to cardiovascular and respiratory morbidity and mortality. Indirect impacts will probably result from increases in certain vector-borne diseases (e.g. malaria,

Box 10.2 Effects of climate change on health – A case study of The United Kingdom

The UK Department of Health performed one of the most comprehensive studies of the potential impact of climate change on human health (UKDH 2001). The report acknowledges that there are considerable uncertainties relating to predictions, but they conclude on the basis of a medium to high climate-change scenario (i.e. 1% per year increase in greenhouse gas concentrations) that in the United Kingdom, by the 2050s

- cold-related winter deaths are likely to decline substantially, by perhaps 20,000 cases annually;
- heat-related summer deaths are likely to increase, by around 2,800 cases annually;
- because of warmer incubation temperatures, cases of food poisoning are likely to increase significantly, by perhaps 10,000 cases annually;
- vector-borne diseases may present local problems, but the increase in their overall impact is likely to be small;
- waterborne diseases may increase but, again, the overall impact is likely to be small;
- the risk of major disasters caused by severe winter gales and coastal flooding is likely to increase significantly;
- in general, the effects of air pollutants on health are likely to decline, but the effects of ozone, during the summer, are likely to increase: several thousand additional deaths and a similar number of hospital admissions may occur each year;
- because of stratospheric ozone depletion and increasing ultraviolet radiation, cases of skin cancer are likely to increase by perhaps 5,000 cases per year and cataracts by 2,000 cases per year;
- measures taken to reduce the rate of climate change by reducing greenhouse gas emissions could produce secondary beneficial effects on health.

dengue, and schistosomiasis), marine-borne diseases (cholera and toxic algae), decreases in food productivity (malnutrition or increased starvation), increased air pollution (asthma and cardiorespiratory disorders), and weather disasters, and sea-level rise (deaths, injuries, and infectious diseases).

The urban poor will be particularly subject to negative health effects of climate change. The percentage of the world's population living in cities is expected to increase to 60% by the latter half of this century. This trend will certainly increase the risks to human health from disease and air pollution. New investments in potable water supplies, air quality, and sanitary waste disposal will be necessary to minimize health risks.

The medical community and the World Health Organization are beginning to recognize the serious public health challenge of global warming. Many physicians and health care professionals now believe that unmitigated

Box 10.3 Physicians' statement on global climate change and human health
(Excerpts from PSR 2002. Physicians for Social Responsibility. Available from *http://www.psr.org*)

More than 1,100 individual physicians and health professionals from 25 countries, including eight Nobel Laureates in Medicine, have signed the following statement:

As physicians and health professionals, we are concerned about the potentially devastating and possible irreversible effects of climate change on human health and the environment. We urge the U.S. and other nations of the word to take prompt and effective actions – both domestically and internationally – to achieve significant reductions in greenhouse gas emissions.

There is mounting evidence that climate change, of the scale currently projected, would have pervasive adverse impacts on human health and result in significant loss of life. Potential impacts include increased mortality and illness due to heat stress and worsened air pollution, and increased incidence of vector-borne infectious diseases, such as malaria, schistosomiasis and dengue, diseases related to water supply and sanitation, and food-borne illnesses. Expanding populations of pest species, impaired food production and nutrition, and extreme weather events such as floods, droughts, forest fires and windstorms would pose additional risks to human health. Infants, children, and other vulnerable populations – especially in already-stressed regions of the world – would likely probably suffer disproportionately from these impacts.

As public health professionals who believe firmly in the wisdom of preventive action, we endorse strong policy measures to stabilize greenhouse gas concentrations in the atmosphere at a level that would prevent dangerous anthropogenic interference with the climate system, as called for in the United Nations Framework Convention on Climate Change.

The time has come for the nations of the world to act. The science is credible, and the potential impacts profound. Prudence – and a commitment to act responsibly on the behalf of the world's children and all future generations – dictate a prompt and effective response to climate change.

human-induced climate change will result in increased death and disease (Box 10.3).

References

Balbus JM and Wilson ML 2000 *Human Health and Global Climate Change: A Review of Potential Impacts in the United States*. Arlington, VA: PEW Center on Global Climate Change, p. 43 Available from: *http://www.pewclimate.org*.

Beniston M and Tol RSJ 1998 Europe. In: Watson RT, Zinyowera MC and Moss RH, eds *The Regional Impacts of Climate Change. Intergovernmental Panel on Climate Change*. Cambridge: Cambridge University Press, pp. 149–185.

Bouma MJ, Sondorp HE and van der Kaay HJ 1994 Climate change and periodic epidemic malaria (letter). *Lancet* **343**: 1440.

Brown KS 1996 Do disease cycles follow changes in weather? *Bioscience* **46**(7): 479–481.

Bufalini JJ, Finkelstein PL and Durman EC 1990 Impact of climate change on air quality. In: Smith JB and Tirpak DA, eds *The Potential Effects of Global Climate Change on the United States*. New York, NY: Hemisphere Publishing Corporation, pp. 485–524.

Cifuentes L, Borja-Aburto VH, Gouveia N, Thurston G and Davis DL 2001 Hidden health benefits of greenhouse gas mitigation. *Science* **293**: 1257–1259.

Colwell RR and Huq A 1994 Environmental reservoir of *Vibrio cholerae*. In: Wilson ME, Levins R and Spielman A, eds Disease in Evolution: Global Changes and Emergence of Infectious Diseases. *Annals of the New York Academy of Sciences* **740**: 44–54.

Epstein PR, Ford TE and Colwell RR 1994 Marine ecosystems. Health and Climate Change. Special Issue. *The Lancet* 14–17.

Githeko AK, Lindsay SW, Confalonieri UE and Patz JA 2000 Climate change and vector-borne diseases: a regional analysis. *Bulletin of the World Health Organization* **78**(9): 1136–1146.

Hay SI, Cox J, Rogers DJ, Randolph SE, Stern DI, Shanks GD, et al. 2002 Climate change and the resurgence of malaria in the East African highlands. *Nature* **415**: 905–909.

Hopp MJ and Foley JA 2001 Global-scale relationships between climate and the dengue fever vector *Aedes aegypti*. *Climatic Change* **48**: 441–463.

Kalkstein LS 1994 Direct impacts on cities. Health and Climate Change. Special Issue. *The Lancet* 26–28.

Kovats RS, Haines A, Stanwell-Smith R, Martens P, Menne B and Bertollini R 1999 Climate change and human health in Europe. *British Medical Journal* **318**: 1682–1685.

Kovats RS, Menne B, McMichael AJ, Corvalan C and Bertollini R 2000 *Climate Change and Human Health: Impact and Adaptation*. WHO/SDE/OEH/00.4, Geneva and Rome: World Health Organization, p. 22.

Last JM 1997 New causes for new diseases. *World Health* **50**: 12–13.

Lindgren E and Gustafson R 2001 Tick-borne encephalitis in Sweden and climate change. *The Lancet* **358**: 16–18.

Listorti JA 1997 Environmental health dimensions of climate change and ozone depletion. *Energy Environment Monitor* **13**: 103–120.

Longstreth JA and US EPA 1990 Human health. In: Smith JB and Tirpak DA, eds *The Potential Effects of Climate Change on the United States*. Washington, DC. New York: Hemisphere Publishing Corporation, pp. 525–556.

Lovinsohn ME 1994 Climatic warming and increased malaria incidence in Rwanda. *The Lancet* **343**: 714–718.

Martens P 1999 How will climate change affect human health? *American Scientist* **87**: 534–541.

Martens WJM, Niessen LW, Rotmans J, Jetten TH and McMichael AJ 1995 Potential impact of global climate change on malaria risk. *Environmental Health Perspectives* **103**: 458–464.

Martens P, Kovats RS, Nijhof S, de Vries P, Livermore MTJ, Bradley DJ, et al. 1999 Climate change and future populations at risk of malaria. *Global Environmental Change* **9**: S89–S107.

Martin P and Lefebvre M 1995 Malaria and climate: sensitivity of malaria potential transmission to climate. *Ambio* **24**: 200–207.

McMichael AJ and Haines A 1997 Global climate change: the potential effects on health. *British Medical Journal* **315**: 805–809.

Montague P 1995 Climate and infectious disease, Part 1. *Rachel's Environment and Health Weekly* **466**: 1–9. Available from: *http://www.monitor.net/rachel/r466.html*.

Murray CJL and Lopez AD 1996 Estimating causes of death: new methods and global and regional applications for 1990. In: Murray CJL and Lopez AD, eds *The Global Burden of Disease*. Cambridge: Harvard School of Public Health, pp. 118–200.

PSR 2002 *Physicians for Social Responsibility*. Available from: *http://www.psr.org*.

Rogers DJ and Packer MJ 1994 Vector-borne diseases, models, and global change. *The Lancet* **342**(8882): 1282.

Rogot E and Padgett SJ 1976 Associations of coronary and stroke mortality with temperature and snowfall in selected areas of the United States 1962–1966. *American Journal of Epidemiology* **103**: 565–575.

Russell RC 1998 Mosquito-borne arboviruses in Australia: the current scene and implications of climate change for human health. *International Journal of Parasitology* **28**(6): 955–969.

Stone R 1995 If the mercury soars, so may health hazards. *Science* **267**: 957,958.

UKDH 2001 *Health Effects of Climate Change in the UK: An Expert Review for Comment*. Report 22452. London: Expert Group on Climate Change and Health in the UK. UK Department of Health, p. 290 Available from: *http://www.doh.gov.uk/hef/airpol/climatechange/index.htm*.

US EPA 2002a *Global Warming*. United States Environmental Protection Agency. Available from: *http://www.epa.gov/globalwarming*.

US EPA 2002b *Air Quality Planning and Standards*. United States Environmental Protection Agency. Available from: *http://www.epa.gov/oar/oaqps*.

Van Dolah FM 2000 Marine algal toxins: origins, health effects, and their increased occurrence. *Environmental Health Perspectives* **108**: 133–141.

Ward MA and Burgess NR 1993 *Aedes albopictus*: a new disease vector for Europe? *Journal of the Royal Army Medical Corps* **139**: 109–111.

Watts DM, Burke DS, Harrison BA, Whitmire RE and Nisalak A 1987 Effect of temperature on the vector efficiency of *Aedes aegypti* for Dengue-2 virus. *American Journal of Tropical Medicine and Hygiene* **36**: 143–152.

Woodward A, Hales S and Weinstein P 1998 Climate change and human health in the Asia Pacific region: who will be most vulnerable? *Climate Research* **11**(1): 31–38.

Chapter 11

Mitigation: Reducing the Impacts

"... a prudent society hedges against potentially dangerous future outcomes, just as a prudent person buys health insurance."

Stephen Schneider
National Center for Atmospheric Research, 1988

Introduction

If actions are not taken soon to reduce (or mitigate) the impacts of human-induced climate change, the consequences could be far reaching. Even though we may be uncertain of the exact magnitude of climate change or its specific effects, the impacts on human society could range from slight to catastrophic. Caution is called for. Many scientists urge application of the "precautionary principle" or more simply stated: better safe than sorry. How can the risk of serious impacts be reduced? How can we avoid or lessen the consequences of our large profligate emission of greenhouse gases? There are several possible approaches to reduce or mitigate human-induced climate change. We could:

- capture or sequester carbon emissions,
- reduce global warming or its effects through geoengineering,
- enhance natural carbon sinks,
- convert to carbon-free and renewable energy technologies,
- conserve energy and use it more efficiently,
- adapt to climate change.

Capture or Sequester Carbon Emissions

If we could reduce CO_2 emissions at the source, we could eliminate much of the greenhouse warming potential, but this is easier said than done. Removing large quantities of CO_2 from the air is difficult. Consider, for example, the self-contained underwater rebreathing apparatus. In this system used by navy divers (and less frequently by sport divers), the diver's exhaled air is filtered through a chemical cartridge containing soda lime (mostly calcium hydroxide, with small amounts of sodium and potassium hydroxide)

Climate Change: Causes, Effects, and Solutions John T. Hardy

ISBNs: 0-470-85018-3 (HB); 0-470-85019-1 (PB)

to remove CO_2, and then the air is rebreathed. Unlike the more popular SCUBA system, no air bubbles are released into the surrounding water. Thus, chemical filters can economically remove small quantities of CO_2.

Mobile sources, that is, motor vehicles, are relatively compact and preclude addition of large bulky CO_2-absorbing filtration equipment. An average US passenger car emits about 14.3 kg (31.4 pounds) of CO_2 per day (US EPA 2000). Applying the current technology of lime $Ca(OH)_2$, as a chemical CO_2 absorber, the average passenger car would require about 8,778 kg of lime per year to absorb its CO_2 emissions. The lime would then need to be disposed of or treated to remove CO_2 before it could be reused. For mobile sources, both the volume of material and the cost are prohibitive.

For stationary sources such as power plants, carbon-sequestering technologies offer more promise of success. Today, capturing emissions at the source is not very efficient (3 to 5% in gas plants and 13 to 15% in coal plants) and the cost of doing so is high. However, new technologies are being developed (Herzog 2001). Agencies such as the US Department of Energy (US DOE 2002a) and the International Energy Agency (IEA 2002), as well as nine of the world's largest energy companies are actively researching new energy technologies, some of which are fossil-fuel–based. For example, in the integrated coal gasification combined cycle plant, coal is gasified to CO and H_2. The CO reacts with steam to form CO_2 and H_2. The CO_2 is removed and the H_2 is used in a gas turbine. The organic solvent monoethanolamine (MEA) effectively absorbs CO_2 (CO_2 Capture Project 2002).

Carbon dioxide can be stored underground in deep geologic formations or in the deep ocean. Under high pressure (e.g. in the deep ocean) one molecule of liquid CO_2 combines with seven molecules of water and forms a solid. At the Sleipner oil and gas field in the North Sea, CO_2 (a by-product of natural gas extraction) is compressed and pumped into a subsurface sandstone layer 1,000 m below the seabed.

An analysis of different carbon capture and sequestration technologies suggests that capturing 90% of the CO_2 from a power plant would add about 2 cents per kilowatt-hour to the cost of electricity, a level that competes favorably with the current costs of renewable and nuclear energy (David 2000).

Despite these promising technologies, large-scale carbon sequestration, even from stationary sources, poses a number of problems. First, the quantity of carbon to be sequestered is enormous. Gas and coal-fired electric power plants emit huge quantities of CO_2 and account for about 33% of worldwide emissions. A typical 500-MW coal-fired power plant emits (depending on efficiency) about 10,560 metric tonnes of CO_2 per day (Stultz and Kitto 1992). Second, there are environmental and safety concerns. Ocean storage involves environmental questions such as acidification of the seawater when the CO_2 is dissolved. Also, the lifetime of solid CO_2 stored in the deep sea is not known. It might form bubbles and return to the surface. Underground terrestrial storage involves safety questions. Leaking CO_2 could, in high enough concentrations, suffocate nearby animal and human populations.

Reduce Global Warming or its Effects by Geoengineering

Many proposed technical solutions to deal with human-induced warming fall under the heading of "earth systems engineering and management" or "geoengineering," that is, large-scale schemes to manipulate the Earth's climate and mitigate the effects of greenhouse

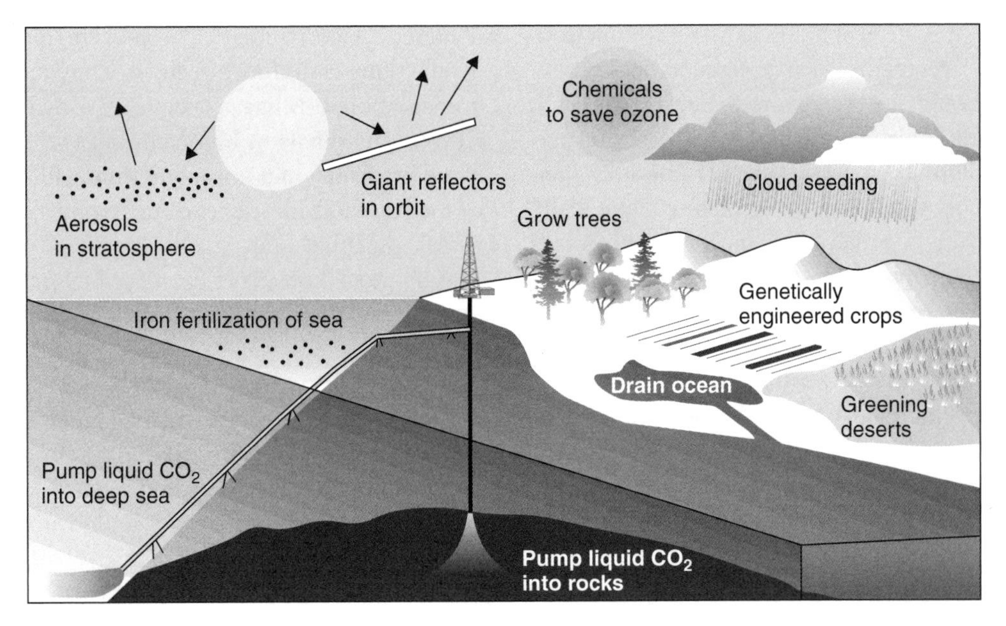

Fig. 11.1 Various geoengineering and technical approaches to mitigating the effects of global climate change (From Keith DW 2001. Geoengineering. *Nature* **409**: 420. Copyright (2001) Macmillan Magazines Limited, with kind permission of Kluwer Academic Publishers).

warming (Figure 11.1) (Begley 1991, Schneider 2001). Proposals include using fleets of large aircraft or large guns to release dust into the lower stratosphere and reflect sunlight back into space. Other proposals to reduce solar input to our planet would send billions of aluminized reflective balloons into the stratosphere or orbit 50,000 mirrors with an area of 100 km^2 (39 mi^2). These schemes raise numerous questions regarding possible harmful effects on ecosystems. For example, reduced solar input, in addition to reducing the greenhouse effect, might reduce photosynthesis in crops and natural vegetation, reducing agricultural and forest productivity.

Even the problem of greenhouse-induced sea-level rise fosters technological solutions. Proposals include greatly expanding freshwater storage in land-locked reservoirs and draining ocean water into below-sea-level continental depressions (Newman and Fairbridge 1986). Several areas of the world are well below the current sea level. These include the Imperial Valley of California, the Qattara Depression of Northwestern Egypt, the Dead Sea rift valley between Israel and Jordan, the Salina Gaulicho of Argentina, and the Eritrea depression in Ethiopia. Also, as the water drops from sea level into these depressions, some could be channeled through turbines to generate large quantities of electricity. In fact, engineering feasibility studies have been conducted for the Dead Sea. "Project Noah" proposes to dig a canal from the Mediterranean Sea to the Dead Sea. About 6,000 km^2 of seawater would then be drained from the Mediterranean into the Dead Sea. Planners estimate this could offset as much as 50% of the projected sea-level rise between 1990 and the year 2050. However, substantial areas of the agriculturally rich and populated Jordan Valley would be flooded.

Large-scale technological fixes to environmental problems such as greenhouse warming should not be ignored. Further research may discover an effective and at least partial solution. However, even if possible, their high cost generally makes them economically unfeasible. Furthermore, technology has created these problems and changed our planet in ways we do not fully understand. In our rush to find a "quick fix" for the problems we have created, we must be careful not to create additional problems. Such technological approaches deal with the end results, rather than the root cause (world dependence on fossil fuel) of global greenhouse warming.

Enhance Natural Carbon Sinks

If natural sinks of CO_2 could be enlarged, they would remove more CO_2 from the atmosphere. The ocean's role as a significant sink for CO_2 might be enhanced (Chapters 1 and 8). Phytoplankton, in the lighted surface layers of the ocean, assimilate dissolved CO_2, and through photosynthesis, convert it to organic carbon (biomass). Also, some species, most notably the coccolithophores, use CO_2 to build exoskeletons of solid $CaCO_3$ (calcium carbonate). When phytoplankton die, they sink, and a proportion of the fixed carbon is removed to the deep ocean for long-term (>200 years) storage. Phytoplankton, like other plants, require nutrients for growth. One nutrient that often limits the growth of oceanic phytoplankton is iron. The experimental addition of iron to small volumes of seawater often stimulates a rapid growth or bloom of phytoplankton.

Some scientists give serious consideration to the idea of dumping large quantities of iron into the ocean to stimulate algal blooms and promote the removal of dissolved CO_2 to the deep ocean. In pilot experiments (IRONEX) in the Pacific and Antarctic Ocean, tons of iron were dumped from ships and the effects on phytoplankton monitored (Monastersky 1995, Boyd et al. 2000). The plankton bloomed, but the effects were transient and disappointing. However, some researchers believe this approach is worth pursuing and patents for ocean fertilization procedures have been filed in anticipation of a global market for carbon mitigation credits.

To be effective on a large scale, such a scheme would need to involve hundreds, if not thousands, of ships dumping iron full time, and would be exceedingly costly. Also, adding iron to the ocean would probably cause a major change in the structure of open ocean ecosystems that support fish, including those fish that are commercially harvested. Nevertheless, the debate continues within the scientific community as to whether this scheme represents a feasible and worthwhile mitigation option (Johnson and Karl 2002) or a misguided and possibly dangerous alteration of ocean systems (Chisholm et al. 2001).

Forest carbon sinks could be enhanced. Trees, through photosynthesis, remove CO_2 from the atmosphere and store it as organic carbon until the tree dies and decays, or is burned, releasing the carbon back into the atmosphere as CO_2. The World's vegetation and forests currently store about 610 Gt of carbon. Young growing forests shift carbon from the atmosphere into temporarily stored organic biomass, but a mature old-growth forest is probably close to equilibrium, giving off as much carbon to the atmosphere (as trees die and decay) as it removes through photosynthesis. Planting a new tree could effectively offset some CO_2 emissions for the life of the tree, often 100 to 300 years, and large-scale reforestation could significantly reduce the rate of CO_2 buildup in the atmosphere.

If we could double our current rate of reforestation each year, we could delay greenhouse warming for a decade or two, possibly long enough to develop alternative sources of energy (Botkin 1989). Estimates for the United States suggest that reforestation of 30 million hectares of economically marginal crop, pasture, and non-Federal forest land (about 3% of US land area) could sequester about 5% of 1990 US CO_2 emissions at a cost of $7 per ton of CO_2. Large-scale reforestation will be difficult and expensive. It may be difficult to find large enough areas suitable for reforestation on this scale (Rubin et al. 1992). However, reforestation should form one component of a broad strategy to address greenhouse warming. Again, caution is called for. Some scientists suggest that more forests could decrease the Earth's albedo so that the darker surface would absorb more heat and add to global warming.

Convert to Carbon-Free and Renewable Energy Technologies

Alternative (renewable nonfossil fuel) energy sources could significantly reduce greenhouse gas (GHG) emissions. Many are already proven and effective sources of energy, but they are dwarfed in magnitude by our consumption of coal, oil, and gas. Hydroelectricity from rivers and streams currently supplies 20% of global electrical demand. However, many streams remain untapped and this source could be expanded, particularly through the use of multiple small-stream generators on local scales. The Three Gorges Dam in China, the largest hydroelectric dam in the world, when completed, will generate 84 billion kWh of electricity – an amount of energy equivalent to 40 to 50 million tons of coal per year. However, dams and hydroelectric plants have their own set of environmental impacts, ranging from human population displacement (1.9 million people in the case of Three Gorges) to interference with migrating fish and decreased downstream water flow and sedimentation.

Wood and plant fiber represent biomass that can be burned and converted to heat to warm buildings, drive steam turbines to generate electricity, or power motor vehicles (US DOE 2002b). Wood burning for cooking and space heating is probably the oldest source of energy used by humans. In the burning process, organic carbon is converted to CO_2 and released into the atmosphere. However, burning simply releases CO_2 that was captured during photosynthesis. If the same number of trees are replanted, there is no net effect on atmospheric CO_2 concentration.

Plants can also be fermented to produce liquid fuels. Wood fiber can be converted to methyl alcohol and burned. Currently, most motor vehicles in Brazil run on 100% ethyl alcohol, mostly derived from the fermentation of maize. Large-scale wood or fiber farms could capture solar energy and convert it into biomass fuel, thus reducing our reliance on fossil fuel. Biodiesel is a fuel made from vegetable oil, for example, soybean oil or from modified used restaurant grease. It can be burned either pure or as a mixture with petroleum-derived diesel fuel in standard diesel engines. In Europe, 34% of all cars currently have diesel engines, and biodiesel made from rapeseed (canola) oil is sold in over 1,000 stations in Germany alone. Converting prime agricultural land from food production to fuel production is probably undesirable. However, large areas of arid or marginal land could be used to raise less-demanding biomass crops.

Wind, essentially a form of solar energy driven by differences in atmospheric pressure between one area and another, represents another promising source of alternative power. Humans have used wind power for millennia.

Sailing ships traveled the world, carrying humans and their cargo from one continent to another. Windmills have been used for centuries to pump water and to drive large stone grinding wheels to mill grain. New technology provides an effective means for windmills to drive electrical generators, and wind generators are being installed at locations around the globe (Box 11.1; Figure 11.2).

Large wind farms, consisting of hundreds of wind generators capable of generating significant quantities of electricity are now fairly common. There are over 35,000 wind turbines worldwide, providing over 12,000 MW of total global generating capacity (AWEA 2002). In the United States, wind-generated electricity currently represents a small fraction of the total electrical capacity, but according to Battelle's Pacific Northwest Laboratory,

Box 11.1 Wind power (From AWEA 2002. American Wind Energy Association. 122C Street NW, Washington DC 2001. Available from: *http://www.awea.org*)

What is a wind turbine and how does it work?

A wind energy system transforms the kinetic energy of the wind into mechanical or electrical energy that can be harnessed for practical use. Mechanical energy is most commonly used for pumping water in rural or remote locations. Wind electric turbines generate electricity for homes and businesses and for sale to utilities. Turbine subsystems include

- a rotor, or blades, which convert the wind's energy into rotational shaft energy;
- a nacelle (enclosure) containing a drive train, usually including a gearbox and a generator;
- a tower to support the rotor and drive train; and
- electronic equipment such as controls, electrical cables, ground support equipment, and interconnection equipment.

How much electricity can one wind turbine generate?

The ability to generate electricity is measured in watts. Watts are very small units, so the terms kilowatt (kW, 1,000 watts), megawatt (MW, 1 million watts), and gigawatt (GW = 1 billion watts) are most commonly used to describe the capacity of generating units such as wind turbines or other power plants.

Electricity production and consumption are most commonly measured in kilowatt-hours (kWh). A kilowatt-hour is one kilowatt (1,000 watts) of electricity produced or consumed for one hour. One 50-W light bulb left on for 20 h consumes one kilowatt-hour of electricity (50 W × 20 h = 1,000 Wh = 1 kWh). Wind turbines being manufactured now have power ratings ranging from about 250 W to 1.8 megawatts (MW). The output of a wind turbine depends on the turbine's size and the wind's speed through the rotor. A 10-m-diameter rotor would generate 25 kW or 45 MWh, while a 71-m rotor would produce 1,650 kW or 5,600 MWh.

A 10-kW wind turbine can generate about 16,000 kWh annually, more than enough to power a typical household. A 1.8-MW turbine can produce more than 5.2 million kWh in a year – enough to power more than 500 households. The average US household consumes about 10,000 kWh of electricity each year.

World Growth

Some 6,500 MW of new wind energy generating capacity were installed worldwide in 2001, amounting to annual sales of about $7 billion. This is the largest increase ever in global wind energy installations, well above the capacity added in 2000 (3,800 MW) and 1999 (3,900 MW). The world's wind energy–generating capacity at the close of 2001 stood at about 24,000 MW.

Germany set a world and national record of more than 2,600 MW of new generating capacity installed during the year. Germany, Denmark, and Spain are demonstrating that wind can reliably provide 10 to 25% and more of a region or country's electricity supply.

In the United States, the wind energy industry exceeded previous national records in 2001, installing nearly 1,700 MW or $1.7 billion worth of new generating equipment. The new installations account for close to a third of the world wind energy generating capacity added in 2001.

The global wind energy market continues to be dominated by the "big five" countries with over 1,000 MW of generating capacity each: Germany, the United States, Spain, Denmark, and India. The number of countries with several hundred megawatts is growing larger, however, and it may be that in the next several years – if the current rafts of proposed projects are developed – Brazil and the United Kingdom will see their own wind-generating capacity exceed the 1,000-MW mark.

wind could potentially supply about 20% of the nation's electricity. Furthermore, wind-generated electricity could be greatly expanded. A recent study performed by Denmark's BTM Consult for the European Wind Energy Association and Greenpeace found that by the year 2017, wind could provide 10% of the world's electricity. However, there are limitations. Desirable wind-generating sites are limited to those areas that have winds of consistent moderate to strong speed without strong maxima and minima.

The huge potential energy of the Sun offers another alternative to fossil fuel. The average annual amount of solar energy reaching the Earth's surface (198 W m^{-2}) far exceeds all current human energy requirements. With proper architectural design, passive solar space heating of buildings could be greatly expanded. The feasibility of this low-tech approach is generally underestimated. It requires no new technology and can usually be accomplished for about the cost of conventional building construction. It only requires an adequate area of south-facing glass windows, proper roof overhang to shade summer sun, good insulation, and some thermal mass such as concrete or water inside the building for heat storage (Kachadorian 1997). Even in northern temperate latitudes,

Fig. 11.2 A 250-kW turbine installed at the elementary school in Spirit Lake, Iowa, provides an average of 350,000 kWh of electricity per year, more than is necessary for the 4,900 m^2 (53,000 ft^2) school. Excess electricity fed into the local utility system has earned the school $25,000 over five years. The school uses electricity from the utility at times when the wind does not blow (Courtesy AWEA 2002. *American Wind Energy Association*. 122C Street NW, Washington DC 2001. Available from: *http://www.awea.org*).

Fig. 11.3 A passive solar home located in Western Washington State, USA, at about 48 °N latitude. The annual cost for supplemental (nonsolar) energy in this cloudy and rainy location is less than $150 (Photo by author).

space-heating requirements from conventional sources (wood, electricity, or fossil fuel) can be reduced to a small fraction with proper passive solar building design (Figure 11.3).

Solar energy can also be converted to electricity using photovoltaic cells. When photons from the Sun strike the surface of these cells (often made of silicon), electrons are released creating an electrical current. The efficiency of solar cells is increasing, while the cost of their manufacture is decreasing. Researchers are investigating thin-film "plastic" solar cells that use organic semiconductors. This new technology promises the possibility of increasing quantum efficiency (conversion of photons to electricity) by five orders of magnitude (Schön et al. 2000).

Currently, photovoltaic systems cost about $5 to $6 per watt output. Today, a rooftop array of solar cells of about 13.2 m^2 (142 ft^2) can generate about 1,500 W of electricity, or enough to meet the needs of many energy-efficient US households. The cost for such a system with a 20-year guaranteed life would be about $9,000 or $428 per year. This does not include the cost of installation or the discount rate, that is, the fact that this is an "up-front" cost and not spread out over time. In the United States, a fixed (nonrotating) solar panel area of 45 m^2 (484 ft^2) can generate an annual 4,899 kWh of electricity at a cost saving of $250 in Seattle, Washington, or 6,786 kWh of electricity at a saving of $461 in Atlanta, Georgia (NCPV 2002). Many states now require utility companies to buy back any excess solar-generated electricity from private households or businesses. The use of solar energy, whether for space heating or electricity generation, does require certain site characteristics. Specifically, the site must have "solar access," that is, not have buildings or vegetation blocking the Sun from the south.

Proponents and detractors argue about the potential role of nuclear energy as an alternative to fossil fuel. Nuclear fission (splitting the atom in a controlled chain reaction) can be used to generate heat, produce steam, and power turbines that generate electricity. It is a technology first developed almost 50 years ago and, unlike fossil-fuel combustion, it generates no greenhouse gases. Nuclear energy now supplies about 17% of the world's electricity. In the United States it accounts for 18% of electric production, but in some countries, for example, France, it is the major (76%) source of electricity (World Bank 2002).

Several problems argue against expansion of nuclear energy. First, the safety of nuclear plants remains in question. Accidents at Three Mile Island in the United States and Chernobyl in the former Soviet Union testify to the risks involved. Second, safe disposal and long-term storage of highly radioactive waste material remains a challenge. Finally, nuclear power generation is very expensive. The costs of construction and operation of nuclear plants make the electricity generated generally more expensive than that of conventional power plants. Probably in response to the combination of these factors, global nuclear power generating capacity grew during the 1990s by only 1% per year compared to annual growth rates of 17% for solar cells (24% in 2001) and 24% for wind power.

Nevertheless, proponents hail nuclear energy as a means of reducing GHG emissions. They argue that damage to human health and the environment from nuclear power plants has been historically small compared to that from fossil-fuel plants that emit not only GHGs but also carcinogenic combustion products (e.g. polyaromatic hydrocarbons). Controlled nuclear fusion of hydrogen promises to produce huge amounts of power with little waste problem. However, effective fusion is still very much in the research arena and will probably not become feasible until late in this century, if ever.

Fuel cells, first developed for spaceship life-support systems, now offer the promise of helping to support life on Earth (Burns et al. 2002). Fuel cells have an energy efficiency of about 55% compared to the less than 30% efficiency of most internal combustion gasoline engines. Most fuel cells rely on one of the two processes. One process uses hydrogen to produce electricity. The hydrogen can be derived from a variety of feedstocks including water (via electrolysis using electricity), natural gas, coal, biomass, or organic waste. No GHGs are produced. The products are water and electricity (Figure 11.4, Box 11.2). As part of a demonstration project in clean technologies, The Global Environment Facility, a multilateral trust fund of UNDP, UNEP, and the World Bank, plans to subsidize the operation of 40 to 50 fuel cell powered buses as demonstrations in Brazil, Mexico, India, Egypt, and China.

Alternatively, fuel cells can use electricity to produce hydrogen fuel. Electricity from photovoltaic cells or other sources is used to split water (H_2O) into oxygen (O_2) and hydrogen (H_2) (photolysis). The hydrogen can then be burned as a fuel to power vehicles, heat buildings, or run appliances like refrigerators.

Technical problems for the worldwide use of hydrogen remain, namely, the development of safe (H_2 is explosive) and compact hydrogen storage systems. Hydrogen can be liquefied, but this requires considerable

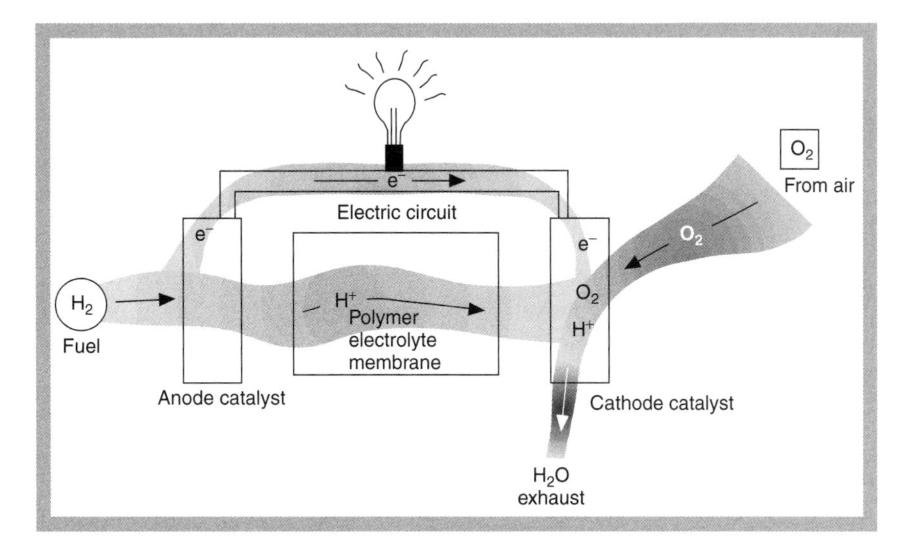

Fig. 11.4 Fuel cells – how they work and what they produce (From BTI 2000. *Breakthrough Technologies Institute/Fuel Cells 2000*. The Online Fuel Cell Information Center. Accessed September 9, 2002 from: *http://www.fuelcells.org/*).

Box 11.2 What is a fuel cell? (From BTI 2000. Breakthrough Technologies Institute/Fuel Cells 2000. The Online Fuel Cell Information Center. Accessed September 9, 2002. Available from: *http://www.fuelcells.org/*)

In principle, a fuel cell operates like a battery. Unlike a battery, a fuel cell does not run down or require recharging. It will produce energy in the form of electricity and heat as long as fuel is supplied. A fuel cell consists of two electrodes sandwiched around an electrolyte. Oxygen passes over one electrode and hydrogen over the other, generating electricity, water, and heat. Hydrogen fuel is fed into the "anode" of the fuel cell. Oxygen (or air) enters the fuel cell through the cathode. Encouraged by a catalyst, the hydrogen atom splits into a proton and an electron, which take different paths to the cathode. The proton passes through the electrolyte. The electrons create a separate current that can be utilized before they return to the cathode, to be reunited with the hydrogen and oxygen in a molecule of water. A fuel cell system that includes a "fuel reformer" can utilize the hydrogen from any hydrocarbon fuel – from natural gas to methanol, and even gasoline. Since the fuel cell relies on chemistry and not combustion, emissions from this type of a system would still be much smaller than emissions from the cleanest fuel combustion processes.

energy. Hydrogen offers great potential if some technological, storage, and safety barriers can be overcome (Ogden 1999). Researchers are also investigating the use of algae and microbes to produce hydrogen.

Ocean thermal energy conversion (OTEC) uses the potential energy represented by the differences in temperature between deep and surface ocean water to produce electricity. It is practical in certain areas where the temperature difference between warm surface water and deep water differs by at least 20 °C. In an open-cycle OTEC system, warm seawater is "flash" evaporated in a vacuum chamber to produce steam. The steam expands through a low-pressure turbine that is coupled to a generator to produce electricity. The steam exiting the turbine is condensed by cold seawater pumped from the ocean's depths. If a surface condenser is used in the system, the condensed steam remains separated from the cold seawater and provides a supply of desalinated water (Figure 11.5). The technology was field-tested in Hawaii in 1993, where an OTEC plant at Keahole Point, Hawaii, produced 50,000 W of electricity. The potential is great, but considerable research needs to be completed before this technology becomes practical. Nevertheless, OTEC could provide power to selected coastal or island communities located near deep water (NREL 2002).

Lunar and solar cycles drive ocean tides, and winds create ocean surface waves. Tidal energy can be exploited in two ways. "Barrages" allow tidal waters to fill an estuary via sluices and to empty through turbines. Tidal streams harness offshore underwater tidal currents with devices similar to wind turbines. Both are limited to areas that experience high tidal ranges. Tidal stream technology is in its infancy, with only one prototype 5-kW machine operational in the world. The exploitable tidal energy potential in Europe is 105 TWh year^{-1} (TWh = terawatt hours = 10^{12} watt hours) from tidal barrages (mostly in France and the United Kingdom) and 48 TWh year^{-1} from tidal stream turbines (mostly around UK shores). A 240-MW barrage has been operational at La Rance in France since 1967. Both technologies are currently not economically competitive with other forms of energy and no further deployments are anticipated before 2010 (E.C. 2002). Also, tidal energy schemes affect the environment, influencing water levels, currents, and sediment transport.

Ocean waves contain huge amounts of energy (Falnes and Lovseth 1991). Waves are another form of stored solar energy, since the winds that produce waves are caused by pressure differences in the atmosphere arising from solar heating. Globally, total wave power generating potential is about 2,000 TWh year^{-1} (E.C. 2002) or more than 150 times more energy than the total global electrical consumption of 13.7 TWh for the year 2,000 (US DOE 2002c). Wave energy potentials in the European Union area alone have been estimated conservatively at 120 to 190 TWh year^{-1} (offshore) and 34 to 46 TWh year^{-1} (nearshore).

A great variety of wave energy devices have been proposed and several have been deployed in the sea as prototypes or demonstration schemes. The pendulum device (Figure 11.6a) consists of a rectangular box that is open to the sea at one end. A pendulum flap is hinged over this opening, so that the action of the waves causes it to swing back and forth. This motion is then used to power a hydraulic pump and generator. Only small devices have been deployed and tested. Another example is the oscillating water column device composed of a partly submerged concrete or steel structure, which has an opening to the sea below the water line, thereby enclosing a column of air above a column of water. As waves impinge on the device, they cause the water column to rise and fall, which alternately compresses and depressurizes the air column (Figure 11.6b). This air is allowed to flow to and from the atmosphere through a turbine, which drives an electric generator. The tapered channel system (Figure 11.6c) consists of a gradually narrowing channel

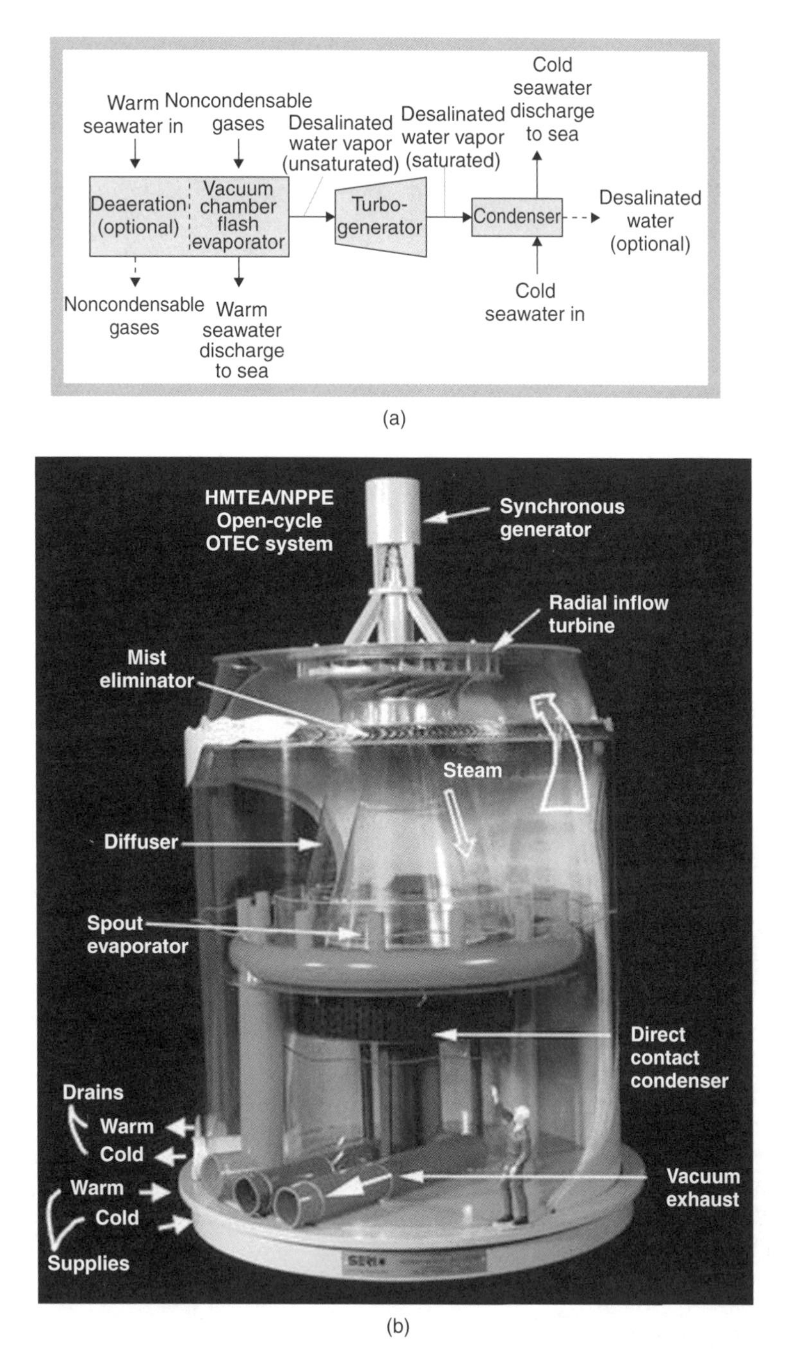

Fig. 11.5 An open-cycle ocean thermal energy conversion system (OTEC). In tropical waters, ocean surface temperatures may reach 80 °F, while at depths of 5,000 feet, temperatures are near freezing. Warm surface water is pumped to a vacuum chamber to produce steam that drives a turbine. At the same time, a heat exchanger uses cold water pumped from depths to condense spent steam into drinkable water (Part (a) reproduced with permission of U.S. Department of Energy, National Renewable Energy Laboratory, Golden, Colorado, USA *http://www.nrel.gov/otec/*).

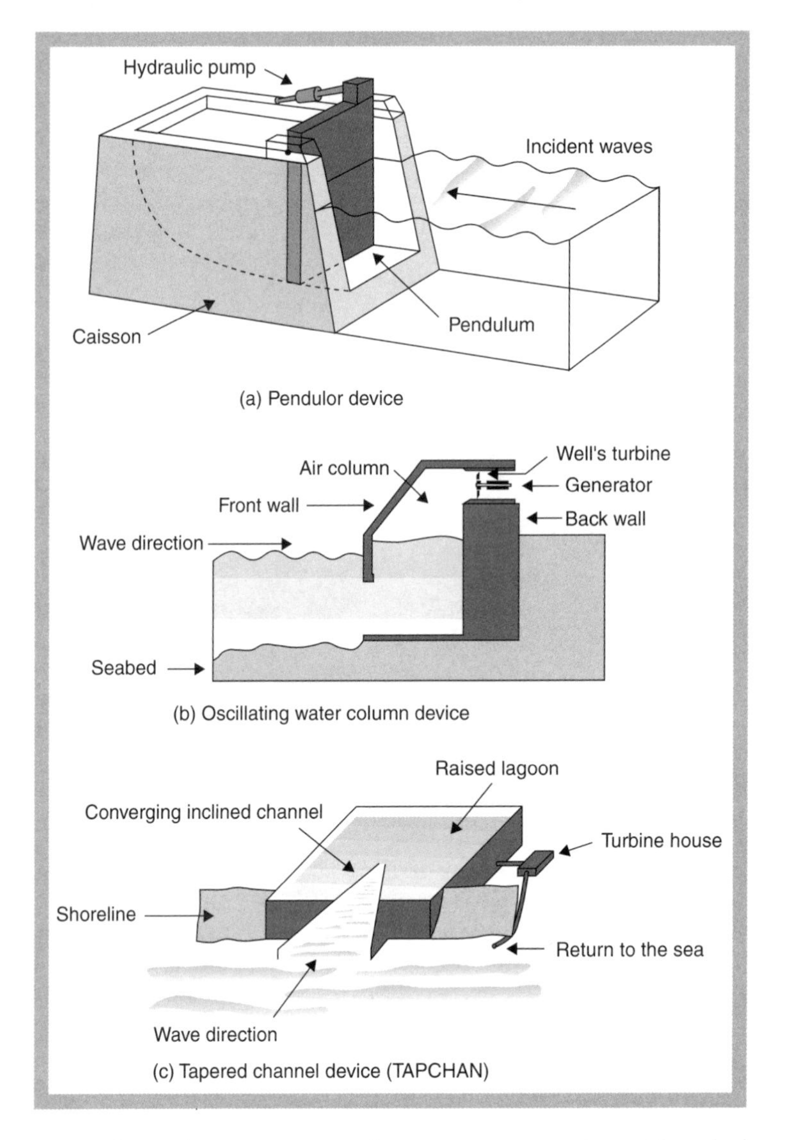

Fig. 11.6 Three approaches to harnessing energy from ocean waves: (a) pendulum device; (b) oscillating water column; (c) tapered channel device (EC 2002. *Atlas Data of Information. The Fourth Framework Programme for Research and Technological Development*. Department of Trade and Industry. Reproduced by permission of the Department of Trade and Industry).

with wall heights typically 3 to 5 m above mean water level. The waves enter the wide end of the channel and, as they propagate down the narrowing channel, the wave height is amplified until the wave crests spill over the walls to a reservoir, which provides a stable water supply to a conventional low head turbine. The requirements of low tidal range and suitable shoreline limit the usefulness of this device.

Most wave energy schemes remain in the research stage, but a significant number have been constructed as demonstration projects. The main countries involved with the development of wave power have been Denmark, India, Ireland, Japan, Norway, Portugal, United Kingdom, and United States. Although the feasibility of wave energy has been demonstrated, many of the current uncertainties of cost and performance will need to be overcome before large-scale development is pursued.

The developing world, as it industrializes, is rapidly increasing its use of fossil fuel. Greenhouse gas emissions from developing nations, as a whole, will soon exceed those of the major industrialized countries. The transfer of energy-efficient and alternative energy technologies from developed to developing countries could help slow the growth rate of CO_2 emissions. Such technology transfer will require international planning and cooperation.

Each energy source has advantages and disadvantages (Dresselhaus and Thomas 2001). Fossil fuels are currently abundant, inexpensive, and efficient, but produce GHGs as well as other pollutants. They are nonrenewable and will eventually be exhausted. Hydroelectric plants may interfere with stream flows and fish migrations as well as flood large land areas. Solar, wind energy, and tidal power generators are limited geographically. Geothermal energy is practical in only a few locations. Ocean thermal energy has low efficiency, is limited geographically, and requires very large heat exchangers. Nuclear plants are expensive and produce toxic radioactive waste.

Currently, none of these alternatives to fossil fuel represent a large fraction of global power generation. However, a concerted research and development effort, perhaps promoted with government incentives, could make at least some of these alternatives more technically and economically feasible (Fulkerson et al. 1989). Collectively, they have the potential to significantly reduce fossil-fuel consumption.

Conserve Energy and Use It More Efficiently

None of the nonfossil energy sources are currently available in the quantity necessary to totally replace fossil fuel. Improving energy efficiency, that is, the amount of energy generated per greenhouse gas emitted, still represents the most viable option for reducing GHG emissions over the next few decades (Fulkerson et al. 1989). Improving energy efficiency will help delay greenhouse warming while alternative fuel sources are implemented.

In the transportation sector, several alternatives to gasoline-guzzling vehicles are available. Walking, bicycling, carpooling, or using mass transit (bus or train) can greatly increase efficiency by decreasing the quantity of CO_2 emitted per passenger mile. Thee-quarters of US commuters drive to work alone. Also, the fuel efficiency of motor vehicles could be greatly improved, and it could be done now. Electric cars offer one alternative, and research on improved batteries for vehicles continues. Currently, two-thirds of the world's electricity is generated from fossil fuel. Therefore, electric cars that require battery recharging ultimately add CO_2 to the atmosphere. However, if electricity is produced from nonfossil-fuel energy, then cars using it will contribute virtually nothing to GHG emissions.

A new generation of gas/electric hybrid vehicles, using both rechargeable electric batteries and gasoline engines with up to $29\,L\,km^{-1}$ (70 miles per gallon) of gasoline, significantly reduce CO_2 emissions per

(a)

(b)

Fig. 11.7 Energy-efficient gas/electric hybrid vehicles now on the market offer high mileage on regular gasoline supplemented with electric power. (a) Honda Insight 68 mpg highway (Reproduced by permission of Honda). (b) Toyota Prius 52 mpg highway (Reproduced by permission of Toyota (GB) Plc).

passenger kilometer (Figure 11.7). Also, compared to conventional gasoline vehicles, those run on compressed natural gas (although generally shorter range) could reduce GHG emissions by 40% (MacLean et al. 1999).

Cogeneration, the simultaneous production of electricity and another form of useful thermal energy (such as heat or steam), is highly efficient. For example, when separate processes are used to produce steam and electricity, roughly two-thirds of the energy produced is wasted. The combined production of steam and electricity from the same energy source makes use of most of this wasted energy, greatly increasing fuel efficiency and reducing emissions. A variety of industries are using different cogeneration schemes to reduce fossil-fuel consumption and save money. These are generally energy-intensive industries such as petroleum and metals extraction and refining, or chemical and paper manufacturing. Examples of cogeneration include electrical generating plants that provide warm water to heat large commercial greenhouses or steel-manufacturing plants that use blast furnace heat to produce steam that in turn drives a turbine to generate electricity.

Incremental steps in energy conservation or efficiency can, in total, make a significant contribution to reducing GHG emissions. Quality of life is somewhat subjective, but the UN has ranked countries on the basis of measures such as living conditions, health care, education, and crime. It is clear that increasing quality of life does not necessarily depend on increasing energy consumption. Indeed, by many of these measures, the quality of life of some European Countries (e.g. Sweden, Denmark, Switzerland) exceeds that of the United States, even though the per capita energy consumption in Europe is much lower. Gross domestic product (GDP) can be compared between countries using purchasing power parity (PPP) rates, where an international dollar has the same purchasing power over GDP as a US dollar has in the United States. On this basis, the quantity of CO_2 emitted per PPP$ of GDP in the United States is 3.5 and 1.8 times that of Switzerland and the United Kingdom, respectively, while that of the Russian Federation is seven times greater than that of Switzerland (World Bank 2002).

Many currently available options could improve our energy efficiency, that is, the amount of useful work produced per energy consumed. This means that the same goods and services could be produced with much smaller expenditures of energy. Options include gasoline vehicles that use fewer gallons per mile, home heating by gas-fired heat pumps rather than older, less-efficient technologies, increased use of fluorescent versus incandescent light bulbs, the use of smart "set back" automatic thermostats in building heating, and better building insulation to reduce winter heat loss and summer heat gain.

The United States emits about 25% of the global GHGs and could, through a variety of energy efficiency and other currently available measures, reduce its emissions by 10 to 40% at either low cost or a net cost savings (Figure 11.8). For example, at least eleven technologies could reduce the use of electrical energy in buildings. Each incremental step, from lighter reflective roofs and shade trees (reducing air-conditioning demand) to more energy-efficient water heaters, could contribute to energy savings. In total, implementation of all measures could reduce US building energy use by 45% and also result in a net cost savings (Rubin et al. 1992). Greenhouse gas emissions (as CO_2 equivalent warming potential) from the US residential–commercial sector could be reduced by 890 Mt year^{-1} at an average cost of \$62 ton^{-1} of CO_2 removed. The average fuel economy of US autos as well as electric power plants could also be significantly increased at a net cost savings.

In the United States, "efficient use" represents the single greatest potential source of rapidly available and inexpensive energy (Lovins and Lovins 2002). In 2001, the people of California, faced with massive electrical shortages, cut peak electrical demand per dollar of GDP by 14% in six months, ending a crisis that some thought would require

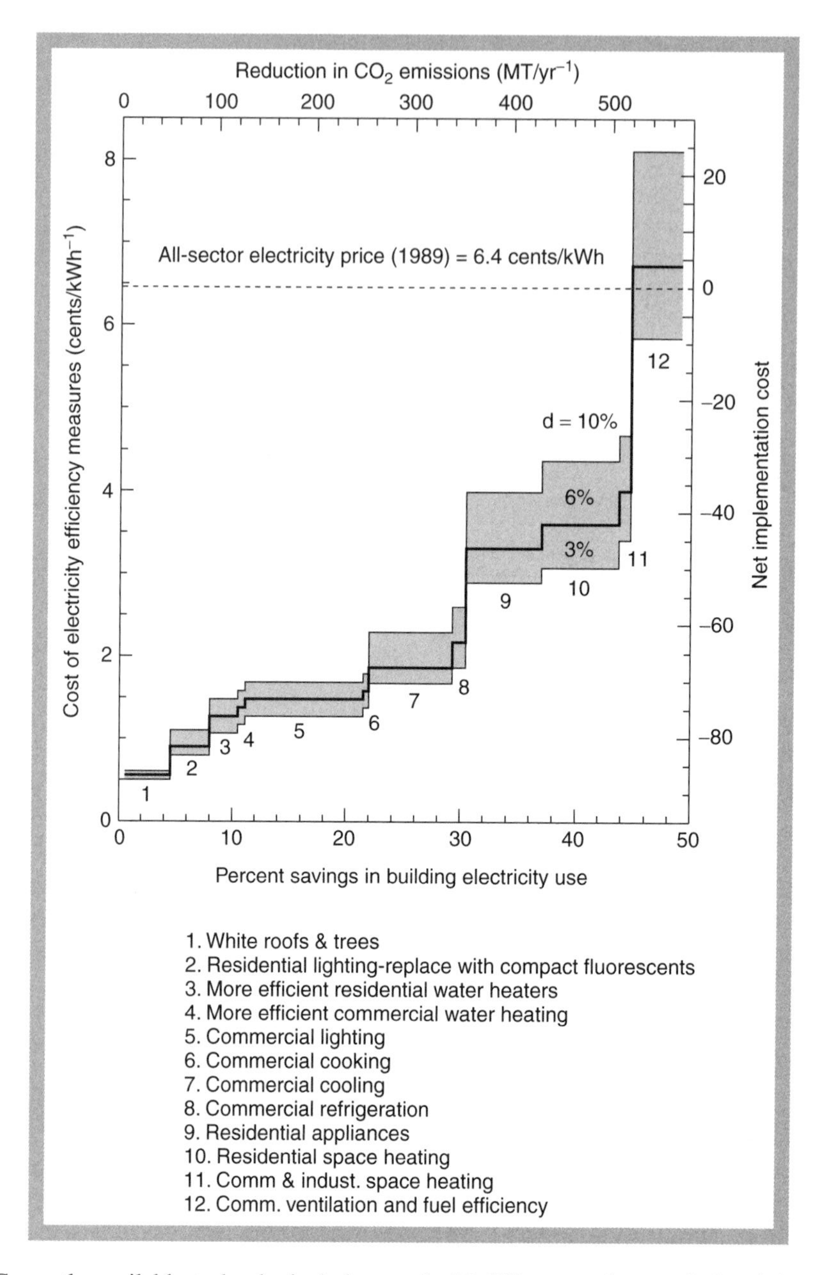

Fig. 11.8 Currently available technological changes in 12 different end uses of electricity could reduce building energy use by 734 billion kWh or 45% of the US demand at a net cost savings. Each step is the annualized investment cost of a given technological option numbered at left. For example, white roofs and trees (1) could decrease air-conditioning requirements, and reduce CO_2 emissions by 40 mt year^{-1}, saving 4% in building electrical use and 0.5 cents per kWh (Reprinted with permission from Rubin ES, Cooper RN, Frosch RA, Lee TH, Marland G, Rosenfeld AH, et al. 1992. Realistic mitigation options for global warming. *Science* **257**: 148–149 & 261–266. Copyright (1992) American Association for the Advancement of Science).

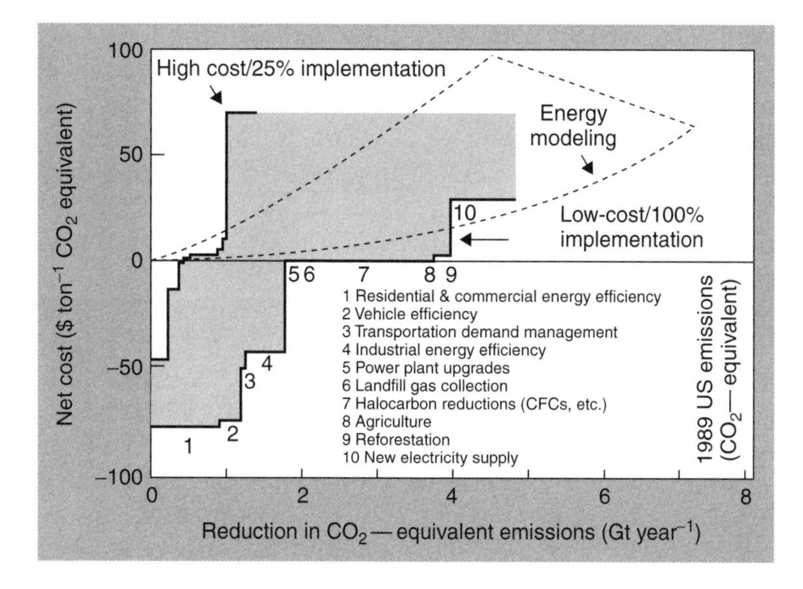

Fig. 11.9 Cost effectiveness versus emission reduction potential for 10 mitigation options. Implementation rates of 25 to 100% of maximum potential characterize the uncertainty range. Energy modeling studies encompass the range from other studies. For example, improvements in residential and commercial energy efficiency could reduce emissions by almost 1 Gt year^{-1} at a net overall savings of about $49 to $59 per ton (Reprinted with permission from Rubin ES, Cooper RN, Frosch RA, Lee TH, Marland G, Rosenfeld AH, et al. 1992. Realistic mitigation options for global warming. *Science* **257**: 148–149 & 261–266. Copyright (1992) American Association for the Advancement of Science).

1,300 to 1,900 more power plants nationwide. Between 1975 and 2,000, the US reduction in "energy intensity" (brought about by such things as better building insulation and lighting, and more efficient vehicles) saved an amount of energy equivalent to three times the total oil imports and five times the domestic oil production. Since 1996, saved energy has been the nation's fastest growing "source." Energy policy can have a direct bearing on fossil-fuel demand and GHG emissions (Chapter 12). Conservation incentives, such as utilities rewarding customers for reducing their energy demands, or allowing businesses to write off energy-saving investments against taxable income, can lead to significant reductions in energy use.

Overall, analysis suggests that improvements in energy efficiency could reduce US GHG emissions by 800 to 3,100 Mt year^{-1}, that is, about 10 to 40% of 1990 emissions. At the same time such measures would produce an annual cost savings of about $10 to $110 billion per year (Figure 11.9). These results assume inflation-adjusted discount rates (rates of return) of 3 to 10%. However, in the United States, individuals and businesses are often reluctant to invest in change unless the payback is immediate or at most 2 to 3 years, not the 5- to 10-year returns expected from most investments in energy efficiency.

Different energy choices will result in different levels of GHG emissions, but overall energy efficiency is the result of complex interactions over the entire energy cycle. The technological and economic factors that influence GHG emissions include the full fuel cycle, that is, conversion from the primary energy source to actual end use including different conversion, transport, and end-use systems (Nakic′enovic′ 1993). In the industrialized countries, energy intensity (watts $year^{-1}$ $dollar^{-1}$) has been increasing about 1% per year since the 1860s. In addition, energy use has undergone "decarbonization," that is, a shift from fuels of high carbon content (coal) to those of lower carbon content such as oil, natural gas, or carbon-free nuclear energy. However, with economic output increasing at 3% per year, the improvement in intensity will not keep pace with the growth in CO_2 emissions. Computer models of the full fuel cycle suggest that choices in how a service (e.g. building lighting) is provided can result in differences in CO_2 emissions of 90%. Thus, CO_2 emissions could be reduced substantially using currently available technology.

Adapt to Climate Change

Even if GHG emissions were drastically reduced today, the Earth's climate would probably continue to warm for some time. Regardless of the rate of fossil-fuel combustion, if all the known fossil-fuel reserves (about 4,000 GtC) are burned, CO_2 will reach about 1,000 ppmv and the Earth will be warmed by >5 °C by the end of the millennium (Lenton and Cannell 2002). Hence, some scientists and politicians argue that we must prepare to adapt to the inevitable climate change. Research funded by the Electric Power Research Institute suggests that a variety of adaptation measures, from better building insulation to altered crop species, could be instituted to lessen the impacts of climate change on a variety of sectors (Table 11.1).

Adaptation can be effective, especially if the value of the resulting benefit is greater than the cost of the adaptation. The net economic impact of climate change may be lessened in the case of efficient adaptation measures, but may be increased by inefficient measures. Building new seawalls to protect against sea-level rise can be very expensive, but if it saves a valuable tourist beach or the homes of many people, then it may be worthwhile. On the other hand, investing billions in fossil fuel–powered air conditioning might bring some reduction in medical costs associated with heatstroke, but it could add even more to a variety of costs related to further GHG emissions and global warming. Adaptation obviously does not solve the long-term problem of damage and increased costs from continued GHG emissions.

Taking Action

Many actions can be taken that would greatly reduce GHG emissions and or reduce the impacts of greenhouse warming. These actions can be undertaken globally, nationally, and individually. Policies related to energy and GHG emissions are discussed more fully in Chapter 12. Globally, we need to strengthen and adhere to international treaties on climate change. The climate problem is global and its mitigation will require international cooperation. Although US delegates signed the 1997 Kyoto Protocol (Chapter 12), Congress never ratified the treaty, and the United States has fallen far short of the emission reduction goals it agreed to in Kyoto.

Nationally, individual countries need to institute good faith efforts of their own, perhaps going beyond international agreements to mitigate as much as they possibly can. Carbon taxes, expanded mass transit systems, and increased use of alternative energy sources may be expensive in the short term, but could have numerous long-term benefits in air quality and human health while, at the same time, mitigating climate change.

Table 11.1 Market sector adaptations to climate change (From Mendelsohn R 2000. Efficient adaptation to climate change. *Climatic Change* **45**: 583–600, original copyright notice with kind permission of Kluwer Academic Publishers).

Sector	Private/public	Adaptation
Agriculture	Private	Alter crop species
–	–	Alter timing
–	–	Irrigation
–	Public	Plant breeding
Sea-level rise	Private	Depreciate vulnerable buildings
–	Public	Seawalls as needed
–	–	Beach enrichment
Forestry	Private	Harvest vulnerable trees
–	–	Plant new trees
–	–	Intensify management
Energy	Private	New cooling capacity
–	–	Changes in insulation
–	–	Cool building designs
–	Public	New building codes
Water	Private	Invest in water efficiency
–	Public	Shift water to high-value uses
–	–	Divert/store more water
–	–	Flood zoning
Biodiversity	Public	Move endangered species
–	–	Manage landscapes
–	–	Plant-adapted species
Health	Private	Prepare for extreme weather
–	–	Avoid insect bites
–	Public	Control disease carriers
–	–	Treat infected people
–	–	Control diseased ecosystems
Aesthetics	Private	Adapt behavior (e.g. recreation)
–	Public	Educate public of adaptive options

Some steps are being taken to prepare for sea-level rise. In the United States, policymakers have stressed the need to review and amend policies and laws surrounding coastal property (Titus 1998). In the meantime, people are elevating their homes, and some communities are requiring coastal structures to be built several feet higher in anticipation of sea-level rise.

Several alternative energy technologies probably have the potential to make significant contributions to the world's energy needs. However, a great deal of research is needed to make improvements in materials, generation capacities, storage, conversion efficiencies, and power transmission from these nonfossil-fuel technologies (Dresselhaus and Thomas 2001). To delay and reduce the effects of greenhouse warming, research and technology development efforts should match, or exceed in scale, previous efforts such as the Manhattan Project to develop the atomic bomb or the Apollo Project to land human beings on the Moon.

Individually, each person must take responsibility for his or her own contribution to global climate change. Small incremental changes, based on individual choices, can add up to substantial change. Individuals should consider, whenever possible, choosing a fuel-efficient car, reducing automobile use, making their home more energy-efficient, planting trees, recycling, and reducing consumption. Individuals have a responsibility to learn more about climate change and educate others. In democratic nations, individuals should insist that political candidates reveal their views on climate change and its mitigation and choose those candidates who are well informed.

Specifically, individuals can

- turn off electric lights and appliances when not in use;
- set home and workplace thermostats lower at night or when gone and higher when air conditioning is used;
- turn water heaters down when gone for a period of time;
- set refrigerators no lower than 5 °C (41 °F) and check the door gasket;
- unplug idle electric appliances;
- replace incandescent light bulbs with energy-efficient fluorescents;
- wash full loads of clothes and dishes using cold-water rinse and preferably after 8 pm or during other off-peak hours;
- use a microwave rather than a conventional oven when possible;
- install improved insulation and seals in your home;
- walk, bicycle, carpool, or use mass transit when possible instead of a single passenger vehicle;
- select an energy-efficient automobile and combine tasks to reduce the number of trips;
- encourage others to adopt energy-saving practices.

The exact impacts of climate change are not, and may not be, totally predictable. So why take mitigative actions that may be expensive in the face of uncertainty? There are several reasons that argue in favor of acting sooner rather than later. First, although exact effects are not predictable and could in some instances be minimal, it is also true that such uncertainty means that effects could be very serious or even catastrophic. Second, we have already committed the Earth to significant climate change; the longer we wait to substantially reduce GHG emissions, the more serious, difficult, and expensive that change will be. Third, many of the suggested mitigative measures, such as increased energy efficiency, will have positive environmental (e.g. improved air quality) and economic benefits. Finally, we have a responsibility to future generations to hand over the planet with a climate that can sustain the quality of life that we ourselves have experienced, rather than a planet with a serious problem requiring huge costs to repair.

Summary

Reducing the negative impacts of human-induced climate change on natural ecosystems and humans represents the greatest environmental challenge of this century. At least five courses of action could potentially slow the rate of greenhouse warming. All are currently the subject of extensive research by government institutions and private industries. Capturing or sequestering the carbon dioxide from fossil-fuel combustion at the source is technically feasible, at least for stationary sources, but currently represents a significant added cost for manufacturing or power generation. Some large-scale geoengineering proposals to reduce warming or deal with its effects may be worth further research. However, many are unproven, probably very expensive, and probably carry high risks of environmental damage. Natural carbon sinks, for example forests, can be enhanced to absorb more anthropogenic CO_2 emissions. However, reforestation is needed on a huge scale to significantly offset growing carbon emissions. New carbon-free and renewable energy technologies could each contribute a share of our growing energy needs while contributing little or nothing to atmospheric CO_2 levels. Research will continue to improve the potential of these sources, while reducing their costs. Energy conservation and increasing energy efficiency, represented by actions of government agencies, industries, and individuals, currently represent the least expensive "supply" of carbon-emission-free energy. However, efficiency increases alone will not provide enough energy to sustain current levels of economic growth. Substantially meaningful reductions in total global GHG emissions and stabilization of atmospheric CO_2 can only be achieved by a concerted worldwide effort combining different mitigation approaches.

References

AWEA 2002 *American Wind Energy Association*. 122C Street NW, Washington, DC 2001. Available from: *http://www.awea.org*.

Begley S 1991 On the wings of Icarus. *Newsweek* **May**: 64,65.

Botkin DB 1989 *Can we plant enough trees to absorb all the greenhouse gases?* Paper presented at the University of California Workshop on Energy Policies to Address Global Warming, September 6–8, Davis, California.

Boyd PW, Watson AJ, Law CS, Abraham ER, Truli T, Murdoch R, et al. 2000 A mesoscale phytoplankton bloom in the polar Southern Ocean stimulated by iron fertilization. *Nature* **407**: 695–702.

BTI 2000 *Breakthrough Technologies Institute/Fuel Cells 2000*. The Online Fuel Cell Information Center. Accessed September 9, 2002, from: *http://www.fuelcells.org/*.

Burns LD, McCormick JB and Borroni-Bird CE 2002 Vehicle of change. *Scientific American* **October**: 64–73.

Chisholm SW, Falkowski PG and Cullen JJ 2001 Dis-crediting ocean fertilization. *Science* **294**: 309–310.

CO_2 Capture Project 2002 *An International Effort Funded by Nine of the World's Leading Energy Companies*. Accessed September 7, 2002, from: *http://www.co2captureproject.org*.

David J 2000 *Economic Evaluation of Leading Technology Options for Sequestration of Carbon Dioxide [Masters Thesis]*. Cambridge: Massachusetts Institute of Technology. Available from: *http://sequestration.mit.edu/bibliography/*.

Dresselhaus MS and Thomas IL 2001 Alternative energy technologies. *Nature* **414**: 332–337.

EC 2002 *Atlas Data of Information. The Fourth Framework Programme for Research and Technological Development*. The European Commission. Accessed September 7, 2002 from: *http://europa.eu.int/comm/energy_transport/atlas/homeu.html*.

Falnes L and Lovseth J 1991 Ocean wave energy. *Energy Policy* **19**(8): 768–775.

Fulkerson W, Reister DB, Perry AM, Crane AT, Kash DE and Auerbach SI 1989 Global warming: an energy technology R&D challenge. *Science* **246**: 868–869.

Herzog HJ 2001 What future for carbon capture and sequestration? *Environmental Science and Technology* **35**(7): 148–153.

IEA 2002 *International Energy Agency*. Available from: *www.ieagreen.org.uk*.

Johnson KS and Karl DM 2002 Is ocean fertilization credible or creditable? *Science* **296**: 467, 468.

Kachadorian J 1997 *The Passive Solar House*. White River Junction, Vermont, USA: Chelsea Green Publishing Company.

Keith DW 2001 Geoengineering. *Nature* **409**: 420.

Lovins AB and Lovins LH 2002 Mobilizing energy solutions. *The American Prospect* **13**(2): 18–21.

MacLean H, Heather L and Lave LB 1999 Environmental implications of alternative-fueled automobiles: air quality and greenhouse gas tradeoffs. *Environmental Science and Technology* **34**: 225–231.

Mendelsohn R 2000 Efficient adaptation to climate change. *Climatic Change* **45**: 583–600.

Monastersky R 1995 Iron versus the greenhouse: oceanographers cautiously explore a global warming therapy. *Science News* **148**: 220–222.

Lenton TM and Cannell MGR 2002 Mitigating the rate and extent of global warming: an editorial essay. *Climatic Change* **52**: 255–262.

Nakic′enovic′ N 1993 CO_2 mitigation: measures and options. *Environmental Science and Technology* **27**(10): 1986–1989.

NCPV 2002 *National Center for Photovoltaics. U.S. Department of Energy*, Washington DC. Accessed September 7. from: *http://www.nrel.gov/ncpv/.*

Newman WS and Fairbridge RW 1986 The management of sea-level rise. *Nature* **320**: 319–321.

NREL 2002 *National Renewable Energy Laboratory. U.S. Department of Energy*, Washington, DC. Accessed September 7, 2002 from: *http://www.nrel.gov.*

Ogden JM 1999 Prospects for building a hydrogen energy infrastructure. *Annual Review of Energy and the Environment* **24**: 227–279.

Rubin ES, Cooper RN, Frosch RA, Lee TH, Marland G, Rosenfeld AH, et al., 1992 Realistic mitigation options for global warming. *Science* **257**: 148, 149, 261–266.

Schneider SH 2001 Earth systems engineering and management. *Science* **409**: 417–421.

Schön JH, Kloc C, Bucher E and Batlogg B 2000 Efficient organic photovoltaic diodes based on doped pentacene. *Nature* **403**: 408–410.

Stultz SC and Kitto JB, eds 1992 *Steam: Its Generation and Use*. (40th Edition). New York, NY: Babcock and Wilcox Company, Barberton, Ohio, pp. 24-1–24-13.

Titus JG 1998 Rising seas, coastal erosion, and the takings clause: how to save wetlands and beaches without hurting property owners. *Maryland Law Review* **57**: 1279–1399.

U.S. DOE 2002a *U.S. Department of Energy, Office of Science, Carbon Sequestration Program*, Washington, DC. Available from: *http://cdiac2.esd.ornl.gov.*

U.S. DOE 2002b *U.S. Department of Energy Biofuels Program*, Washington, DC. Accessed September 7, 2002b from: *http://www.ott.doe.gov/biofuels/.*

U.S. DOE 2002c *U.S. Department of Energy*, Washington, DC. *http://www.energy.gov/sources/index.html.*

U.S. EPA 2000 *Average Annual Emissions and Fuel Consumption for Passenger Cars and Light Trucks.* Air and Radiation, Office of Transportation and Air Quality. EPA420-F-00-013.

World Bank 2002 *World Development Indicators 2001.* Available from: *http://www.worldbank.org.*

Chapter 12

Policy, Politics, and Economics of Climate Change

"The economy is a wholly-owned subsidiary of the environment."

Timothy Wirth, Former US Senator

"We discredit the White House claim that protecting the climate will harm the economy."

Heidi Wills, Chair of the City Council Energy and Environmental Policy Committee, Seattle, Washington, February 21, 2002

Introduction

Humans are altering the Earth's climate. The problem is global. The greenhouse emissions of one country impact all countries. The solution to such a global problem can only come through international cooperation. Cooperative efforts to solve the problem are under way. However, modern economies depend on fossil-fuel energy, and reducing this dependence and greenhouse gas (GHG) emissions is likely to take considerable time. Also, policy makers, even if they agree on the severity of the problem, often disagree on the best approach to solve the problem.

The science of climate change, like all science, contains some degree of uncertainty. Conclusions are always subject to modification as new information is discovered. However, most climate scientists agree that if recent rates of GHG emissions continue, the result will be widespread serious damage to ecosystems and human enterprises.

In the face of uncertainty, many policymakers, as well as scientists and individuals, advocate the precautionary principle. Thus, they argue, reducing GHG emissions involves some costs, but it will also produce benefits. Assuming worst-case climate change predictions, reducing GHG emissions now is an insurance policy against a possible global environmental catastrophe. Taking action now will be less costly than waiting until later.

Climate Change: Causes, Effects, and Solutions John T. Hardy
 ISBNs: 0-470-85018-3 (HB); 0-470-85019-1 (PB)

To achieve GHG emission reductions, governments have formulated national policies and signed international agreements. Almost 100 countries have agreed to an international treaty – the 1997 Kyoto Protocol – to reduce GHG emissions and lessen the rate of climate change. However, some argue that science has not yet proven beyond a doubt that climate change will result in significant damage to ecosystems or economies. They believe that reducing fossil-fuel use will place too large a burden on industry and the economy.

Thus, the climate-change debate has moved into the political and policy arena. The economic costs and benefits of different policy options are an active ongoing area of research and debate by opposing groups. On the one hand, most conservation groups and some scientists, economists, and politicians, through reports, presentations, and lobbying actively, promote policies to mitigate climate change. On the other hand, many fossil fuel and energy trade groups and a few scientists, economists, and politicians argue for unrestricted fossil-fuel emissions or policies of minimal climate-change action.

Meanwhile, one thing remains clear – unless the largest contributing countries sharply reduce their emissions, atmospheric GHG concentrations will continue to increase rapidly. The consequences of not achieving a global agreement to reduce GHG emissions, although not certain, are likely to be costly.

International Cooperation – From Montreal to Kyoto

Climate change is a global problem and can only be solved through international cooperation (Luterbacher and Sprinz 2001). International cooperative efforts to study the Earth's climate have grown since the founding of the International Meteorological Organization (IMO) in 1873, culminating in international agreements to mitigate anthropogenic climate change (Table 12.1).

The 1987 Montreal Protocol regarding stratospheric ozone depletion by chlorofluorocarbons was a landmark in international governmental cooperation on the environment (UNEP 2001). For the first time, countries from around the world approved an international agreement on environmental protection. They agreed to reduce the production and consumption of 96 chemicals (mostly chlorofluorocarbons and halons) that deplete the stratospheric ozone layer and lead to increases in damaging ultraviolet radiation. One hundred and eighty countries have ratified the Montreal Protocol, but ratification of subsequent amendments that accelerate the chlorofluorocarbon phase-out with stronger control measures lag far behind. The agreement significantly reduced the rate of ozone depletion and avoided tens of millions of cases of human cancer that would have otherwise resulted from increased ultraviolet radiation (UNEP 2001). Perhaps as important, the Protocol provided an example of effective international cooperation to solve a global environmental problem. This example would subsequently be applied in addressing global greenhouse warming. In particular, Montreal demonstrated the usefulness of the "precautionary principle," that is, waiting for complete scientific proof can delay action to the point at which damage is irreversible.

June 1992 marked an important first step in international efforts to address climate change. One hundred and ninety six countries met at Rio de Janeiro, Brazil and through negotiation agreed on a "Framework Convention on Climate Change" (UNFCCC 2003). The goal of the convention was to "stabilize greenhouse gas concentrations in the atmosphere at a level that will prevent a dangerous anthropogenic interference with the climate

Table 12.1 Climate conferences and treaties.

Conference	Organizer	Location and description	Conclusion and principal recommendations
Conference on the Assessment of the Role of Carbon Dioxide and Other Greenhouse Gases	WMP & UNEP, ICSU	Villach, Austria *http://unfccc.int/resource/ccsites/senegal/fact/fs214.htm*	Significant climate change highly probable States should initiate consideration of developing a global climate convention
Montreal Protocol	UNEP	Montreal *http://www.unep.org/ozone/mp-text.shtml*	International treaty on substances that deplete the ozone layer
Toronto Conference	Government of Canada	Toronto *http://www.unep.ch/iucc/fs215.htm*	Global CO_2 emissions should be cut by 20% by 2005 States should develop a comprehensive framework convention on the law of the atmosphere
Ministerial Conference on Climate Change	Netherlands	Noordwijk, Netherlands *http://www.unep.ch/iucc/fs218.htm*	Industrialized countries should stabilize greenhouse gas emissions as soon as possible Many countries support stabilization of emissions by 2000
IPCC First Assessment Report	WMO & UNEP	*http://www.ipcc.ch/pub/reports.htm*	Global mean temperature likely to increase by about 0.3 °C per decade, under business-as-usual emissions scenario
Second World Climate Conference	WMO & UNEP	Geneva *http://unfccc.int/resource/ccsites/senegal/fact/fs221.htm*	Countries need to stabilize greenhouse gas emissions Developed states should establish emissions targets and/or national programs or strategies
United Nations Conference on Environment and Development	UN	Rio de Janeiro *http://www.un.org/esa/sustdev/agenda21.htm*	FCCC opened for signature
Conference of the Parties (COP 1)	UNFCCC	Berlin *http://unfccc.int/*	Authorized negotiations to strengthen FCCC
Conference of the Parties (COP 3)	UNFCCC	Kyoto *http://unfccc.int/resource/convkp.html*	Kyoto Protocol signed by many countries

(*continued overleaf*)

Table 12.1 (*continued*)

Conference	Organizer	Location and description	Conclusion and principal recommendations
Conference of the Parties (COP 6)	UNFCCC	The Hague *http://cop6.unfccc.int/*	European and US negotiators fail to agree
World Summit on Sustainable Development	UN	Johannesburg *http://www.johannesburgsummit.org/*	

system." This goal was to be achieved within a time frame "sufficient to allow ecosystems to adapt naturally to climate change, ensure food production is not threatened, and enable economic development to proceed in a sustainable manner" (UNFCCC 2003).

The Framework Convention was the first step. Countries continued to work together to formulate an international binding agreement to effectively reduce GHG emissions.

The 1997 meeting in Kyoto, Japan, drafted the "Kyoto Protocol" (UNFCCC 2003). The Protocol calls for industrialized countries to reduce their combined GHG emissions to 5.2% below 1990 levels within the period 2008 to 2012. For the United States, the world's largest emitter of GHGs, Kyoto mandated a 7% reduction in emissions. Annex I (developing) countries were not required to reduce GHG emissions. The Protocol includes six gases aggregated into CO_2 equivalents on the basis of their individual greenhouse warming potential (GWP), that is, their radiative forcing (heat-trapping) potentials. Countries joining agreed to

- make available national inventories of greenhouse gas emissions and sinks;
- formulate national programs to mitigate climate change;
- promote technologies to reduce emissions;
- promote and sustain sinks, for example, forests;
- cooperate in information and research.

However, The Kyoto Protocol does not specify how countries might achieve the targeted reduction. Many subsequent meetings have attempted to fill in details and forge a consensus for a strong international agreement.

Meeting Kyoto Targets

The process of negotiating and implementing the Kyoto Protocol illustrates the differences in philosophy, politics, and approach between nations. Some countries, such as the United States, argue that the Kyoto Protocol goes too far and that the goals are unrealistic. Others, such as several small island states threatened with a rising sea level, agreed to the Protocol, but specifically stated their belief that it does not go far enough to meet the goal of preventing a dangerous anthropogenic interference with the climate system.

Emission reduction targets and approaches to meeting those targets differ by country. The Kyoto Protocol requires participant countries to formulate and publish their individual national implementation plans, that is, to state their goals and how these are being achieved. Periodic national reports are available through the UNFCCC (2003). For example, under the Protocol the United Kingdom agreed to reduce GHG emissions 12.5% below 1990 levels by 2008 to 2012. Under an even more ambitious domestic goal, the United Kingdom plans to reduce CO_2 emissions 20% below 1990 levels by 2010. The approaches of some countries are more likely to achieve GHG stabilization than those of others (Table 12.2). Generally, approaches that provide economic incentives for emissions reductions are likely to be both cost-effective and successful (IPCC 2002).

Different approaches to meeting the emissions reductions mandated by the Kyoto Protocol include tax incentives (both negative and positive), voluntary incentives, emissions

Table 12.2 Country plans to curb greenhouse gas emissions (Adapted and updated from Stone R 1994. Most nations miss the mark on emission-control plans. *Science* **266**: 1939).

Country	Target	Key measure	Critique[a]
Denmark	Reduce annual CO_2 emissions to 80% of 1988 levels by 2005	Carbon tax, improve energy efficiency, increase use of alternative energy sources	Likely to succeed
France	Limit annual CO_2 emissions to 2 tons per person by 2000	Greater use of public transportation and nuclear power, proposed carbon tax	Emissions will increase as population grows
Germany	Reduce annual greenhouse gas emissions to 50% of 1987 levels by 2005	Close eastern factories, promote wind and photovoltaic energies, tax relief to alternative energy consumers	Little impact on use of coal
Japan	Stabilize per capita annual CO_2 emissions at 1990 levels by 2002	Improve vehicle fuel efficiency, build more nuclear power plants	More coal plants, few alternative sources
Spain	Cap annual CO_2 emissions at 125% of 1990 levels by 2000	Rely more on natural gas and less on coal	CO_2 emissions will increase
United Kingdom	Reduce annual CO_2 emissions to 1990 levels by 2000	Taxes on fuel and power consumption, convert from coal to natural gas	Likely to achieve target
United States	Reduce annual CO_2 emissions to 1990 levels by 2000	Increase efficiency of utilities, clean car initiative, industry incentives	Relies heavily on voluntary measures

[a] An international coalition of environmental groups critique of available plans.
Sources: Individual National Plans, US Climate Action Network, Climate Network Europe.

trading, carbon, carbon sink credits, and clean development mechanisms.

Taxes

A reduction in fossil-fuel use can be promoted using either the carrot or the stick approach. That is, positive incentives could include government tax credits for the development or installation of alternative energy sources, energy conservation measures such as additional building insulation, or carpooling. However, many individuals and groups argue for also including negative incentives. This approach, based on the "polluter pays" principle, proposes that the cost of pollution control should be included in the goods and services produced. It is easier to assess the quantity of pollutants emitted than the resultant ecological damage. These measures include a carbon tax. How carbon tax revenues are used will determine the effectiveness of this approach. An additional gasoline tax, although currently unpopular, would reduce gasoline consumption. Also, investing the gas tax revenues in alternative energy systems would contribute further to reducing GHG emissions.

Many European countries have adopted this concept. Denmark, for example, intent on reducing GHG emissions 20% below 1988 levels by 2005, assesses a tax on CO_2 of $16 per tonne for households and half that for industries, while renewable energy is not taxed. Denmark has also instituted major tax and economic incentives to implement alternative energy (e.g. wind power) and waste recycling, and they plan to double their forested area over the next 80 to 100 years (UNFCCC 2003).

Voluntary Incentives

US negotiators stress a free market approach to reducing GHG emissions, for example, voluntary market-based incentives that include labeling household appliances with energy star ratings, so consumers can choose energy-efficient models. Under the Voluntary Reporting of Greenhouse Gases Program of the US Department of Energy, companies may submit reports of their actions to reduce GHG emissions, but no independent verification of the documentation is required. Reported emissions reductions in 2,000 were 2.7% of total US GHG emissions (UNFCCC 2003).

Emissions Trading

In 1990, the United States proposed a global market of free trading in carbon emissions (Sun 1990). Each country would be allocated a CO_2 or GHG emissions quota or fixed number of carbon credits. If country X can show that it has emitted less than its quota, then it can sell its excess credits to another country. Conversely, if country X really needs to emit more, it can purchase credits on an open global market. A similar system is now used within the United States where SO_2 and NO_X, allowances established by EPA, can be bought or sold. These allowances can be bought at EPA's annual auction, through a commodities broker or through environmental groups that "retire" allowances, so they cannot be used (US EPA 2002). To actually reduce total global emissions of GHG, the overall global allocation would need to be frozen and then reduced from its current level. US negotiators and even some large coal-fired utilities have favored this approach to emissions reduction.

Emissions trading has promise, but many difficult political and social questions challenge implementation. Using gross domestic product (GDP) as a basis for initial credit allocations would be advantageous to large industrialized countries. Population as a base for credit allocation would favor countries with large populations such as China and

India, while countries such as the United States would need to greatly reduce emissions. Should energy-efficient nations like Japan, or nations that rely heavily on nuclear energy, like France, be awarded with extra credits? Perhaps, some combination of national population and GDP could form an acceptable basis for global-emissions credits. Finally, how will the "value" of carbon credits be determined? Some economists suggest that the cost of emissions permits should "float" on the free market until they hit a predetermined ceiling (Pizer 2002).

Authors at the US Brookings Institute call an international permit trading system *flawed* and *doomed to failure* because it would focus on emissions stabilization rather than reduction, involve large international transfers of wealth, and lead to major changes in exchange rates and patterns of international trade. They propose a more modest step, that is, an international agreement to set up a system of national permits and emissions fees (McKibbin and Wilcoxen 1997).

Carbon Sink Credits

If carbon credits are considered as part of an international emissions allocation strategy, then some countries (most notably the United States) argue that a country's current forested area or annual planting of trees should be subtracted as a carbon sink and that amount should be added to their carbon allocation. In a November 1999 meeting, US negotiators suggested that forest and soil sinks should count toward about half of the US carbon-reduction target. Other industrialized nations objected to this strategy and to the fact that the world's largest single source of GHGs should be allowed to meet its obligation without substantially reducing its emissions. A report by the UK Royal Society suggests that carbon sinks provide only a limited and transient answer to sequestering the large quantities of carbon released by fossil-fuel combustion (Pickrell 2002).

Clean Development Mechanisms (CDMs)

Another method for meeting emissions targets for the Kyoto Protocol allows an industrialized country to receive credits by joining with a developing country (which under Kyoto has no obligation to reduce emissions) in an emission-reducing project in the developing country. Even this approach raises questions. For example, what types of projects would qualify – construction of nuclear power plants, hydroelectric dams, mass transit systems, or only alternative energy projects such as solar or wind power? (Box 12.1).

Post-Kyoto Developments

Success in progressing toward GHG emission reduction targets differs between countries. In March 2000, the European Commission launched the European Climate Change Programme (ECCP) to develop policies including an emissions trading scheme, and ensure that the EU achieves the 8% cut in emissions by 2008 to 2012 to which it committed under the Kyoto Protocol. By 2000, European Union average GHG emissions were 3.5% below 1990 levels (Figure 12.1), but in North America, CO_2 emissions continued to rise to levels more than 13% higher than that in 1990.

To enter into force, the Kyoto treaty needs to be ratified, accepted, or acceded to by 55 countries that emitted, in total, 55% of the industrialized world's carbon dioxide emissions in 1990. As of February 2003, 104 governments, totaling 43.9% of global emissions, had accepted or ratified the Protocol. In spring 2002, the Russian government signaled that they would move ahead and possibly ratify the Protocol.

Box 12.1 Get a higher return on your investments in sustainable energy and energy efficiency

Senter International, an agency of the Netherlands Ministry of Economic Affairs, helps companies investing in renewable energy and energy efficiency in developing countries or in Central and Eastern Europe improve the return on their investments. The Dutch government buys the reduction in greenhouse gas emissions (*carbon credits*) that these projects generate, thus creating an additional source of income to boost the economic feasibility of projects and accelerate their implementation (Carboncredits.nl 2002).

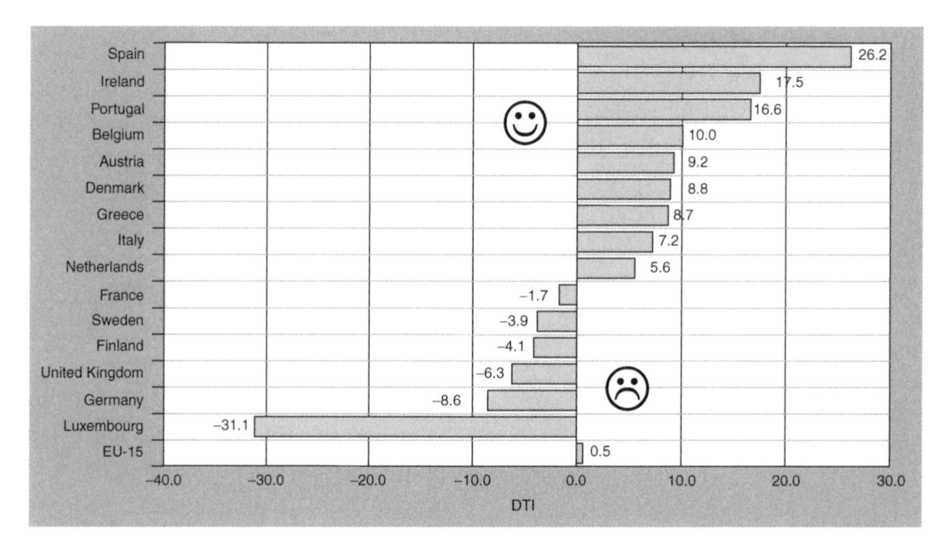

Fig. 12.1 In 2000, total EU greenhouse gas emissions were 3.5% below their 1990 level. The distance-to-target indicator (DTI)[1] is a measure of the deviation of actual greenhouse gas emissions in 2,000 from the linear target path between 1990 and the Kyoto Protocol target for 2008 to 2012, assuming that only domestic measures will be used (From EEA 2002. *European Environment Agency. http://www.eea.eu.int/*).

The United States initially signed the treaty, but later President George W. Bush announced that the US administration would neither support ratification of the Kyoto Protocol nor enforce emission reduction measures in the United States. This policy apparently stems from concerns about negative economic impacts due to costs of emissions reductions. The United States also objected to the fact that mandatory reductions were limited to industrialized nations and that developing countries would not need to meet Protocol emission reduction targets. President George W. Bush called for more research to reduce

[1] The Danish DTI is 0.7 index points if Danish greenhouse gas emissions are adjusted for electricity trade in 1990. This methodology is used by Denmark to monitor progress toward its national target under the EU "burden sharing" agreement. For the EU emissions, total nonadjusted Danish data have been used.

the uncertainties about global warming. In response, the National Academies of Science of 17 countries reaffirmed the science already conducted by the IPCC. They stated that, "The balance of the scientific evidence demands effective steps now to avert damaging changes to Earth's climate" (Science 2002).

In Bonn in July 2001, 180 governments agreed to revisions of the Kyoto rules. Revised mechanisms for implementation included emissions trading and limited allowances for carbon sinks such as forests (Giles 2001). However, environmental groups argued that, by subtracting carbon sinks from emissions allowances, targets are effectively reduced from the 5.2% of the 1990 level agreed on at Kyoto to something more like 1.8% (WWF 2002).

Also, in Bonn, the European Union placed climate change among its top environmental priorities, and established the ECCP to implement policies to meet the goals of Kyoto (EU 2002). Several cooperative European efforts are under way. For example, the European Automobile Manufacturers Association has committed itself to reducing CO_2 emissions from automobiles 25% by 2008. Studies by the European Union estimate that their cost of meeting Protocol emission reduction goals would be reduced 25 to 30% by adoption of a regional EU-wide emissions trading scheme. The EU intends to implement a cap-and-trade system covering 46% of all CO_2 emissions in 2010 (EU 2002).

On August 1, 2001, the US Senate Committee on Foreign Relations voted 19 to 0 in favor of US participation in future climate negotiations. The committee stated that the United States should not abandon "... its shared responsibility to help find a solution to the global climate change dilemma." Perhaps in response, on February 14, 2002, President Bush announced a new climate-change strategy that would set a voluntary "greenhouse gas intensity" target for the United States. The greenhouse gas intensity is the ratio of greenhouse gas emissions (GHG) to economic output (GDP). This ratio has already been decreasing for decades as energy efficiency improves (see Chapter 11). However, even though this ratio is decreasing, because of economic and population growth, total GHG emissions continue to rise.

Some have called this US strategy ineffective "... blowing smoke," and emphasized that, at the very least, any effective reporting scheme must be mandatory (The Economist 2002) (Figure 12.2). The US administration's target of an 18% reduction in energy intensity between 2002 and 2012 will actually allow total emissions to increase 12% over the same period – about the same as usual. Current projections of GHG emissions for the United States indicate an increase of 39% (or with intensity improvements about 33%) over 1990 levels by 2012 (EU 2002).

On June 11, 2002, President George W. Bush spoke about global climate change and hinted at a more active role for the United States in solving the climate-change problem. He noted that the recent warming trend was "due in large part to human activity."

In May 2002, the European Commission and its 15 member nations ratified the Kyoto Protocol. This raised the total number of participating countries to 69 and met the first requirement for the treaty to become international law, that is, ratification by at least 55 countries. However, ratification by additional countries will be needed to meet the second criteria – inclusion of countries emitting 55% of the global 1990 emissions. As of February 2003, countries representing an additional 11.1% of 1990 emissions were needed. The 55% goal could be met through

Fig. 12.2 The attitude of some governments to climate mitigation. Auth © 2002 Philadelphia Inquirer. Reprinted with permission of Universal Press Syndicate.

ratification by the United States or Russia (36.1 and 17.4% of global emissions, respectively) or by a combination of other countries (UNFCCC 2003).

The Politics of Climate Change

As scientific research increasingly confirms the potential serious consequences of human-induced climate change, the debate on climate-change mitigation policies has widened into the political arena. The continuing policy debate has engaged individuals, nongovernmental organizations, national governments, and international organizations (Hecht and Tirpak 1995).

Opposing organizations attempt to influence climate-change policy (Box 12.2). Conservation groups such as The World Wildlife Fund (*www.worldwildlife.org*), Sierra Club (*www.sierraclub.org*), and Earth Policy Institute (*http://earth-policy.org*) advocate rapid implementation of agreements to reduce GHG emissions. On the other hand, industry groups such as the Global Climate Coalition (GCC) (*http://www.globalclimate.org/*) and the American Petroleum Institute (*http://www.api.org/globalclimate/*) coordinate and promote strong and outspoken opposition to controls on GHG emissions. For example, Citizens for a Sound Economy Foundation (*http://www.cse.org/*) reported that in their survey of 36 state climatologists, most believe global warming "is a largely natural phenomenon," so reducing GHG emissions would not affect climate. The more moderate International Climate Change Partnership (*http://www.iccp.net/*) stands for constructive engagement by industry in formulating policy responses to climate change, and includes leading industries such as 3 M Company, AlliedSignal, AT&T, Boeing, Chevron, Dow, Dupont, Eastman Kodak, Enron, and General Electric. Other industry groups involved in the climate-change debate include the

Box 12.2 Opposing views

The following excerpts are from the websites of two nongovernmental organizations.

The Global Climate Coalition

"... the coalition... opposed Senate ratification of the Kyoto Protocol that would assign such stringent targets... that economic growth in the US would be severely hampered and energy prices for consumers would skyrocket. The GCC also opposed the treaty because it does not require the largest developing countries to make cuts in their emissions."

World Wildlife Fund

"Climate change and global warming... threaten the survival of nature and the well-being of people around the world... reducing CO_2 emissions ... can help prevent unnecessary loss of life, reduce human suffering and economic disruption from global warming and save plants and animals from being wiped out over huge areas in the future."

World Business Council for Sustainable Development (*http://www.wbcsd.ch/*) and the Business Council for Sustainable Energy (*http://www.bcse.org/*) that promote environmentally responsible development and clean energy technologies. All groups, through publications and talks, attempt to influence energy and climate-change policy.

Kyoto Without the United States

US politics surrounding the climate-change issue represent a tug-of-war between those who fear potential economic damage from restrictions on fossil-fuel combustion and those who fear for the health of the planet and the longer-term economic consequences resulting from climate change. Policies have swung from one pole to another with each new US administration (Box 12.3). In March 2001, US President George W. Bush announced that the US administration would not support ratification of the Kyoto Protocol and on June 11, 2001, he called the treaty "fatally flawed." "No one can say for certainty what constitutes a dangerous level of warming, and therefore what level must be avoided." He also attacked the agreement for not including developing countries, like China, the world's second-largest source of greenhouse gases. Thus, Protocol participants were left with two options: (1) adjust the Protocol to meet US objections, that is, force developing nations to also reduce emissions or (2) ratify the Protocol without the United States. Some European groups point to the huge investment in intellectual and research effort that has gone into achieving the Kyoto Protocol. Renegotiating a new agreement would not necessarily improve the agreement and would delay implementation for years.

In a July 2001 survey of four European countries, 80% of respondents agreed that European governments should begin making GHG emission cuts in line with the Kyoto Protocol whether the United States does or not (WWF 2002). An analysis of the economic implications of unilateral European ratification suggests several things (Harmelink et al. 2002). First, without emissions trading, different countries would experience a 0.3 to

Box 12.3 History of US climate and energy policy

Climate change is a global problem. However, as the world's largest single emitter of greenhouse gases (GHGs), the policies and politics surrounding this issue in the United States are particularly critical. In the post-WW II era, US energy consumption exploded. By 1973, energy consumption was growing three times faster than the population. Autos were getting 10% less gas mileage than in 1961 and refrigerators used five times more energy than in the 1940s.

Then, in 1973, one event sparked an abrupt halt to this energy binge. The Organization of Petroleum Exporting Counties (OPEC) reduced oil production. Because of the high rate of consumption in the United States, shortages quickly developed, resulting in long lines at auto gasoline stations. Some in the US Congress suggested that a major commitment to alternative energy development and energy conservation would need to be undertaken. The scope of this commitment would need to match that of the Manhattan Project (to develop the atom bomb in WW II) or the Apollo Project (to land a human on the Moon). The federal government (Carter administration) instituted major energy conservation measures, including tax benefits for upgrading residential insulation, installation of solar water heaters, wind power generators, and so on. However, by 1980, with oil supplies again abundant, the Reagan Administration put any idea of energy conservation on hold. By the mid 1980s, all alternative energy sources together added up to less than 3% of US energy needs. As one politician stated, the entire alternative energy budget "... is not enough to buy a booster rocket on the space shuttle."

In 1987, the US Congress approved, and President Ronald Reagan signed the US Climate Protection Act (PL 100–204). This law directed the EPA and the Department of State to prepare a report describing the scientific understanding of climate change and the policy options for dealing with it. This legislation recognized the importance of protecting the Earth's climate from human-induced change. Unfortunately, it fell far short of any actual actions or regulations that might mitigate such change.

1988 was a critical year in the politics of climate change. In response to a severe drought in Central North America and high summer temperatures almost everywhere, many people began to ask the question – is a radical shift in climate actually under way? The whole issue of human-induced climate change was thrown into the forefront of public debate. US Senator Timothy Wirth called the greenhouse effect "the most significant economic, political, environmental and human problem facing the 21st Century." Senator Albert Gore studied the climate issue, traveled to the Antarctic to observe ozone depletion research and spoke at meetings and conferences on climate change.

During the Presidential candidate debates and the campaign leading up to the election of 1988, candidate George Bush (senior) said, if elected, "... the greenhouse effect will be matched by the White House effect." In July 1988, 16 senators unveiled a comprehensive plan to combat global warming. It would have forced reductions in CO_2 emissions of 20% by the year 2000 and required the administration to draft a national energy plan emphasizing ways to reduce fossil-fuel use. The bill authorized $1 billion through 1992 for research and development on solar energy and safer nuclear reactors. It also authorized \$1.5 billion to provide birth control information internationally, but Congress defeated the bill.

Even individual state governments began a flurry of climate-related research and legislation. For example, the 1988 Washington State Legislature held workshops and heard expert testimony on climate change. On February 2, 1989, the state approved a resolution advising the US Congress to support research on climate change.

In 1989, the US Congress held a hearing chaired by Senator Joseph Lieberman on "Responding to the Problem of Global Warming" (CEPW 1989). The committee gathered information and took testimony from experts about the problem of global warming. However, the new Republican administration suggested that much more research was needed before any action to stem GHG emissions could be justified. Indeed, through the Department of Energy, the Republican Administration requested a 50% reduction in the national energy conservation program – a budget that had already been cut almost 70% from 1980 levels.

By 1990, many in the United States recognized the need to increase detailed research on climate change. A National Academy of Sciences Panel on Global Change was formed and chaired by former Washington State Governor Daniel Evans. Also, the US Global Change Research Program was formed to coordinate research among 10 federal agencies. Overall, US funding for research on climate change has grown substantially since 1990. By fiscal year 2001–2002, this program had a budget of \$1.64 billion to study climate change, almost a tenfold increase since 1989–1990 (USGCRP 2002).

However, in terms of energy consumption, little has changed in the past few decades; refrigerators use 600 kWh $year^{-1}$ when 300 kWh $year^{-1}$ is possible and the United States lags far behind Japan and Europe in energy efficiency. Automobiles achieve an average 10 km L^{-1} (24 mpg) of gas when 30 km L^{-1} (70 mpg) is possible.

The new administration of George W. Bush, in May 2001, announced a new US energy plan. The plan contained some elements of energy efficiency (e.g. tax credits for alternative energy sources and funds for fuel cell research), but emphasized expanding supplies of fossil fuels. Among other things, it recommended that the United States

- study impediments to federal oil and gas exploration and development;
- consider economic incentives for offshore oil and gas development;
- reexamine the current federal, legal, and policy regime regarding siting of energy facilities in the coastal zone and Outer Continental Shelf;
- consider oil and gas exploration and development in the Arctic National Wildlife Reserve;
- expedite renewal of the Trans-Alaska Pipeline System rights-of-way;
- invest \$2 billion over 10 years in research on coal technology;
- reduce the time and cost of the hydropower licensing process;
- support the expansion of nuclear energy as a major component of our national energy policy.

Energy policy has become a hot topic for debate in the United States. Generally, Republicans favor expansion of fossil-fuel supplies by such measures as opening the Arctic National Wildlife Refuge to oil drilling, while Democrats favor increasing renewable energy supplies and increasing auto gasoline efficiency requirements. Republicans charge that Democrats "... completely ignored harsh realities in favor of their distant dreams" (Adams 2001).

On March 13, 2002, the US Senate debated raising the Corporate Average Fuel Economy (CAFÉ) standards for passenger cars from 24 to 36 miles per gallon over a 13-year period. It would have been the first increase in standards in 25 years. The Bush administration opposed the change and the house rejected it.

In contrast to this Republican "supply-side" energy policy, many European and other countries, as well as some parties in the United States, advocate policies that promote energy efficiency and conservation (see Chapter 11). These policies include taxes on fossil-fuel consumption (i.e. a carbon tax), greatly expanding mass transit, conducting large-scale reforestation programs, and providing tax incentives for alternative energy research or installation (hydropower, biomass fuel, wind energy, solar energy, fuel cells).

1.9% lower economic growth by 2010 as a result of implementing the Kyoto Protocol CO_2 reductions. Second, if emissions trading, along with reductions in other GHGs were considered, most industrialized countries will experience only a 0.1% (or for the United States about 0.2%) reduction in GDP in 2010. Third, the majority of implementation costs would be compensated for by overall reductions in air pollution control costs, that is, by reducing CO_2 emissions, other air pollutants are reduced at the same time. Finally, the study concluded that 85 to 95% of the reduction target for the European Union could be achieved without affecting its economic competitiveness. In some scenarios, the European Union, because of increased energy efficiency and trade competitiveness, could actually benefit economically from US rejection of the Kyoto Protocol.

A growing number of city and local governments around the world are addressing climate change. Cities for Climate Protection (CCP) is a campaign of the International Council for Local Environmental Initiatives (ICLEI 2002). The CCP offers a framework for local governments to develop a strategic agenda to reduce emissions and global warming. Five hundred local governments, representing 8% of global GHG emissions, are participating in the Campaign. For example, in the United States, the Seattle City Council unanimously passed a resolution to fully eliminate or mitigate fossil fuels associated with the City's electric supply. They also passed a "Kyoto" resolution to inventory all city GHG emissions and reduce them, through a variety of actions, to 7% below 1990 levels, as proposed for the United States by the Kyoto Protocol.

Benefits and Costs of Mitigating Climate Change

How much we benefit economically from mitigating climate change depends on what the costs would be if we do not. However, climate models are somewhat uncertain to begin with, and when their output is coupled to economic models and regional differences, the uncertainties loom even larger. Thus, the total global cost of climate change is not known. We must rely on individual sector-by-sector (e.g. agriculture, fisheries, etc.) and national estimates of damage.

One of the direct benefits of implementing an agreement like the Kyoto Protocol is reduced fossil-fuel consumption. This alone would have several consequences. For many industrialized countries, benefits would include increased energy security resulting from a decreased need for oil imports, new business opportunities for companies producing alternative energy technologies, and increased energy efficiency and economic competitiveness.

The benefits of policies to mitigate climate change extend beyond direct effects on climate to what are termed *ancillary benefits* (OECD 2000). These include health, ecological, economic, and social benefits. If these benefits can be given monetary value, they can then be subtracted from the costs of mitigation. For example, sulfur dioxide (SO_2), a common air pollutant from fossil-fuel combustion, causes acidic rain damage to habitats, and as a lung irritant impacts human health. Therefore, money spent reducing fossil-fuel emissions not only has the benefit of reducing global warming but also of reducing other air pollutants such as SO_2. Economically, this means a reduction in SO_2 abatement costs as well as a reduction in health costs related to air quality. In fact, a 15% reduction in CO_2 emissions would result in reductions in many harmful pollutants including ozone, particulate matter, and SO_2 (Figure 12.3).

Another benefit of emissions reductions will be a decrease in costs for repairing the ecological and health damages of climate change (Chapters 5–10). An international insurance agency estimates that more frequent tropical storms, sea-level rise, and damage to food and water supplies resulting from a doubling of atmospheric CO_2 will cost more than $300 billion per year (Munich Re 2002).

The costs to meet Kyoto emission reduction targets are difficult to estimate. Numerous studies using computer models that incorporate economic and energy parameters predict a variety of outcomes. Modeling approaches including bargaining and game theory (Nash 1953) are numerous, complex, and beyond the scope of this chapter (see Mabey et al. 1997). However, many estimated costs for individual economic sectors have been published. For example, climate-change mitigation may involve a shift from other fossil fuels to cleaner and more efficient natural gas, requiring tens of thousands of new kilometers of gas pipelines and tripling the cost of electric generation (Anon 2000).

Costs for the United States to meet Kyoto targets range from about 0.2 to 4% of GDP per year with typical estimates of about 1 to 2.5% of GDP. The GCC, the industry group that generally opposes Kyoto, estimates that US economic losses from the Protocol by 2010 would range from $120 billion to $440 billion dollars per year. If US growth in GDP continues to grow at rates similar to the past 20 years, this would mean a 0.9 to 3.4% decrease in 2010. In other words, the GCC estimate of costs agrees with those of other studies. In the United States, an innovative energy plan, exceeding Kyoto Protocol goals by reducing emissions 10% below 1990 levels, could be achieved at a net cost of $530 per household (Alliance to Save Energy et al. 1997).

Thus, the costs of meeting Kyoto targets could be significant. However, the costs of not meeting the targets may be just as great. For example, The American Solar Energy Society estimates that continued use of fossil fuels will cost the United States about $100 billion (1989 dollars) a year in environmental and

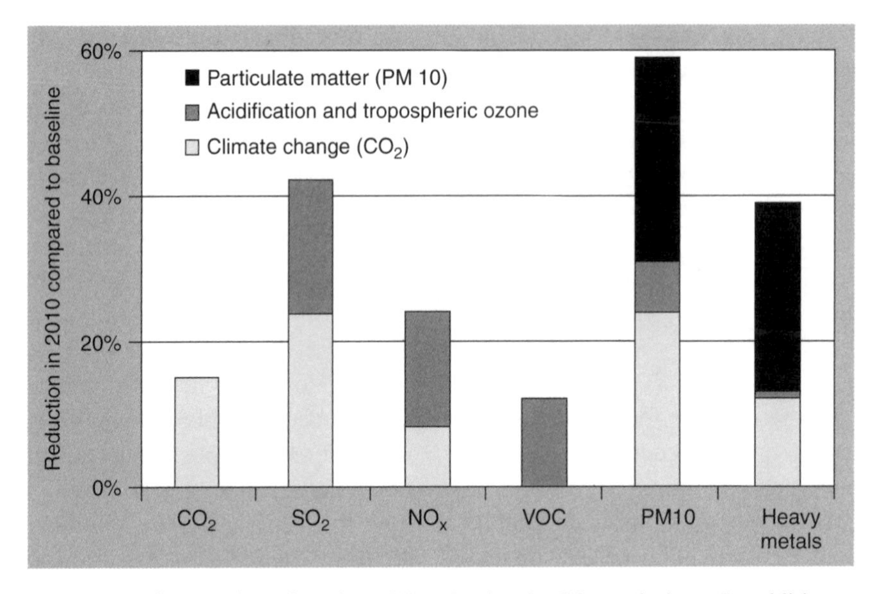

Fig. 12.3 Ancillary (spillover) benefits of a 15% reduction in CO_2 emissions. In addition to reductions brought about by acidification and tropospheric ozone reduction policies and PM10 reduction measures, CO_2 reduction leads to additional emission reductions of 24% for SO_2, 8% for NO_x, 24% for PM10. If no climate-change policies were implemented, an additional investment of 6 billion euros per year would be needed to reach acidifying emissions reduction targets. SO_2 = sulfur dioxide, NO_x = nitrogen oxides, VOC = volatile organic compounds, PM10 = particulate matter (Reproduced from Wieringa K 2001. *European Environmental Priorities: An Integrated Economic and Environmental Assessment.* RIVM Report 481505010, National Institute of Public Health and the Environment (Rijksinstituut voor volksgezondheid en milieu) Bilthoven, The Netherlands, *http://arch.rivm.nl/ieweb/ieweb/*).

health costs alone (CEPW 1989). Gains in energy efficiency of 10 to 30% can be accomplished along with net economic gains (Bruce et al. 1996).

Emission reduction targets of the Kyoto Protocol are based on an aggregate of the warming potential of each of the GHGs. However, each gas has a different atmospheric lifetime, so their radiative forcing or warming potential alone cannot predict overall economic costs. The impact of one additional ton of a specific gas will depend on the mixture of other gases already present at that time, and when and over what time period it is added. In principle, substantial savings would be possible if emissions reductions focused on countries and gases that yield the largest reduction in warming potential per dollar expended. In practice, programs such as the Kyoto Clean Development Mechanisms will be needed to ensure appropriate capital flows and technology transfers between countries (Bruce et al. 1996).

The risk of ecological, human, and economic damage from human-induced climate change is substantial and well documented. Any costs of reducing GHGs as insurance against climate-change damage is like bearing the small cost of inoculating our children against diseases even though we are not certain they will be exposed to them. It has also been likened to a game of Russian roulette with negligible short-run benefits (unconstrained fossil-fuel consumption) weighed against huge potential losses (from climate-related disasters). "Even worse, the gun is pointed not so much at us as at our children" (DeCanio 1997).

The Future – What is Needed?

Many now consider the Kyoto Protocol only as a first step in what may be a negotiating process lasting decades. Even if the Protocol were fully implemented soon, and industrialized countries held their emissions at the targeted levels for the rest of the twenty-first century, GHG concentrations in the atmosphere will continue to rise. The time at which global CO_2 exceeds twice the preindustrial level would only be delayed by about a decade (Dooley and Runci 2000). Long-term GHG stabilization will require application of new energy technologies; yet in many industrialized countries, investment in publicly and privately funded energy research and development is actually declining (Dooley and Runci 2000). For example, between 1985 and 1998, the United States, the European Union, and the United Kingdom collectively reduced their public sector investments in energy R&D by 35% in real terms, and US private sector investments fell 53%.

To understand climate change and its effects requires large multidisciplinary research programs. Many nations are indeed expanding their research efforts to meet the challenge. For example, in the United Kingdom, the National Research Councils have come together to create the multibillion-dollar Tyndall Centre for Climate Change Research. The Centre integrates scientific, social, and technological research and develops sustainable solutions to the societal challenges of climate change. In the United States, the US Global Change Research Program, within the Executive Office of Science and Technology Policy, coordinates research among federal agencies to provide a sound scientific basis for national and international decision making on global change issues.

Developing countries will contribute an increasing share of global GHG emissions. For example, China's contribution to total global carbon emissions is expected to increase from 7% in 1990 to 25% in 2020. For Southeast Asia, this percentage will increase from 7% in 1990 to 25% in 2065. In all, the developing world will account for about 75% of future energy growth and most of this will come from fossil fuel. This means that participation of developing countries in international agreements and capacity building in terms of non-fossil-fuel technology is critical to stabilizing GHGs. Some authors argue for a century-long plan with a target of reducing emissions to 50% of their 1990 levels by 2100 and thus stabilizing atmospheric CO_2 at about 550 ppmv while accommodating significant growth in per capita income (Dooley and Runci 2000). Future agreements might include a "graduation" or "ability to pay" clause in which non-Annex 1 countries (developing nations) agree that once their economy reaches a certain per capita income, they automatically become bound by agreed-on emission reduction targets. By 2100, most countries could potentially "graduate" to this economic level (Edmonds and Wise 1999).

In 1997, a group of 2,500 US economists, including 8 Nobel Laureates, signed a public pronouncement calling for preventative steps to deal with the threats of global warming (Box 12.4).

Summary

Climate change has emerged as a major political issue. Some industry groups point to the uncertainties of climate-change science and the economic costs involved in reducing

Box 12.4 Statement endorsed by over 2,500 economists including 8 Nobel Laureates

1. The review conducted by a distinguished international panel of scientists under the auspices of the Intergovernmental Panel on Climate Change has determined that "the balance of evidence suggests a discernible human influence on global climate." As economists, we believe that global climate change carries with it significant environmental, economic, social, and geopolitical risks, and that preventive steps are justified.
2. Economic studies have found that there are many potential policies to reduce greenhouse gas emissions for which the total benefits outweigh the total costs. For the United States in particular, sound economic analysis shows that there are policy options that would slow climate change without harming American living standards, and these measures may in fact improve US productivity in the longer run.
3. The most efficient approach to slowing climate change is through market-based policies. In order for the world to achieve its climatic objectives at minimum cost, a cooperative approach among nations is required – such as an international emissions trading agreement. The United States and other nations can most efficiently implement their climate policies through market mechanisms, such as carbon taxes or the auction of emissions permits. The revenues generated from such policies can effectively be used to reduce the deficit or to lower existing taxes.

The original drafters of this statement are Kenneth Arrow, Stanford University; Dale Jorgenson, Harvard University; Paul Krugman, MIT; William Nordhaus, Yale University; and Robert Solow, MIT.

The Nobel Laureate signatories are Kenneth Arrow, Stanford University; Gerard Debreu, University of California, Berkeley; John Harsanyi, University of California, Berkeley; Lawrence Klein, University of Pennsylvania; Wassily Leontief, New York University; Franco Modigliani, MIT; Robert Solow, MIT; and James Tobin, Yale University.

All signatories endorse this statement as individuals and not on behalf of their institutions.

fossil-fuel consumption. Others, especially conservation groups, respected climate scientists, and economists, argue that the potentially high cost of climate change to ecosystems and humans necessitates reduction in GHG emissions now, even in the face of uncertainty.

The 1992 Framework Convention for Climate Change formed the basis for continuing international cooperative efforts to reduce GHG emissions. The subsequent Kyoto Protocol proposes to reduce overall global GHG emissions below 1990 levels through a variety of mechanisms. It forms the basis for continuing negotiations that may eventually lead to an effective binding international treaty.

Estimates of the costs of mitigating climate change range from moderately substantial to negative (i.e. a cost benefit). Predictions that actions similar to those proposed by the Kyoto Protocol will result in economic disaster seem unscientific. In fact, many of the world's most notable economists believe that actions to reduce GHG emissions will have a long-term net economic benefit.

References

ACEA 2002 *European Automobile Manufacturers Association http://www.acea.be/ACEA/index.html.*

Adams R 2001 *Senate's Energy Policy Rift Grows as Democrats Unfurl their Bill and GOP Loses ANWR Vote*. Congressional Quarterly Dec. 8, p. 2907. Available from: *http://www.cq.com.*

Alliance to Save Energy, American Council for an Energy-Efficient Economy, Natural Resources Defense Council, Tellus Institute, and Union of Concerned Scientists, 1997 *Energy Innovations: A Prosperous Path to a Clean Environment*. Washington, DC, p. 172, *http://www.ase.org/.*

Anon 2000 Greenhouse gas reduction news. *Electric Perspectives* **25**(2): 4–6.

API 2002 *American Petroleum Institute* Available from: *http://www.api.org/globalclimate/.*

Bruce JP, Lee H and Haites EF, eds 1996 *Climate Change 1995: Economic and Social Dimensions of Climate Change. Contribution of Working Group III to the Second Assessment Report of the Intergovernmental Panel on Climate Change*. Cambridge: Cambridge University Press.

Carboncredits.nl 2002 *Senter International, Netherlands Ministry of Economic Affairs. http://www.senter.nl/asp/page.asp?id=i001003&alias=erupt.*

CEPW 1989 *Responding to the Problem of Global Warming*. Committee on Environment and Public Works, United States Senate. Hearing before the Subcommittee on Environmental Protection; August 10. Superintendent of Documents, Washington, DC, p. 122.

DeCanio SJ 1997 *The Economics of Climate Change*. San Francisco: Redefining Progress. Available from: *www.rprogress.org.*

Dooley JJ and Runci PJ 2000 Developing nations, energy R&D, and the provision of a planetary public good: a long-term strategy for addressing climate change. *Journal of Environment and Development* **9**(3): 215–239.

Edmonds J and Wise M 1999 Exploring a technology strategy for stabilizing atmospheric CO_2. In: Carraro C, ed. *International Environmental Agreements on Climate Change*. Boston: Kluwer Academic Publishers, pp. 131–154.

EEA 2002 *European Environment Agency, http://www.eea.eu.int/.*

EU 2002 *European Union*, Available from: *http://www.europa.eu.int/comm/environment/climat/home_en.htm.*

Giles J 2001 "Political fix" saves Kyoto deal from collapse. *Nature* **412**: 365.

Harmelink M, Phylipsen D, de Jager D and Blok K 2002 Kyoto without the U.S.: costs and benefits of EU ratification of the Kyoto Protocol. ECOFYS Energy and Environment, Utrecht, The Netherlands. A report to the World Wildlife Fund. Available from: *www.ecofys.com/climate.*

Hecht AD and Tirpak D 1995 Framework agreement on climate change: a scientific and policy history. *Climatic Change* **29**: 371–402.

ICLEI 2002 *The International Council for Local Environmental Initiatives*. Available from: *http://www.iclei.org/.*

IPCC 2002 *Intergovernmental Panel on Climate Change. United Nations Environment Programme and World Meteorological Organization*, Available from: *http://www.ipcc.ch/.*

Luterbacher U and Sprinz DF, eds 2001 *International Relations and Global Climate Change*. Cambridge: MIT Press.

Mabey N, Hall S, Smith C and Gupta S 1997 *Argument in the Greenhouse: The International Economics of Controlling Global Warming*. London: Routledge.

McKibbin WJ and Wilcoxen PJ 1997 *A Better Way to Slow Global Climate Change*. Policy Brief 17, June 1997. Washington, DC: The Brookings Institution. Available from: *http://www.brook.edu.*

Munich Re 2002 *Munich Re Group, Munich, Germany*, Available from: *www.munichre.com.*

Nash JF 1953 The bargaining problem. *Econometrica* **21**: 128–140.

OECD 2000 Ancillary Benefits and Costs of Greenhouse Gasmitigation. Organization for Economic Cooperation and Development, *Proceedings of an IPCC Co-Sponsored Workshop*. 27–29 March, Washington, DC, p. 583, Available from: *http://www.oecd.org/env/cc.*

Pickrell J 2002 Scientists shower climate change delegates with paper. *Science* **293**: 200.

Pizer W 2002 *Resources for the Future*, Available from: *http://www.rff.org.*

Wieringa K 2001 *European Environmental Priorities: An Integrated Economic and Environmental*

Assessment. RIVM Report 481505010, National Institute of Public Health and the Environment (Rijksinstituut voor volksgezondheid en milieu) Bilthoven, The Netherlands, *http://arch.rivm.nl/ieweb/ieweb/*.

Science 2002 The science of climate change. *Editorial* **292**: 1261.

Stone R 1994 Most nations miss the mark on emission-control plans. *Science* **266**: 1939.

Sun M 1990 Emissions trading goes global. *Science* **247**: 520–521.

The Economist 2002 United States: blowing smoke. *Climate Change* **362**(8260): 27–28.

UNEP 2001 *United Nations Environment Programme. Backgrounder: Basic Facts and Data on the Science and Politics of Ozone Protection. http://www.unep.org/ozone/mp-text.shtml.*

UNFCCC 2003 *United Nations Framework Convention on Climate Change*, Available from: *http://unfccc.int/*.

U.S. EPA 2002 *http://www.epa.gov/airmarkets/index.html*.

USGCRP 2002 *Our Changing Planet: The FY 2002 US Global Change Research Program. Report by the Subcommittee on Global Change Research, Committee on Environment and Natural Resources of the National Science and Technology Council*, Available from: *http://www.usgcrp.gov/*.

WWF 2002 *World Wildlife Fund. Policy News*, Available from: *www.worldwildlife.org*.

Appendix A

Units

Symbol	Quantity	Unit	Equal to	English equivalent
BkW	Billion kilowatts	Energy	10^{12} W	
°C	Degrees Celsius	Temperature		°F = (9/5) °C + 32
cal	Calorie	Energy	4.19 J	
d	Dyne (see Newton)	Pressure		
g	Gram	Mass	10^{-3} kg	
Gt	Gigatons	Mass	10^{9} tonnes	
ha	Hectare	Area		2.47 acres
J	Joule	Energy	$kg\,m^2\,s^{-2}$	
ka or ky	Thousand years	Time		
kcal	Kilocalories	Energy	10^3 calories	
kg	Kilogram	Mass	10^3 g	2.2 pounds
km	Kilometers	Length	10^3 m	0.62 miles
km^2	Square kilometers	Area		0.4 square miles
kW	Kilowatt	Energy	10^3 W	
kWh	Kilowatt hours	Energy	1 kW acting over 1 h	
L	Liter	Volume		0.3 gallon
m	Meter	Length		3.3 feet
Mt	Million tonnes	Energy	10^6 tonnes	
MW	Megawatt	Energy	10^6 W	
N	Newton	Force	$kg\,m\,s^{-2}$	
$N\,m^{-2}$	Newton m^{-2}	Pressure		10 dynes cm^{-2}
nm	Nanometer	Length	10^{-9} m	
Pg	Petagrams	Mass	10^{15} g	
s	Second	Time		
t	Tonne (metric)	Mass	10^3 kg	1.1 ton (short or US)
TW	Terawatt	Energy	10^{12} W	

Climate Change: Causes, Effects, and Solutions John T. Hardy
 ISBNs: 0-470-85018-3 (HB); 0-470-85019-1 (PB)

Symbol	Quantity	Unit	Equal to	English equivalent
μm	Micron	Length	10^{-6} m	
W	Watt	Power	$J\,s^{-1} = kg\,m^2\,s^{-3}$	
y	Year	Time		

Appendix B

Abbreviations and Chemical Symbols

AOGCM	Atmosphere-ocean general circulation model
BP	Before the present time
^{12}C	Carbon 12 – most abundant stable isotope
^{14}C	Carbon 14 – rarer radioactive isotope
$CaCO_3$	Calcium carbonate (limestone)
$Ca(OH)_2$	Calcium hydroxide (lime)
CFC(s)	Chlorofluorocarbon
CH_4	Methane
CO	Carbon monoxide
CO_2	Carbon dioxide gas
Delta Δt_{2x}	Change in temperature from a doubling of atmospheric CO_2
DMS	Dimethyl sulfide
ENSO	El Niño Southern Oscillation
GAT	Global average temperature
GCM	General circulation (climate) model
GDP	Gross domestic product
GDPA	Gross domestic product agriculture
GWP	Greenhouse warming potential
HadCM3	Hadley Center for Climate Research
HCFC(s)	Hydrochlorofluorocarbon
IPCC	Intergovernmental Panel on Climate Change
N_2O	Nitrous oxide
NAO	North Atlantic Oscillation
NO_x	Nitrogen oxide(s)
^{16}O	Oxygen – most abundant isotope
^{18}O	Oxygen – less abundant heavy isotope
O_3	Ozone
OH^-	Hydroxyl radical
PFC(s)	Perfluorocarbon
ppb	Parts per billion by mass
ppbv	Parts per billion by volume

Climate Change: Causes, Effects, and Solutions John T. Hardy
 ISBNs: 0-470-85018-3 (HB); 0-470-85019-1 (PB)

pphm	Parts per hundred million
ppmv	Parts per million by volume
RCM	Regional climate model
SF6	Sulfur hexafluoride
SRES	Special Report on Emissions Scenarios by IPCC
UNESCO	United Nations Education, Scientific, and Cultural Organization
UNFCCC	United Nations Framework Convention on Climate Change
UV-b	Ultraviolet-b radiation (280- to 320-nm wavelength)

Appendix C

Websites on Climate Change

General

US Global Change Information Office – Links to a great variety of information on greenhouse warming and climate change, especially for the United States. See, for example, "Ask Dr Global Change" and your questions will be answered in a few days.
http://www.gcrio.org/

Intergovernmental Panel on Climate Change (IPCC) – The authoritative UN science group under the auspices of WMO and UNEP. A good site for reviewing current research on climate change. Publications and reports are available. See, for example, full text of "Regional Effects of Climate Change."
http://www.ipcc.ch/

US EPA Global Climate Website – A good source of basic information and references, especially focusing on the United States.
http://www.epa.gov/globalwarming/

US Global Change Research Program – Describes agency initiatives and budgets for research and development related to climate change.
http://www.usgcrp.gov/

Energy Information Agency – US DOE link to many sources of information on both fossil and nonfossil energy.
http://www.eia.doe.gov

European Union – Climate Change Research and Programs in Europe.
http://www.europa.eu.int/comm/environment/climat/home_en.htm

Government of Canada Global Climate Change
http://www.climatechange.gc.ca/english/actions/action_fund/index.shtml

Natural Resources Canada
http://climatechange.nrcan.gc.ca/english/index.asp

United Nations Development Fund and Global Environment Facility – Description of project funding on climate change issues.
http://www.undp.org/gef/portf/climate.htm

United States Geological Survey (USGS) – Global Climate Change
http://GeoChange.er.usgs.gov/

Climate Network Europe – Provides many links to other climate information sites.
http://sme.belgium.eu.net/~mli10239/links.htm

Climate Change: Causes, Effects, and Solutions John T. Hardy
 ISBNs: 0-470-85018-3 (HB); 0-470-85019-1 (PB)

NOAA Global Warming Home Page
http://lwf.ncdc.noaa.gov/oa/climate/globalwarming.html

Journal Articles and Literature on Climate Change

NOAA Links to Climate Products and Services on the WWW
http://www.pmel.noaa.gov/constituents/climate-services.html

Renewable Energy World Magazine
http://www.jxj.com/magsandj/rew/2001_01/index.html

The New Scientist – Journal web page with brief articles and links on all aspects of climate change.
http://www.newscientist.com/hottopics/climate/

Nature The journal Nature – article abstracts and some full-text articles covering the breadth of science with some on climate change.
http://www.nature.com/nature/

Environmental Science & Technology – Journal of the American Chemical Society. On-line abstract and full-text articles, some dealing with climate change.
http://pubs.acs.org/journals/esthag/index.html

Global Change – Electronic version of research journal with articles on many aspects of climate change.
http://www.globalchange.org/default.htm

Global Change Master Directory – NASA Searchable databases on many aspects of climate change.
http://gcmd.nasa.gov/

Climate Change Education

UNEP Vital Climate Graphics
http://www.grida.no/climate/vital/index.htm

US Department of Energy Global Change Education – Information on graduate fellowships and undergraduate summer internships dealing with climate change research.
http://www.atmos.anl.gov/GCEP/

NASA Globe website for Earth Images
http://viz.globe.gov/viz-bin/home.cgi

Websites by Chapter Subject Area

Section I Climate Change – Past, Present, and Future

Chapter 1 – Earth and the Greenhouse Effect

University of Reading – Atmospheric Radiation and Climate
http://www.met.rdg.ac.uk/~radiation/

NOAA Pacific Marine Environmental Laboratory – Atmospheric Chemistry and Global Change
http://saga.pmel.noaa.gov/atmochem.html

Carbon Dioxide Information and Analysis Center
http://cdiac.esd.ornl.gov/

Chapter 2 – Past Climate Change: Lessons From History

Civilization and past effects of climate
http://www.msnbc.com/news/564306.asp

Chapter 3–Recent Climate Change: The Earth Responds

Climate Trend – Data for specific locations worldwide from the IPCC Data Center
http://www.climatetrend.com/

NASA – Earth Observatory
http://earthobservatory.nasa.gov/Study/GlobalWarm1999/

NOAA Data Center – Portal for searching and downloading climate data of the National Oceanic and Atmospheric Administration
http://www.epic.noaa.gov/cgi-bin/NOAAServer?stype=HTML3

Chapter 4 – Future Climate Change: The Twenty-First Century and Beyond

IPCC Data Distribution Center – A source of recent research and data on climate change with links to specific reports and other sites.
http://ipcc-ddc.cru.uea.ac.uk/

Climatic Research Unit University of East Anglia – Information on climate change and links to other sites.
http://www.cru.uea.ac.uk/

NASA Climate Modeling Goddard Institute for Space Studies
http://www.giss.nasa.gov/research/modeling/

Hadley Center for Climate Prediction UK – Part of the UK Meteorological Office. This is an excellent site to review global climate model predictions.
http://www.met-office.gov.uk/research/hadleycentre/index.html

National Center for Atmospheric Research (USA)
http://www.ncar.ucar.edu/ncar/index.html

Population Action International – This site particularly highlights the links between world population growth and recent and future climate change.
http://www.cnie.org/pop/CO2/intro.htm

Section II Ecological Effects of Climate Change

Climate Change Impact on Species and Ecosystems – A bibliographic list of research articles.
http://eelink.net/~asilwildlife/CCWildlife.html

Chapter 5 – Effects on Freshwater Systems

US Geological Survey Website on Climate Change – Information on US climate-change research, particularly that dealing with soils and water resources.
http://GeoChange.er.usgs.gov/

NASA Global Hydrology and Climate Center
http://wwwghcc.msfc.nasa.gov

Chapter 6 – Effects on Terrestrial Systems

US Forest Service – How the distribution of trees will be affected by climate change
http://www.fs.fed.us/ne/delaware/4153/global/global.html

Canadian Forest Service Climate Change Information
http://www.nrcan.gc.ca/cfs-scf/science/resrch/climatechange_e.html

Chapter 7 – Effects on Agriculture

US Department of Agriculture – NRCS, Agriculture, and Climate Change
http://www.nrcs.usda.gov/technical/land/pubs/ib3text.html

Global Climate Change and Agricultural Production – FAO Book edited by F. Bazzaz.
http://www.fao.org/docrep/W5183E/W5183E00.htm

Two Essays on Climate Change and Agriculture. UN FAO Economic and Social Development Paper *http://www.fao.org/DOCREP/003/X8044E/X8044E00.HTM*

Chapter 8 – Effects on the Marine Environment

US National Oceanic and Atmospheric Organization – NOAA Office of Global Programs *http://www.ogp.noaa.gov/enso/*

Section III Human Dimensions of Climate Change

Chapter 9 – Effects on Human Infrastructure

Cities for Climate Protection *http://www.iclei.org/co2/index.htm*

Electric Power Research Institute *http://www.epri.com/*

US Department of Energy – National Energy Policy and other energy-related information. *http://www.energy.gov/*

International Energy Agency – Information on world energy supplies and policies. *http://www.iea.org/*

Population Action International – This site particularly highlights the links between world population growth and climate change *http://www.cnie.org/pop/CO2/intro.htm*

Chapter 10 – Effects on Human Health

Program of the Health Effects on Global Climate Change of Johns Hopkins University – Lists of publications and links to information on health effects. *http://www.jhu.edu/~climate/*

Health Canada, Climate Change and Health Office – Information on health effects of climate change. *http://www.hc-sc.gc.ca/hecs-sesc/hecs/climate/*

World Health Organization – Information on health effects of climate change, lists of reports and conferences, downloadable articles. *http://www.who.int/peh/climate/climate_and_health.htm*

Chapter 11 – Mitigation: Reducing the Impacts

American Forests – A nonprofit conservation group. This website allows individuals to enter information on of their own energy consumption and estimate the number of trees that need to be planted to offset their greenhouse gas emissions. *http://www.americanforests.org/resources/ccc/*

US Department of Energy – Sustainable Development *http://www.sustainable.doe.gov/*

Carbon sequestration bibliography *http://sequestration.mit.edu/bibliography/*

US Department of Energy – Office of Energy Efficiency and Renewable Energy *http://www.eren.doe.gov/*

Canada – Transportation Initiatives *http://www.ec.gc.ca/press/clim4_b_e.htm*

Photovoltaics – US Dept of Energy, National Center for Photovoltaics. Provides information and interactive

models to calculate energy output and cost savings for any selected area of the United States.
http://www.nrel.gov/ncpv/

European Commission – Energy Site – Contains information on renewable sources of energy.
http://europa.eu.int/comm/energy_transport/atlas/homeu.html

The Online Fuel Cell Information Center – Information on all aspects of fuel cells as an energy source.
http://www.fuelcells.org/

Biofuels – US Department of Energy information on biofuel sources and research.
http://www.ott.doe.gov/biofuels/

Carbon Dioxide Sequestration – Description of the carbon dioxide capture project and other aspects of CO_2 sequestration.
http://www.co2captureproject.org

Energy Information Agency – Information on renewable and fossil and alternative fuels. Good links to other sources.
http://www.eia.doe.gov

American Wind Energy Association – Great site for information on wind energy in general and wind energy projects in the United States.
http://www.awea.org

Renewable Energy – The US Department of Energy's laboratory for renewable energy and energy efficiency research and development.
http://www.nrel.gov/

Danish Wind Industry Association – All about wind energy.
http://www.windpower.org/core.htm

US Department of Energy – Wind Energy Program
http://www.eren.doe.gov/wind/

American Wind Energy Association
http://www.awea.org/default.htm

Alternative Fuels Data Center – US Department of Energy Website with information on a variety of renewable and noncarbon energy technologies and links to related sites.
http://www.afdc.doe.gov/altfuels.html

Renewables – European Commission site on energy.
http://europa.eu.int/comm/energy_transport/atlas/htmlu/renewables.html

Chapter 12 – Policy, Politics, and Economics

United Nations Framework Convention on Climate Change (UNFCCC) – The site for current information on UN and international efforts at global treaties to reduce greenhouse gas emissions and climate-change effects.
http://www.unfccc.de/

US Department of State – site for policy aspects of climate change.
http://www.state.gov/g/oes/climate/

Resources for the Future
http://www.rff.org/books/descriptions/climatechange_anthology.htm

Indiana University – A Welfare Economics Perspective
http://www.spea.indiana.edu/richards/Welfare_Economics/

Ohio State University – Climate Change: Science, Policy, and Economics
http://ohioline.osu.edu/ae-fact/0003.html

Resources for the Future – Weathervane, A Digital Forum on Climate Change Policy
http://www.weathervane.rff.org/

World Summit on Sustainable Development Johannesburg 2002
http://www.johannesburgsummit.org/

The Pew Center – Climate Policy
http://www.pewclimate.org/policy/

Climate Action Network Europe
http://www.climnet.org/

Conservation and Environmental Action Groups

World Wildlife Campaign Climate Change
http://www.panda.org/climate/

Sierra Club *www.sierraclub.org*

Earth Policy Institute
http://earth-policy.org

Hot Earth – An environmental action group dedicated to reducing potentials for human-induced climate change
http://environet.policy.net/warming/index.vtml

Greenpeace – Climate Change Campaign
http://www.greenpeace.org/campaigns/intro?campaign_id=3937

Industry Groups

Global Climate Coalition
http://www.globalclimate.org/

American Petroleum Institute
http://www.api.org/globalclimate/

Citizens for a Sound Economy Foundation
http://www.cse.org/

International Climate Change Partnership
http://www.iccp.net/

World Business Council for Sustainable Development
http://www.wbcsd.ch/

Business Council for Sustainable Energy
http://www.bcse.org/

Global Environmental Management Initiative – What business and private industry is doing related to global climate change.
http://www.businessandclimate.org/

Index

Climate Change: Causes, Effects, and Solutions John T. Hardy

ISBNs: 0-470-85018-3 (HB); 0-470-85019-1 (PB)